AF656207

Bacterial Wilt of Solanaceous Vegetables

Bacterial Wilt of Solanaceous Vegetables

Jai Prakash Sharma
Principal Scientist (Plant Pathology)
ICAR Research Complex for Eastern Region
Research Centre, Plandu, Ranchi
E-mail: jaibina_05@yahoo.co.in

Shivendra Kumar
Principal Scientist and Head
ICAR Research Complex for Eastern Region
Research Centre, Plandu, Ranchi
E-mail: sk_harp@yahoo.co.in

Dinesh Singh
Principal Scientist
Division of Plant Pathology
Indian Agricultural Research Institute
New Delhi-110012
E-mail: dinesh_iari@rediffmail.com

I.K. International Publishing House Pvt. Ltd.
NEW DELHI

Published by
I.K. International Publishing House Pvt. Ltd.
S-25, Green Park Extension
Uphaar Cinema Market
New Delhi–110 016 (india)
E-mail: info@ikinternational.com
Website: www.ikbooks.com

ISBN 978-93-84588-22-9

© 2016 I.K. International Publishing House Pvt. Ltd.

All rights reserved. No part of this publication may be reproduced, stored in a retrieval system, or transmitted in any form or any means: electronic, mechanical, photocopying, recording, or otherwise, without the prior written permission from the publisher.

Published by Krishan Makhijani for I.K. International Publishing House Pvt. Ltd. S-25, Green Park Extension, Uphaar Cinema Market, New Delhi–110 016 and Printed by Rekha Printers Pvt. Ltd., Okhla industrial Area, Phase II, New Delhi–110 020.

भारतीय कृषि अनुसंधान परिषद
कृषि अनुसंधान भवन-II
पूसा, नई दिल्ली -110 012
INDIAN COUNCIL OF AGRICULTURAL RESEARCH
KRISHI ANUSANDHAN BHAWAN-II
PUSA, NEW DELLHI-110 012

डॉ एन. के. कृष्ण कुमार
उप महानिदेशक (बाग. वि.)
Dr. N. K. Krishna Kumar
DEPUTY DIRECTOR GENERAL (HORTICULTURAL SCIENCE)

FOREWORD

Global agricultural production of the world sustains an annual loss of about 20 to 30% on an average due to the occurrence of plant diseases on different crops. Occasionally, crop loss is up to 100% under most favorable conditions or when no prophylactic measures are undertaken. Plant diseases are one of the major bottlenecks in agricultural production particularly in irrigated crops. Solanaceous vegetables such as potato, brinjal, tomato, chilli and capsicum which constitute a half of the total vegetables grown, suffer severely in rainy months from bacterial wilt disease especially in costal & inland plateau. Understanding ways to avoid such losses is necessary for increasing the production, productivity and profitability.

I am happy that the authors in the present book entitled "**Bacterial wilt of solanaceous vegetables**" have sourced information of over one century (1886-2014) at one place. Facts detailed in seven chapters encompassing the causal bacterium, symptoms, distribution, detection and diagnosis, genomics and pathogenesis, host range, epidemiology, losses and management have made the book useful for proper understanding of this yield limiting factors.

I take this opportunity to complement Dr. J. P. Sharma, Principal Scientist (Retired), ICAR Research Complex for Eastern Region, Research Centre, Plandu, Ranchi, Dr. Shivendra Kumar, Head, (Retired), ICAR Research Complex for Eastern Region, Research Centre, Plandu, Ranchi and Dr. Dinesh Singh, Principal Scientist, Division of Plant Pathology, Indian Agricultural Research Institute, New Delhi for their effort in compiling valuable information. It would help the researchers, teachers, students, extension workers and progressive farmers to access the information at one source and wish that the readers take advantage.

(N. K. Krishna Kumar)

Preface

Vegetables are important food and highly beneficial for the maintenance of human health and prevention of diseases. They contain valuable food ingredients, which can be successfully utilized to build up and repair the body. They are valued for their high carbohydrate, vitamin, mineral and fibre contents. Vegetables make up a significant proportion of the diet of most of the people and the production of vegetables is a significant factor in ensuring that people has an adequate intake of many essential vitamins, minerals and carbohydrates every day.

China is the leading country in terms of area (21.05 m ha), production (473.0 6 mt) and the productivity (22.5 t/ha) of vegetable crops as per 2010-11 data (FAO website: March 2012). India is the second leading vegetable producing country in the world. Presently, it occupies 8.49 m ha area with the annual production of 146.55 m t (Indian Horticulture Database 2011) and productivity is 17.3 t /ha against world total area 53.97m ha with production 1012.52 m t and productivity 18.8 t/ha. The country being blessed with the unique gift of nature of diverse climate and distinct seasons, make it possible to grow an array of vegetable crops the number exceeding more than hundred. Globally potato is produced in 18.63 m ha area with production of 330.04 m t and productivity 17.7 t/ha, however, the area in China is 50.77 m ha, production 74.79 m t with productivity 14.7 t/ha whereas in India it is grown in 10.86 m ha area, production 42.34 m t and productivity 22.7 t/ha. However, potato being the staple food and easy to mix in several preparations ranks first (26.6%) in total production of vegetables followed by other important solanaceous vegetables like tomato (8.6%) and brinjal (8.0%). Globally tomato is grown in 4.58 m ha area with production of 150.51 m t and productivity of 32.8 t/ha whereas in India it grown in about 0.87 m ha area with production of 16.83 m t with productivity 19.5 t/ha. In China, brinjal is grown in 0.73 m ha with production of 24.5 m t and with productivity of 33.5 t/ha whereas India being the origin of brinjal it is grown in 0.68 m ha with production of 11.89 m t and the productivity is 17.5 t/ha against world productivity of 25.0 t/ha. As regards to global chilli production, it is grown in 17.76 m ha with total production of 71.82 m t and the productivity of 4.04 t/ha and in India it is about 0.79 m ha area with production of 1.22 mt and the productivity 1.5 t/ha. India contributes for around 25% of world chilli production. Andhra Pradesh is the largest producer of chilli in India, which contributes to about 30% to the total area under chili, followed by Karnataka (20%), Maharashtra (1%), Orissa (9%), Tamil Nadu (8%) and other states contributing 18% to the total area under chilli.

India has to go a long way to accelerate the vegetable production, considering the increasing national and international demand and the unfortunate annual loss of about 20 to 30% on an average due to plant diseases in different crops. The projected global climate change also pose threat by way of uncertainty. In this backdrop on the other side, there is a need to achieve target of 225 m t vegetable productions by the end of 2020 and 350 m t by 2030. Bacterial wilt caused by *Ralstonia solanacearum* is one of the most important yields limiting disease of major solanaceous vegetables and need attention for containment at all levels.

A large number of researchers have contributed in understandings the nature of pathogen, the response of the host and ways to contain the disease and mitigate losses in India and abroad. This information is available at different sources. In this book an attempt to bring them at one point has been made to help users. It is expected that the book will be useful for the post-graduate students, researchers, extension workers as well as educated farmers to enhance their knowledge about bacterial wilt of solanaceous crops and means of their management to reduce crop losses at field level.

We are thankful to the scientists, colleagues and friends in sharing their help in preparation of manuscript.

Jai Prakash Sharma
Shivendra Kumar
Dinesh Singh

Authors

Jai Prakash Sharma obtained his Ph.D. in 1987 from Rajendra Agricultural University, Pusa, Bihar. He was a recipient of ICAR merit scholarship, junior research fellowship (ICAR) and senior research fellowship (ICAR). He is a Fellow of Phytopathological Society of India since 1988. He has served in different capacities as senior research assistant, and scientists at Nagaland Centre of ICAR Research Complex for NEH Region, Shillong from 1977 to Feb. 1990 and worked on rice blast, *Helminthosporium* blight on maize, bacterial wilt of solanaceous vegetable crops, and mushroom cultivation. Thereafter as scientist (SS), and senior scientist at CHES, Ranchi (IIHR, Bangalore) from March 1990 and as principal scientist (plant pathology) worked on bacterial wilt disease of solanaceous vegetables. His experience in the field of bacterial wilt has been recognized by Indian Phytopathological Society and awarded him Prof. M. J. Narasimhan Medal (2011) for best research paper. He has published (42) research papers in reputed journals, (8) book chapters, (54) research abstract in symposia (23) technical bulletins and (18) popular articles. He has delivered invited lectures to various universities, talks in All India Radio and Doordarsan, Ranchi and made a documentary film on oyster mushroom cultivation with Bharat Shodh telecast by Doordarsan, New Delhi.

Shivendra Kumar obtained his Ph.D. from Utkal University in 1987. He worked at Central Potato Research Institute and ICRISAT before joining Agricultural Research Service in 1978. At Central Rice Research Institute (1978-84) he worked on physiology of blast disease resistance. After joining Indian Institute of Horticulture Research, Ranchi Centre in 1984, he started working on wilt diseases in solanaceous vegetable crops and guava and continues working in these study areas. He assumed the positions of Principal Scientist in 1997 and Head of the Research Centre, Ranchi in 2003. Dr Kumar was admitted as Fellow of Indian Phytopathological Society in 1992 and was elected to Zonal President (East Zone) in 2004 and received Prof. M J Narasimhan Medal (2011). He has published 40 research papers in journals of high repute.

Dinesh Singh obtained his Ph.D. in Mycology and Plant Pathology from Institute of Agricultural Sciences, Banaras Hindu University, Varanasi. He joined ICAR in January, 1997 at IIVR, Varanasi and in February 1997 at CIPHET, Abohar (Punjab) as scientist and worked on post-harvest diseases of fruits and vegetables. He as senior scientist joined Indian Agricultural Research Institute, New Delhi as senior scientist in 2006. Presently he is working in the field of plant bacteriology particularly bacterial diseases of vegetable crops. He assumed the positions of Principal Scientist in 2012. Dr. Singh was awarded SPPS Fellow (2007), Dr. M. M. Alam Medal (2008), Fellow of Indian Phytopathological Society (2010), Young Scientist Associate Award (2010) and Late Shri P. P. Shinghal Memorial Award (2013). He has published (81) research papers in reputed journals, (3) books, (26) book chapters, (64) research abstract in symposia (5) technical bulletins, (4) training manual, (2) practical manuals and (28) popular articles. He has delivered invited lecture to various universities, various trainings, talks on Doordarsan.

Contents

1
Introduction

1.1 SOLANACEOUS VEGETABLES

Solanaceae is a family of flowering plants that includes several important agricultural crops. The name comes from the Latin word *Solanum* which is perceived as resemblance that the flowers bear to the sun and its rays. The family is also known as "the nightshade or potato family" which includes *Datura, Mandragora* (mandrake), *Atropa belladonna* (deadly nightshade), *Lycium barbarum* (wolfberry), *Physalis philadelphica* (tomatillo), *Physalis peruviana* (Cape gooseberry flower), *Capsicum* (chili pepper, bellpepper), *Solanum* (potato, tomato, eggplant), *Nicotiana* (tobacco), and *Petunia.* Except tobacco (Nicotianoideae) and petunia (Petunioideae), most of the economically important genera are of the subfamily Solanoideae. Many members of the Solanaceae family are used by humans, and are important sources of food, spice and medicine. The solanaceae species are often rich in alkaloids and phenolics (Dauny, 2008) whose toxicity to humans and animals ranges from mildly irritating to fatal in small quantities. Several fruits of the family Solanaceae; especially of the genera *Solanum, Capsicum*, and *Lycopersicon* are used as vegetables. This family has 84 genera and 3000 species, widely distributed in tropical and temperate regions of both the hemispheres. The solanaceous vegetables are of following kinds.

Hypernyms: ("solanaceous vegetable") is a kind of vegetable or veggie (edible seeds or roots or stems or leaves or bulbs or tubers or nonsweet fruits of any of numerous herbaceous plants).

Hyponyms: ("solanaceous vegetable") each of the following is a kind of Iris potato; Murphy Potato; Spud; Tater; White Potato.

Aubergine: Eggplant; Mad Apple (egg-shaped vegetable having a shiny skin typically dark purple but occasionally white or yellow).

Pepper: Sweet and hot varieties of fruits of plants of the genus *Capsicum.*

Tomato: Mildly acid red or yellow pulpy fruit eaten as a vegetable. Husk tomato: Mexican husk tomato; tomatillo (small edible yellow to purple tomato-like fruit enclosed in a bladder like husk).

1.1.1 Tomato

Tomato is a New World plant which was unknown to Europeans before Columbus brought it back at the end of the 15th century. Tomato had a variety of American and Indian names and it is one of these Indian names that give us the English word "tomato." The German name for "wolf peach" gives us the scientific name "*Lycopersicum*" (Jo Pick). The word "tomato" is botanically known as *Solanum lycopersicum* which is, typically a red fruit, originated in South America, and spread around to Spanish colonization of the Americas, and its many varieties are widely grown. The tomato fruit is consumed as raw, or an ingredient in many dishes and sauces, and in drinks. The fruit is rich in lycopene, which is beneficial for health. The tomato is herbaceous green plants with small green fruit and a center of diversity in the highlands of Peru (Cox, 2000; Smith,1994b). The species, *Solanum lycopersicum*, was transported to Mexico and grown and consumed by Mesoamerican civilizations. The first domesticated tomato has been a little yellow fruit, similar in size to a cherry tomato, grown by the Aztecs of Central Mexico (Chester, 2008). The word "tomato" comes from the Nahuatl word *tomatl*, literally "the swelling fruit" Smith (1994). Linnaeus (1753), Luckwill, (1943), (http://tgc.ifas.ufl.edu/vol 56/html/vol56featr.htm) placed the tomato in genus *Solanum lycopersicum*. Miller (1768) moved it in to own genus *Lycopersicon*, which is commonly in use and technically known as *Lycopersicon esculentum* L. Two members of the genus *Solanum* have been successfully hybridized with members of the *Lycopersicon* genus (Rick *et al.,* 1990). Those are *S. lycopersicoides* and *S. pennellii*. Wild tomato species have tiny fruits, and only the red ones are edible. Tomato plants do not tolerate frost, and grow as annuals in colder regions. In warmer regions, they are perennial, and flower regardless of day length. All members of the genus have perfect flowers (hermaphroditic). Cultivated tomato is self-fertile, whereas all other members of the genus are self-incompatible (Simpson and Ogorzaly, 1986) with the exception of *L. pimpinellifolium*, which undergoes various degrees of self-fertilization. Sevral germplasms of wild species of tomato *L. pimpinellifolium* has been found in NEH region (Asati and Yadav, 2004). Abdullahi and Bitrus Choji (2009) have defined tomatoes, as fruity edible plants of a vine just as are cucumbers, squashes, beans and peas that belong to the genus *Lycopersicon esculentum* and solanaceae or nightshade family (Rick *et al*., 1990).

1.1.2 Brinjal

Brinjal is known as aubergine, eggplant, brinjal, garden egg, Jew's apple, mad apple, pea apple, egg apple, guinea squash, melongene, come under solanaceae family and botanically known as *Solanum melongena*. The fruits are commonly used in cooking and is closely related to the tomato and potato. It is a native of India (Valilov, 1931; Doijode, 2001; Tsao and Lo, 2006) and widely consumed as nutritious vegetable. It has been cultivated in southern and eastern Asia since prehistory but appears to have become known to the Western world no earlier than ca.1500. The name *aubergine* [derived from French (Minsheu, 1623, http://www.oed.com/view/Entry/12985?redirectedFrom=aubergine#, Fuchsia Dunlop, 2006)], is the name of the purple color resembling that of the fruit. It is the most common, popular and widely grown vegetable crop of both tropical and sub-tropical world, *viz*, India, Bangladesh, Pakistan, China, Philippines, France, Italy, Balkan area, France, Indonesia, Japan, Mediterranean, Turkey and the United States (Bose and Som, 1986) and the consumers preference is region specific (Kanthaswamy *et al*., 2003) and large number of open pollinated cultivars is grown throughout the country. Brinjal or eggplant, locally known as 'terung' is a popular fruit vegetable in Bangladesh. The eggplant or brinjal or aubergine (*Solanum melongena* L.) represents the non-tuberous group of *Solanum* species (Narasimha Rao, 1979). Dauny (2008) reported that *S. melongena* L. (known as eggplant in the United States and aubergine in France, England and brinjal in India) is one of the few cultivated solanaceous species originating from the Old World, is widely grown in temperate and tropical Asian countries and also a common crop in the Middle East and around the Mediterranean basin and is now cultivated worldwide. The ethno-botanical history of eggplant is quite fascinating given its uses as food crop, medicine, and ornamental by Ancient (Indian) and Medieval (Arabic and European) civilizations, and the diverse beliefs surrounding its use including aphrodisiacal properties and various malevolent and benevolent effects. In the beginning of the 21st century, the health values of eggplant are being revived once again based on phenolic and alkaloid content.

1.1.3 Chilli

Chilli or pepper (also chile pepper or chilli pepper, from Nahuatl chilli) is the fruit of plants from the genus *Capsicum*, members of the nightshade family, Solanaceae. The term in British English and in Australia, New Zealand, India, Malaysia and other Asian countries is just chilli, not pepper. Chilli peppers originated in the Americas. After the Columbian Exchange, many cultivars of chilli pepper spread across the world brought back to Europe by Columbus; Spanish and Portugese introduced peppers to other parts of the world, including India and Asia, used in both food and medicine. *Capsicum annuum* includes the green, bell-shaped, sweet peppers and hot peppers used for spices, including paprika, chilli, and cayenne. In American English it is commonly known as the chili pepper, In British English, the sweet varieties are called whereas in Australian English the name *Capsicum* is commonly used for bell peppers exclusively and chilli is often used to encompass the hotter varieties. Its forms are varied, from large to small, sweet to sour, very hot to bland. Chilli or peppers is a part of the human diet in the Americas since at least 7500 BC. There is archaeological evidence at sites located in southwestern Ecuador that chilli peppers were domesticated more than 6000 years ago (Perry *et al.*, 2007; BBC News Online. 2007) and is one of the first cultivated crops in the Central and South Americas (Bosland, 1996) that is self-pollinating. Columbus was one of the first Europeans to encounter them (in the Caribbean), and called them "peppers" because they, like black and white pepper of the *Piper* genus known in Europe, have a spicy hot taste unlike other foodstuffs. The monks discovered the pungency in black pepper corns. Chillies were cultivated around the globe after Columbus. Diego Álvarez Chanca, a physician on Columbus' in 1493, brought the first chilli peppers to Spain, and first wrote about their medicinal effects in 1494. From Mexico, at the time the Spanish colony that controlled commerce with Asia, chili peppers spread rapidly into the Philippines and then to India, China, Indonesia, Korea and Japan. They were incorporated into the local cuisines. Another account for the spread of chilli peppers is that the Portuguese got the pepper from Spain, and cultivated it in India.The chilli pepper figures heavily in the cuisine of the Goan region of India, which was the site of a Portuguese colony (e.g., vindaloo, an Indian interpretation of a Portuguese dish). Chilli peppers journeyed from India, through Central Asia and Turkey, to Hungary, where it became the national spice in the form of paprika.The hotness of the fruit is indicated in the Scoville scale, which ranges from no heat at zero to pure capsaicin at 16,000,000SHU, which are recorded as the major species. The scale is named after its creator, American pharmacist Scoville devised a method, known as the Scoville Organoleptic Test (Tainter and Grenis, 2001; Anonymous, 1912).

Note: Scoville in 1912 created what is known today as the Scoville Scale. The Scale determines how hot a chilli pepper is based on its capsicin rating (Anonymous, 1912). The Scoville scale is a measurement of the spicy heat (or piquance) of a chilli pepper. The number of Scoville heat units (SHU) (Peter, 2001) indicates the amount of capsaicin present. Capsaicin is a chemical compound that stimulates chemoreceptor nerve endings in the skin, especially the mucous membranes; pure capsaicin has a Scoville rating of 16,000,000 SHU.

1.1.4 Potato

Potato is a starchy, tuberous crop from the perennial of the Solanaceae family (also known as the nightshades) and botanically known as *Solanum tuberosum*. Potatoes were introduced outside the Andes region four centuries ago, and have become an integral part of the world's cuisine. It is the world's fourth-largest food crop, following rice, wheat and maize (U N FAO, 2009). The potato was first domesticated in the region of modern-day southern Peru and extreme northwestern Bolivia (Spooner *et al.*, 2005) between the years 8000 BC and 5000 BC (OIA, 1989). Potatoes provided the principal energy source for the Inca Empire, its predecessors, and its Spanish successor. In Bolivia and Peru above 10,000 feet altitude, tubers exposed to the cold night air turned into chuño (freeze-dried potato, Johns, 1990; Anonymous, 2012) when kept in permanently frozen underground storehouses, chuño can be stored for years with no loss of nutritional value. The Spanish fed chuño to the silver miners who produced vast wealth in the 16th century for the Spanish government. 200 species of wild potatoes are found in the Americas. The potato contains a fragment of the genetic diversity found in the four recognized potato species and 5000 potato varieties are grown in the Andes.

1.2 NUTRITIONAL VALUE OF CROPS

The nutritive values of different solanaceous vegetables are summarized in Table 1.1.

Table 1.1 Nutritional values of solanaceous vegetables (per 100 g) (Source: USDA Nutrient data)

Nutritions	*Tomato*	*Brinjal*	*Chilli*	*Potato*
Energy (kj)	74.0	102.0	166.0	322.0
Carbohydrates (g)	3.9	5.7	8.8	17.0
Sugars (g)	2.6	2.35	5.3	0.78
Dietary fiber (g)	1.2	3.4	1.5	2.2
Fat (g)	0.2	0.19	0.4	0.09
Protein (g)	0.9	1.01	1.9	2.0
Water (g)	94.5	–	88	79.0
Vitamin A equiv.	42 μg (5%)	–	48 μg	2.0(IU)
- beta-carotene (μg)	–	–	534.0	1.0

Contd...

Contd...

- lutein and zeaxanthin	123 μg	–	–	8.0
Thiamine (vit. B1) (mg)	–	0.039	–	0.08
Riboflavin (vit. B2) (mg)	–	0.037	–	0.03
Niacin (vit. B3) (mg)	–	0.649	–	1.05
Pantothenic acid (B5) (mg)	–	0.281	–	0.30
Vitamin B6 (mg)	–	0.084	0.51	0.30
Folate (vit. B9) (μg)	–	22.0	–	16.0
Vitamin C (mg)	14.0	2.2	144 .0	19.7
Vitamin E (mg)	0.54	–	–	0.01
Vitamin K (μg)	–	–	–	1.9
Calcium (mg)	–	9.0	–	12.0
Phosphorus (mg)	–	25.0	–	57.0
Iron (mg)	–	0.24	1 .0	0.78
Manganese (mg)	–	0.25	–	0.15
Magnesium (mg)	–	14.0	23.0	23.0
Potassium (mg)	237.0	230.0	322.0	421.0
Sodium (mg)	–	–	–	6.0
Zinc (mg)	–	0.16	–	0.29
Capsaicin (g)	–	–	0.01 – 6	–
Nocotine (mg)	–	0.01	–	–

1.2.1 Tomato

Tomato is eaten throughout the world and its consumption is believed to benefit the heart, among other organs. Natural genetic variation in tomatoes and their wild relatives has given a genetic plethora of genes that produce lycopene (Fig. 1.1a) (http://upload.wikimedia.org/wikipedia/commons/e/e9/, carotene (Fig. 1.1b), (http://en.wikipedia.org/wiki/Carotene), anthocyanin (Fig. 1.1c) (http://en.wikipedia.org/wiki/). Among them contain the lycopene, (Fig. 1.1a) and carotene (Fig. 1.1b) is the most powerful natural antioxidants. Lycopene especially in cooked tomatoes has been found to help prevent prostate cancer and improved the skin's ability to protect against harmful UV rays (Anonymous, 2008). Tomato varieties are available with double the normal vitamin C (Double rich), 40 times normal vitamin A (97L97), high levels of anthocyanin (blue tomatoes), and two to four times the normal amount of lycopene (cultivars with the high crimson gene). The richest source of lycopene in the diet is tomato and tomato products (Evangelia *et al.* 2005). It also decreases risk of breast cancer (Zhang *et al.*, 2009) head and neck cancers (Freedman *et al.*, 2008) and might be strongly protective against neurodegenerative diseases (Rao and Balachandran, 2002; Suganuma *et al.*, 2002). Tomatoes and tomato sauces and puree lower urinary tract symptoms (BPH) and may have anticancer properties (Fall *el al.*, 1999; Zdeňka *et al.*, 2010) and beneficial for reducing cardiovascular risk associated with type 2 diabetes (Shidfar *et al.*, 2011).

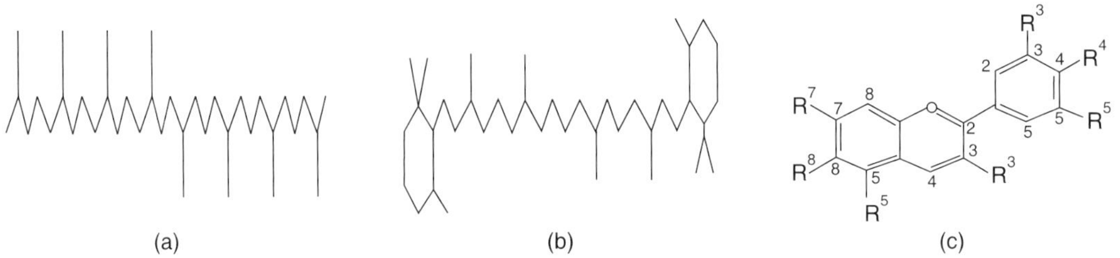

Fig. 1.1 Structural formula of lycopene, carotene and anthocynine, (a) Lycopene (Skeletal formula), (b) β-carotene, (c) Anthocyanin

1.2.2 Brinjal

Brinjal or eggplant fruits are consumed as cooked vegetable in various ways (Rai *et al.,* 1995). The nutritive value of brinjal is comparable with tomato (Chaudhary, 1976) and fruits are rich source of minerals like Ca, Mg and P and fatty acids. Besides, it is used as fresh vegetable and known to have some medicinal properties in curing diabetic patients, asthma, cholera, bronchitis, diarrhea and liver complaints (Tomar and Kalda, 1998). Its fruits and leaves are reported to lower the blood cholesterol (Darlington, 2003). The eggplant is effective in reducing of high blood cholesterol as reported by Institute of Biology of São Paulo State University, Brazil, whereas another study from Heart Institute of the University of São Paulo found no effects at all and does not recommend eggplant as a replacement to statins. It helps to block the formation of free radicals and is also a source of folic acid and potassium (Anonymous, 2007, Health 24.com – Aubergine). Eggplant is richer in nicotine (http://en.wikipedia.org/wiki/Nicotine) (Fig. 1.2a), with a concentration of 100 ng/g (or 0.01 mg/100 g). However, the amount of nicotine from eggplant or any other food is negligible compared to passive smoking (Edward *et al.*, 1993). On average, 9 kg of eggplant contains about the same amount of nicotine as a cigarette. It has some allergies such as itchy skin and/or mouth; mild headache and stomach upset after handling and/or eating eggplant have been reported anecdotally and published in medical journals. The study on some allergic symptoms after consuming eggplant from sample of 741 people taken and found nearly 10% showed allergic symptoms after consuming eggplant, while 1.4% showed symptoms in less than 2 hours reported in India (Harish Babu *et al.*, 2008). Eggplant fruit has induced IgE-mediated immediate-type hypersensitivity reactions, such as rhinorrhoea, urticaria, food allergy and asthma (Muthiah *et al.*, 1996). Contact dermatitis from eggplant leaves (Kabashima and Miyachi, 2004) and allergy to eggplant flower pollen (Gerth van Wijk *et al.*, 1989) have been reported. An individual that is atopic (genetically predisposed to hypersensitivity, such as hay fever) are more likely to have a reaction to eggplant, which may be because eggplant is high in histamines. A few proteins and at least one secondary metabolite have been identified as potential allergens (Pramod and Venkatesh, 2008). Cooking eggplant thoroughly seems to preclude reactions in some individuals, but at least one of the allergenic proteins survives the cooking process. Health experts and nutritionists believe that a well-balanced diet that includes at least five portions of fruit and vegetables per day will help to considerably improve our lifestyles and ultimately extend our lives (http://www.healthy-eating-foods.info/vegetables/aubergine.htm).

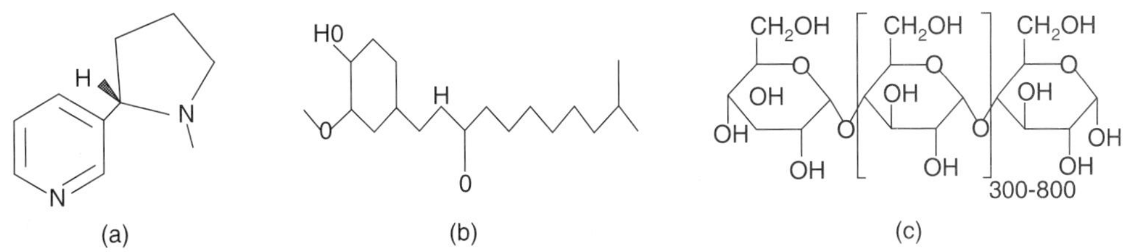

Fig. 1.2 Structural formula of Nicotine, Capsaicin and Starch, (a) Nicotine (b) Capsaicin (c) Amylose

1.2.3 Chilli

Red chillies contain high amounts of vitamin C and carotene (provitamin A). Yellow and especially green chillies (which are essentially unripe fruit) contain a considerably lower amount of both substances. In addition, peppers are a good source of most B vitamins, and vitamin B_6 in particular. They are very high in potassium, magnesium, and iron. Their high vitamin C content can also substantially increase the uptake of non-heme iron from other ingredients in a meal, such as beans and grains. Chilli, cayenne pepper and red pepper contain between 0.1 mg to 60 mg capsaicin per gram. One tablespoon (8 g) of ground chilli pepper contain between 0.8 mg and 480 mg of capsaicin (http://en.wikipedia.org/wiki/File:Kapsaicyna.svg.) (Fig. 1.2b). The hotter the pepper contains more capsaicin and red chillies contain high amounts of vitamin C and carotene (provitamin A) whereas yellow and especially green chillies contain a considerably lower amount of both substances (Roach, 2008). Anandaraj (2009) reported from Kerala that chilli is a very remunerative, widely grown indispensable spice as well as cash crop to the world food industry in Kerala.

1.2.4 Potato

Potato is a starchy stable food. They contains vitamins and minerals, as well as an assortment of phytochemicals, such as carotenoids and natural phenols. Chlorogenic acid constitutes up to 90% of the potato tuber natural phenols. Others found in potatoes are 4-O-caffeoylquinic acid (crypto-chlorogenic acid), 5-O-caffeoylquinic (neo-chlorogenic acid), 3,4-dicaffeoylquinic and 3,5-dicaffeoylquinic acids (Friedman, 1997). A medium-size 150 g (5.3 oz) potato with the skin provides 27 mg of vitamin C (45% of the Daily Value (DV)), 620 mg of potassium (18% of DV), 0.2 mg vitamin B_6 (10% of DV) and trace amounts of thiamin, riboflavin, folate, niacin, magnesium, phosphorus, iron, and zinc. The fiber content of a potato with skin (2 g) is equivalent to that of many whole grain breads, pastas, and cereals.The potato is best known for its carbohydrate (http://en.wikipedia.org/wiki/File:Amylose2.svg) content Fig. 1.2c (26 g in a medium potato). A small but significant portion of this starch is resistant to digestion by enzymes in the stomach and small intestine, and so reaches the large intestine essentially intact. This resistant starch is considered to have similar physiological effects and health benefits as fiber: It provides bulk, offers protection against colon cancer, improves glucose tolerance and insulin sensitivity, lowers plasma cholesterol and triglyceride concentrations, increases satiety, and possibly even reduces fat storage (Raben *et al.*, 1994; Englyst *et al.*, 1992; Cummings *et al.*, 1996; Hylla *et al.*, 1998; Pawanexh, 2009). The amount of resistant starch in potatoes depends much on preparation methods. Cooking and then cooling potatoes significantly increases resistant starch. In cooked potato starch contains about 7% resistant starch, which increases to about 13% upon cooling (Englyst *et al*., 1992).

1.3 INTRODUCTION TO BACTERIAL WILT

Bacterial wilt is one of the most important diseases of solanaceous vegetables, *viz*., potato (Cappel, 1892), tomato and eggplant (Smith, 1896) and tobacco (Smith, 1908). Park and Fernando (1938) studied about the relative resistance of some tomato varieties to bacterial wilt (*Bacterium solanacearum* E.F. Smith). The causal organism was identifed first as a bacterium and described as *Bacillus solanacearum* by Smith in 1896 in the USA. Now its name has changed to *Ralstonia solanacearum* (Smith) Yabbuuchi *et al.* (1995), which is synonyms of *Pseudomonas solanacearum*; *Burkholderia solanacearum*. It was first reported in potato and its bacterial nature was described by Coleman (1909), known by various names, *viz*, Rangpur wilt (Hutchinson, 1913b), tomato, brinjal, chillies (Anonymous,1989), bacterial slime, potato brown rot, Granville wilt of tobacco, moko disease of banana, and southern bacterial wilt of tomato (Buddenhagen and Kelman,1964). Kelman (1953) has published its origin since 19th century in Asia, South America, USA and Australia. The disease was dessiminated in tropical, subtropical and warm-temperate zones (Kelman, 1953; Smith, 1920). Bacterial wilt was a serious problem in pepper in Florida around 1940, when peppers followed potatoes. Today, bacterial wilt on pepper is not a problem anywhere in the southeastern United States (Momol *et al.* 2008). The pathogen has been detected in geraniums (*Pelargonium* spp.) in Wisconsin, USA that causes severe wilting of many important plants (Smith *et al*., 1995) and commonly coexists with polyspecific nematode populations as reported by Deberdta *et al.* (1999). The disease was traced back to the import of geranium cuttings to North America and Europe from the highland tropics Hudelson (2005) where race 3 biovar 2 is endemic (Milling *et al.,* 2009, Tahat and Sijam, 2010). The soil-borne diseases of solanceous vegetables have been compiled (Singh and Sharma, 2012) and shown in Table 1.2. A number of different organisms, *viz*, fungi, bacteria, viruses, nematodes or parasitic plants pathogens cause plant infectious disease (Agrios, 2005). Among those diseases, bacteral diseases are, *viz*, bacterial canker, bacterial speck, bacterial spot, bacterial wilt, damping off, *Verticillium* wilt, are found in India and world. Among them the bacterial wilt disease of solanaceous vegetables is an important soil-borne disease. Moraes (1947) reported that a bacterial wilt of potato due to *Bacterium solanacearum* Ranchi (Jharkhand, Ghufran and Chakravarti, 1960), India (Kishore and Pushkarnath, 1963), Kenya (Muthoni *et al.*, 2012) and Northeastern India (Ram, 1989).

Table 1.2 Soil-borne diseases of solanaceous vegetables in the world (Singh and Sharma, 2014)

Name of Disease	*Causal Organism*	*Host*
A Fungal diseases		
Damping off	*Pythium aphanidermatum, P. arrhenomanes, P. debaryanum, P. myriotylum, P. ultimum, Phytophthora, Fusarium oxysporum, F. solani* f. sp. *minus, Phomopsis vexans, Colletotrichum capsici*	Tomato, *Capsicum* spp., Eggplant
Rhizoctonia damping off and fruit rot	*Rhizoctonia solani* (Teleomorph: *Thanatephorus cucumeris*)	Tomato, *Capsicum* spp., Eggplant,

Contd...

Contd...

Fusarium crown and root rot	*Fusarium oxysporum* f. sp. *radicis- lycopersici*	Tomato
Fusarium wilt	*F. oxysporum* f. sp. *lycopersici, F. oxysporum* f. sp. *capasici*	Tomato, *Capsicum* sp. and *Lycopersicum pimpinellifolium.*
Verticillium wilt	*Verticillium albo-atrum, V. dahliae*	Tomato, pepper and eggplant
White mould/ Sclerotinia blight	*Sclerotinia sclerotiorum, S. minor*	Tomato, *Capsicum* sp., eggplant, pea, cauliflower, beans
Charcoal rot	*Macrophomina phaseolina*	Tomato, *Capsicum* sp.
Corky root rot	*Pyrennochaeta lycopersici*	Tomato
Didymelia stem rot	*Didymelia lycopersici*	Tomato
Black root rot	*Thielaviopsis basicola*	Tomato
Phoma rot	*Phoma destructiva*	Tomato
Buckeye fruit and root rot	*Phytophthora capsici, P. dreshsleri, P. nicotianae* var. *parasitica, P. Parasitica*	Tomato, eggplant, *Capsicum* sp.
B Bacterial diseases		
Bacterial canker	*Clavibacter michiganensis* subsp. *michiganensis*	Tomato, *Capsicum* spp.
Bacterial speck	*Pseudomonas syringae* pv. *tomato*	Tomato
Bacterial spot	*Xanthomonas vesicatoria*	Tomato, *Capsicum* spp.
Bacterial stem rot and fruit rot	*Erwinia carotovora* subsp. *carotovora,* E. *carotovora* subsp. *atroseptica,* E. *chrysanthemi, Pseudomonas* spp.	Tomato, *Capsicum* spp.
Black leg and Bacterial soft rot	*Pectobacterium carotovora* sub sp. *atrospectrum*	Potato
Bacterial wilt	*Ralstonia solanacearum*	Tomato, eggplant, *Capsicum* spp., Potato
Pith necrosis	*Pseudomonas corrugata*	Tomato
Syringae leaf spot	*Pseudomonas syringae* pv. *syringae*	Tomato, *Capsicum* spp.
Ring not	*Clavibacter michiganesis* subsp. *sepodonicus*	Potato

1.4 PATHOGEN

Ralstonia solanacearum is considered a heterogeneous species or a "species-complex" (Fegan and Prior, 2005), supporting the concept of "species-group" already proposed in 1964 (Buddenhagen and Kelman, 1964), following the hypothesis that strains of this species "are the product of long evolution occurring independently in various areas on different hosts" (Buddenhagen and Kelman, 1964). *R. solanacearum* belongs to the β-proteobacteria and is considered a "species complex" Alvarez *et al.* (2010a). *R. solanacearum* is a "species complex" (*R. syzygii* blood disease bacterium (BDB) termed as *R. solanacearum* species complex" "a cluster of closely related isolates whose individual members may represent more than one species" provided its high heterogeneity (Smith, 1896; Yabuuchi *et al.*, 1995, 1996; Fegan and Prior, 2005), and possesses several pathogenicity determinants controlled by a density dependent regulatory network (Schell, 2000a). It enters the plant via the roots, moves through the xylem and causes lethal disease

(Kelman, 1953). Afterwards, the bacterium can survive in the environment, mainly in reservoir plants, soil and water. *R. solanacearum* has not always been the name of this species. Smith was the first to publish a scientific description and classify it in the genus *Bacillus* as *B. solanacearum* in 1896 (Smith, 1896; Smith, 1908; Kelman, 1953). However, afterwards it was moved to the genus *Bacterium*, and to the genus *Pseudomonas* with the name of *P. solanacearum* (Kelman, 1953), temporary reclassified in the genera *Phytomonas* and *Xanthomonas* and eventually transferred back to the genus *Pseudomonas* in 1948 (Kelman, 1953). In 1992, it was placed in the genus *Burkholderia* (Burkholder, 1948; Yabuuchi *et al*., 1973) but, more recent phylogenetic and polyphasic phenotypic analyses pointed out that it would rather be accommodated in the new established genus of *Ralstonia*, (Yabuuchi *et al*., 1995). Since then, the bacterium is named *Ralstonia solanacearum* and belongs to the family Ralstoniaceae included in the β-subdivision of the Proteobacteria (Stackebrandt *et al*., 1988). Peeters *et al.* (2013) reported that *Ralstonia solanacearum* is a widespread bacterial plant pathogen in the post-genomic era.

1.4.1 Characteristics

The single cell of the pathogen is a small rod shaped with rounded ends (Smith, 1896; Kelman, 1953) with an average size of 0.5 to 0.7 by 1.5 to 2.5 μm. Cell wall structure is that of Gram-negative bacteria (Fig. 1.3a,b,c), and flagella when present are polar (Fig. 3d). The bacterium has an oxidative metabolism and is generally considered a strict aerobe. However, under some circumstances, it is able to limited, slow growth when cells are not in direct contact with the air (Kelman, 1953; Kelman and Jensen, 1951). *R. solanacearum* produces poly-β-hydroxybutyrategranules as cell energetic reserve. Its physiological and biochemical characteristics have been extensively reviewed (Kelman, 1953; Anonymous. 1998; Denny and Hayward, 2001; EPPO, 2004; Anonymous. 2006). With respect to growth temperature, *R. solanacearum* strains from tropical areas all over the world have a high temperature optimum (35°C), whereas that of strains occurring at higher altitudes in the tropics and in subtropical and temperate areas is lower (27°C); no growth has been observed at 40°C or 4°C (EPPO. 2004). Approximative minimal and maximal growth temperature values would be 8-10°C and 37-39°C respectively (Kelman, 1953). Regarding pH requirements, in general *R. solanacearum* growth is inhibited in acid media but favoured in alkaline conditions (Kelman, 1953). *R. solanacearum* can grow in 1% NaCl liquid media but, little or none in 2% NaCl (EPPO. 2004). Two morphological types of *R. solanacearum* colonies can be typically observed on agar plates, fluidal or mucoid and afluidal or non-mucoid (*Photo Courtsey of P. Champoiseau, University of* Florida) (Fig. 1.4a,b,c)

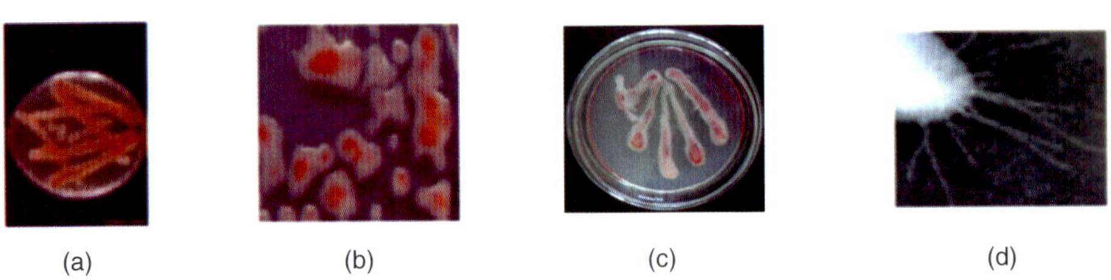

(a) (b) (c) (d)

Fig. 1.3 Morphology of *R. solanacearum*, (a) Appearance of *R. solanacearum*, (b) Virulent culture of *R. solanacearum* on TTC, (c) *R. solanacearum* on TTC in tomato, (d) Flagella of *R. solanacearum*

(Smith, 1920, Kelman, 1953, Denny and Hayward, 2001, EPPO. 2004); and other media (Fig. 1.4d, e, f) in tomato and brinjal.

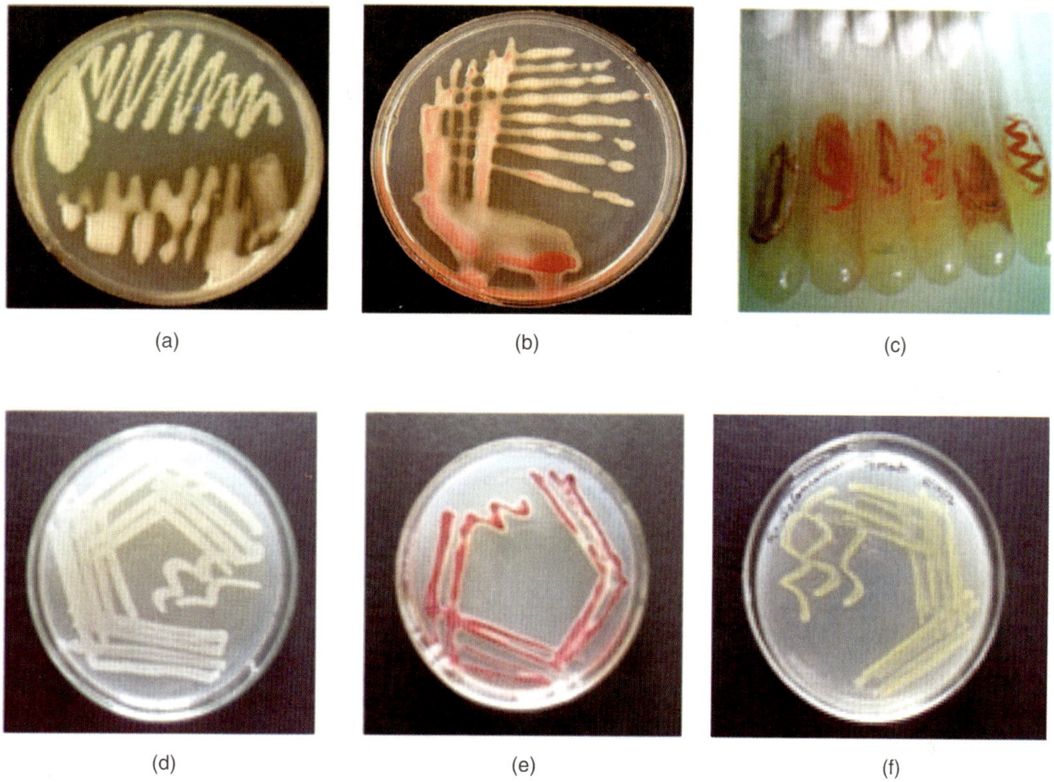

Fig. 1.4 Types of *R. solanacearum* colonies in different media, (a) Virulent (bottom) and non-virulent (top) colonies of *R. solanacearum* on growth medium, (b) Virulent colonies of *R. solanacearum* on TTC agar medium, (c) *R. solanacearum* colonies in tubes on TTC, (d) on Nutient Agar in brinjal, (e) on TTC in brinjal, (f) on Nutient Agar in Tomato

1.4.2 Colony Characters of *Ralstonia solanacearum*

On solid medium, colonies of *R. solanacearum* usually are visible after 2-5 days of incubation at 28°C. On the general nutrient media such as CPG, virulent isolates of *R. solanacearum* develop pearly cream-white, flat, irregular and fluidal colonies often with characteristic whorls in the centre. Avirulent forms of *R. solanacearum* form small round non-fluidal, butyrous colonies which are entirely cream-white. On Kelman's tetrazolium (TTC) and SMSA media, the whorls are blood red in colour. Avirulent forms of *R. solanacearum* form small round non-fluidal, butyrous colonies which are entirely deep red (Fig. 1.5).

The mucoid substance is produced by the accumulation of an exopolysaccharide, which causes these mucoid colonies to exhibit a typical irregularity of their surfaces (Smith, 1920),

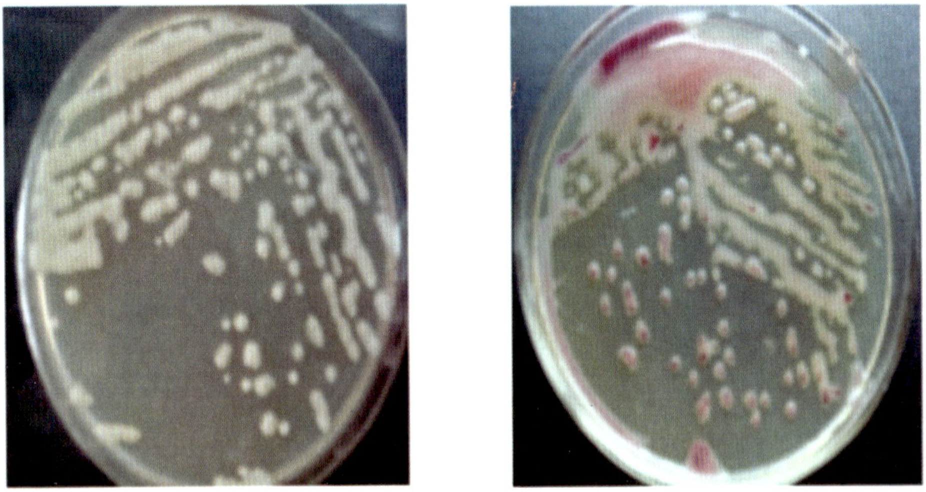

(a) (b)

Fig. 1.5 Colony characters of *solanacearum* on different media. (a) on CPG medium (b) on TTC medium

often with characteristic whorls in the centre. All *R. solanacearum* colonies are non-fluorescent, although a diffusible brown pigment can be produced on some media (Denny and Hayward, 2001). Under certain conditions, *R. solanacearum* colonies spontaneously undergo a change from fluidal to afluidal morphology, linked to a great reduction in disease-inducing capacity of these cells (Kelman, 1954; Buddenhagen and Kelman, 1964; Brumbley and Denny, 1990). This phenomenon is known as "Phenotypic Conversion" (PC) and occurs in most of the strains (Brumbley and Denny, 1990). PC-type variants can be easily observed by prolonged culture on agar plates (Kelman, 1954; Buddenhagen and Kelman, 1964), and when the organism is grown in a non-aerated liquid medium with glucose and an organic source of nitrogen (Kelman and Hruschka, 1973). Motility is transiently expressed during growth (Clough *et al*., 1997). The bacterium can produce one to several polar flagella (Smith, 1920; Kelman, 1953; Clough *et al*., 1997), which provide it with swimming motility to this motility. *R. solanacearum* cell density should be $>10^6$ and $<10^9$ cfu/ml, which is related to the culture age, so that high numbers of motile cells could be obtained in exponential phase, whereas in stationary phase there would be a majority of non-motile (Clough *et al*., 1997). The motility was needed to effectively invade and colonize the host, although cells were non-motile in host xylem vessels (Tans-Kersten *et al*., 2001). The cells from wilted plants were non-motile but, became motile after a few hours in fresh medium (Mao and He, 1998). *R. solanacearum* exhibits twitching motility over solid surfaces (Liu *et al*., 2001; Kang *et al*., 2002), and race 1 and 2 strains show generally more twitching activity than those of race 3 (Liu *et al*., 2001). This form of motility consists of a flagella-independent translocation requiring retractile type IV pili (fimbriae), which is present on *R. solanacearum* (Kang *et al*., 2002) being related to twitching motility (Liu *et al*., 2001), adherence to surfaces and natural transformation (Kang *et al*., 2002). *R. solanacearum* variants having mutations in type IV pili were markedly less virulent on host plants, and consequently motility, adherence and/or type IV pili appear to have a significant role in *R. solanacearum* pathogenesis (Liu *et al*., 2001; Kang *et al*., 2002).

1.4.3 Taxonomy

Olson (2005) has classified the *Ralstonia solanacearum* mentioned in Table 1.3. Taxonomic groups should follow the natural divisions that are apparent after characterizing the phenotype and genotype of related organisms and determining their relationship to the known texa. Until recently, however, the knowledge necessary to discern natural groups was often unavailable and many groups were incorrectly classified in phylogenetic terms. This was certainly true for organisms now considered members of the *Burkholderia* groups, which were long classified as non-fluorescent *Pseudomonas* species (Anzai *et al.* 2000; Palleroni, 1984). However, recent genetic analyses have revealed many new relationships and prompted renaming many bacteria in this and other groups (one Website with official nomenclature in http//www.bacterio.cict.fr.) have been summerised in Table 1.4 (Tahat and Sijam, 2010). De Baere *et al.* (2001) have classified *Ralstonia picketii.* biovar 3/*'thomasii'* strains (Pickett 1994) and of new isolates related to nosocomial recurrent meningitis as *Ralstonia mannitolytica* sp. nov.

Table 1.3 Classification of *Ralstonia solanacearum*

Domain	*Bacteria*
Phylum	Proteobacteria
Class	Betaproteobacteria
Order	Burkholderiales
Family	Ralstoniaceae
Genus	*Ralstonia*
Species	*solanacearum*
Synonyms	*Bacillus solanacearum* (Smith, 1896), *Pseudomonas solanacearum* (Smith) Smith (1914), *Burkholderia solanacearum*(Yabuuchi *et al.*, 1992)
Approved name	Bacterial wilt (pepper, potato, tomato), southern wilt (geranium), brown rot (potato)
Common name	Bacterial wilt of potato, bacterial wilt of solanaceous crops, brown rot of potato, Brown rot of solanaceous crops, southern bacterial.

Table 1.4 Taxonomic list of *Burkholderia* groups

Rank	*Name*	*Comments*
Class	B protobacteria	One of four classes of 'purplebacteria'
Family	*Burkholderia* group	Has nine genera
Genus	*Ralstonia*	Has five species
Species	*pickettii*	Human pathogen; common out group
	solanacearum complex	Plant pathogen, wide host range
	syzygii	Plant pathogen of cloves in indonesia
	Burkholderia	Has human and plant pathogens and symbioants
	Cupriavidus(Wautersia) metalllidurans	Many orthologous genes in *R. Solanacearum*
	necator	Originally *Alcaigenes eutrophus*
	taiwanensis	Nodulating, nitrogen-fixing Mimosa symbiont

Prior and Fegan (2005) reported that *solanacearum* strains are highly variable and adaptable as attested by worldwide distribution and a large and expanding host range. *R. solanacearum* (Rs) and its close relatives, the blood disease bacterium and *R. syzygii*, constitute a species complex, a diverse group of related isolates that represent more than one species. Biovar (Bv) typing and race assessment are methods commonly used for assessing the diversity of Rs strains. However, recent genetic evidence has indicated that these phenotypically-based schemes are not sufficient to encompass the diversity of strains represented in the species Rs. They presented a classification system based upon phylogenetic analysis of sequence data generated from the 16S-23S internal transcribed spacer (ITS) region, the endoglucanase gene and the *mutS* gene. From the phylogenetic analysis of the sequence data the Rs species complex can be subdivided into four monophyletic cluster of strains, termed phylotypes. Strains within each phylotype broadly originate from the same geographic location. Phylotype I includes all strains belonging to biovars 3, 4, and 5 and strains are isolated primarily from Asia. Phylotype II includes strains belonging to biovar 1, 2 and 2T isolated primarily from America. The *R. solanacearum* race 3 potato pathogen, which has a worldwide distribution and the race 2 banana pathogen are both members of phylotype II. Phylotype III contains strains primarily isolated from Africa and surrounding islands. Strains in this group belong to biovar 1 and 2T. Phylotype IV contains strains isolated primarily from Indonesia belonging to biovars 1, 2 and 2T. Each phylotype is composed of a number of groups of strains with a highly conserved sequence termed sequevars. Some sequevars contain strains which are pathogenic on the same hosts or strains of common geographic origin. The phylotyping scheme is highly discriminatory, flexible, additive, and should allow better prediction of the properties of strains.

Wicker *et al.* (2007) investigated a destructive pathogenic variant of the plant pathogen *R. solanacearum* that was consistently isolated in Martinique (French West Indies). Since the 1960s, bacterial wilt of solanaceous crops in Martinique has been caused primarily by strains of *R. solanacearum* that belong to either phylotype I or phylotype II. Since 1999, anthurium shade houses have been dramatically affected by uncharacterized phylotype II strains that also affected a wide range of species, such as *Heliconia caribea*, cucurbitaceous crops, and weeds. From 1989 to 2003, a total of 224 *R. solanacearum* isolates were collected and compared to 6 strains isolated in Martinique in the 1980s. The genetic diversity and phylogenetic position of selected strains from Martinique were assessed (multiplex PCRs, *mutS* and *egl*DNA sequence analysis) and compared to the genetic diversity and phylogenetic position of 32 reference strains covering the known diversity within the *R. solanacearum* species complex. Twenty-four representative isolates were tested for pathogenicity to *Musa* species (banana) and tomato, eggplant, and sweet pepper. Based upon both PCR and sequence analysis, 119 Martinique isolates from anthurium, members of the family Cucurbitaceae, *H. caribea*, and tomato, were determined to belong to a group termed phylotype II/sequevar 4 (II/4). While these strains cluster with the Moko disease-causing strains, they were not pathogenic to banana (NPB). The strains belonging to phylotype II/4NPB were highly pathogenic to tomato, eggplant, and pepper, were able to wilt the resistant tomato variety Hawaii 7996, and may latently infect cooking banana. Phylotype II/4NPB constitutes a new pathogenic variant of *R. solanacearum* that has recently appeared in Martinique and may be latently prevalent throughout Caribbean and Central/South America.

1.4.4 Biovar

Ralstonia solanacearum strains have been traditionally classified into biovars, on the basis of the pathogen ability to utilize and/or oxidize several hexose alcohols and disaccharides (Buddenhagen and Kelman, 1964; Schaad, 1988; Hayward, 1991). Various works on race and biovars are reported by various worker (French, 1985; CABI, 2003; James *et al.*, 2003, 2006; Sagar *et al.,* 2014). Five races and biovars have been reported and described so far (Buddenhagen and Kelman, 1964; Hayward, 1964 & 1991; Denny and Hayward, 2001; Daughtrey, 2003a,b; EPPO, 2004, Denny, 2006a,b). Africa is the continent where the highest diversity can be found (Elphinstone, 2005, 2006). Those races and biovars are summarised in Table 1.5. Biovars are defined by the ability of strains to oxidize three sugar alcohols and three disaccharides (Hayward, 1964). Recently more phylogenetically meaningful system has classified *R. solanacearum* into four major genetic groups called phylotypes that reflect the geographical origin and ancestral relationships between strains (Fegan and Prior, 2005) and shown in Table 1.5. Race 1 groups of the pathogen are established in the southern United States, where they cause bacterial wilt of tomato and other crops, such as pepper, tobacco, and rarely, potato. This race has not spread throughout the United States, in large part due to the seed potato system being isolated from the production system and strains of race 1 not being able to overwinter in northern environments and/or soils with low water holding capacity. In contrast, R3b2 can survive in temperate climates including highland tropics, the UK, and the Netherlands (Elphinstone, 2005), usually in plant tissue (Milling *et al.*, 2009). Race 3 is highly pathogenic on potato and tomato (primary hosts) and can also infect and eventually kill other solanaceous plants (e.g., eggplant, nightshade weeds) as well as geraniums. The host range of this group is not as wide as race 1. It is a particular concern to the potato industry of the United States and could extend the geographic range of *R. solanacearum* in the United States for other crops (tomato). The 8X draft sequence of UW551 (R3b2 strain) genome was published and led to the identification of R3b2-specific genes (Gabriel *et al.*, 2006). Recently taxonomic studies done by Cook *et al.* (1989a,b). By using restriction fragment length polymorphism (RFLP) analysis and sequence analysis of various regions of the genome, race 3, biovar 2 strains belong to RFLP Division II (where strains from America are grouped) and Phylotype II (mainly sequevar 1, Prior and Fegan, 2005a,b) (Table 1.6). Race 3 biovar 2 is the dominant race in mountainous regions in S. America (Andes), where the potato and possibly this race of *R. solanacearum* originates at higher altitudes in other tropical areas and also in the Mediterranean basin, including Egypt and more recently even in NW Europe. Both classification schemes constitute informal groupings at

Table 1.5 Distribution of races and biovars of *R. solanacearum*

Race	*Biovar*	*Host*	*Geographical Distribution*
1	1, 3 or 4	Wide	Asia, Australia, AmericasEurope except Europeon Union (EU)
2	1	*Musa* spp. and *Heliconia* spp.	Caribbean Islands, Brazil, Philippines
3	2	Potato, other Solanaceae, *Geranium*, others	Worldwide except United States and Canada
4	3 or 4	Ginger and related plant species Mioga and Platemma	Asia, widespread
5	5	Mulberry	China

the intra-subspecific level (Hayward, 1991). The relationship between them is only evident with race 3 strains, which generally correlate with biovar 2 phenotype (Buddenhagen and Kelman, 1964; Hayward, 1991). The diversity with *R. solanacearum* species complex shown by Alvarez *et al.*, 2010a (Fig. 1.5). Ramesh *et al.* (2014) reported genome sequences of *R. solanacearum* Indian strains Rs-09-161 and Rs-10-244 were isolated from the coastal region of Goa and from the Andaman and Nicobar Islands.

Table 1.6 Classification of strain based on molecular technique using restriction fragment length polymorphism (RFLP) analysis and sequence analysis of various regions of the genome (Fegan and Prior, 2005a)

Genotype Groups	*Strain/biovars*	*Geographical Origin*
Phylotype I	All Strains of biovar 3, 4 & 5	Asia
Phylotype II	All Race 3 of potato 1, 2 & 2T (Sub group of biovar 2)	America
	Race 2 Banana biovar 1, 2 & 2T (Sub group of biovar 2)	
Phylotype III	Biovar 1 & 2T	Africa, Surrounding Island
Phylotype IV	More heterogenous with Biovar 1, 2 and 2T	Indonesia, Australia, Japan

Wicker *et al.* (2012) also used coalescent genealogy reconstruction to deduce the order in which phylotypes likely evolved, and along with known ecotypes they described eight clades superimposed on the phylotypes (Table 2). When combined with ancestral state reconstruction, these results suggested that the pathogen originated in the Oceanina/Indonesia region, migrated to Africa and from there to South America and to Asia. Analysis of the genomic DNA sequences of six *R. solanacearum* strain, and one each of *R. syzygii* and BDM strains, representing all the phylotypes provided additional insight into the phylogeny of the *R. solanacearum* species complex (Salanoubat *et al.*, 2002; Guidot *et al.*, 2009b; Remenant *et al.*, 2010). Most significantly, based on pairwise comparison of average nucleotide identity, Remanant *et al.* (2011) concluded that the *R. solanacearum* species complex has three groups of strain that exceed the accepted threshold for speciation. Considering these and the extensive genetic results noted above, they proposed two new species and three subspecies (Table 1.7). This conclusion is supported by the limited DNA-DNA hybridization data (Palleroni and Doudoroff, 1971; Vaneechoutte *et al.* 2004) that indicated the RSSC is polyphletic. Unfortunately, the sequence analyses provided no obvious explanations regarding the molecular basis of host preference or ecological niche adaption. Furthermore, two extensive tests examining the pathogenic potential of diverse *R. solanacearum* starins on four solanaceous hosts revealed no good correlation with phylotype or sequevars (Cellier and Prior, 2010); Lebeau *et al.*, 2011; Genin and Denny, 2012).

Genetic diversity of *R. solanacearum* Singh *et al.* (2014a) reported biovars 3 and 4 strains obtained from northern parts of India, which causes a serious wilt disease of solanaceous crops using different techniques. The biovars could not be differentiated by colony, morphological character and pathogenicity test and *hrp* gene sequence analysis, as both the biovars are able to cause wilt disease in solanaceous crop. Ten isolates of *R. solanacearum* were classified into two clusters at 20 per cent similarity coefficient and cluster 1 represented all isolates of biovar 3 (UTT-23, UTT-10, UTT-26, HPT-11, UTT-9, UTT-32, HPC-3) and cluster 2 comprised the biovar 4 (UTT-22, HPT-3, UTT-24). BOX-PCR fingerprint of *R. solancearum* clearly distinguished biovars 3 and 4 grouped into two distinct clusters at 40% similarity coefficient. Cluster 1 represented all isolates of *R. solanacearum* biovar 3 and cluster 2 comprised the biovar 4 isolates. The isolates of *R. solanacearum* have genetic diversity in *hrp*B gene, but it could not differentiate the isolates of biovars 3 and 4. However, the biovars 3 and 4 of *R. solanacearum* can be genetically distinguished by using BOX-PCR and specific primer of RAPD. However, the biovars 3 and 4 of *R. solanacearum* can be genetically distinguished by using BOX-PCR and RAPD.

Table 1.7 Genetic, geographic, and ecotype diversity within the *Ralstonia solanacearum* species complex

Phylotype[a]	*Origin*	*Clade*[b]	*Ecotype information*[c]	*Proposed species*[d]
I	Asia	1	Bacterial wilt of solanaceae, ginger, mulberry	*Ralstonia sequeirae*
IIA	Americas	2	Bacterial wilt of solanaceae, *Musa*	*Ralstonia solanacearum*
IIA[T]		3	Bacterial wilt of solanaceae (south eastern United Sates and Caribbean)	
IIB	Americas	4	Moko disese of *Musa*, new NPB variants	*Ralstonia solanacearum*
IIB		5	Brown rot of potato, bacterial wilt of tomato and geranium (R3bv2), Moko disease; distributed worldwide	
III	Africa	6	Bacterial wilt of solanaceae	*Ralstonia sequeirae*
IV	Indonesia	7	BDB of banana	*Ralstonia haywardii* subsp. *celebenis*
			Bacterial wilt of solanaceae	*Ralstonia haywardii* subsp. *solanacearum*
IV		8	*R. syzygii*, Sumatra disease of clove	*Ralstonia haywardii* subsp. *syzygii*

[a]T indicates that the *R. solanacearum* type strain K60 is in phylotype IIA.

[b]Clades were defined by Wicker *et al.* (2012).

[c]BW, bacterial wilt; R3bv2, previously classified as Race 3 biovar 2; BDB, blood disease bacterium.

NPB, not pathogenic to banana: designates the new pathogenic variants found on Martinique that willt *Anthurum*, *Heliconia*, and cucurbits but not cultivated *Musa*[d]. Although genetically similar, *R. syzygii* is phenotypically quite different from *R. solanacearum* or other proposed *Ralstonia* species and subspecies (Ramenant *et al.*, 2011).

Stefani *et al.* (2005) have developed a protocol to detect latent infections of *R. solanacearum* biovar 2, race 3 in field tomatoes. For two growing seasons, healthy tomato plants were inoculated with a virulent strain, and monitored for between 7 and 55 days. Final extracts used for direct isolation on Kelman's medium and for indirect immunofluorescence staining were prepared from 1 cm segments collected from the base of the lowest sideshoots. Reisolations were identified by colony morphology, PCR, IFAS, and pathogenicity tests on tomato direct isolation . Reisolation was successful from 18 days onwards, with a frequency that constantly increased in the following weeks. Samples prepared by mixing one latently infected segment with increasing numbers of healthy segments revealed a sensitivity threshold of 1:999. This non-destructive protocol was shown to be appropriate for monitoring in the open field. Shekhawat *et al.* (1978) have studied about the distribution of bacterial wilt and races and biotypes of the pathogenin India whereas Chandrashekhara *et al.* (2012) reported that bacterial wilt caused by

R. solanacearum is the most destructive disease of plants. Fifty-seven isolates of *R. solanacearum* causing wilt on different host plants, *viz*., tomato (*Solanum lycopersicum*), brinjal (*S. melongena*), potato (*S. tuberosum*), bird of paradise (*Strelitzia reginae*), ginger (*Zingiber officinale*), chilli (*Capsicum annuum*), capsicum (*C. annuum*), davana (*Artemisia pallens*) and coleus (*Coleus forskohlii*) collected from the different agroclimatic zones of Karnataka and other parts of India. They have isolated 57 isolates and differentiated into race on the basis of their pathogenicity and their ability to infect different host. The isolates were established as race 1. None of the isolates infected mulberry and banana. Fifty-four isolates oxidized and utilized both the disaccharides and sugar alcohols. These isolates were positioned as biovar 3 according to Haywards classification system. Three isolates from Kerala, two ginger, and one tomato strain were not able to utilize dulcitol and lactose. Hence, they were categorized into a new taxo group within the system and designated as biovar-3B for the first time in India. There were 54 isolates which were confirmed as race 1, biovar 3, and 3 isolates were confirmed as race 1, biovar 3B by morphological, physiological, biochemical and pathogenicity studies.

1.4.5 Geographical Distributions

Bacterial wilt caused by *R. solanacearum* is a soil-borne bacterial disease of solanaceous vegetables, found in countries within latitude of 45°N and S, over-wintering in Sweden at 59° and in Peru at an altitude high as 328 m as early as twentieth century, it was recognized in the Southeastern United States, Java, Australia, India and Indonesia, it has since been reported from Portugal, Italy, Central America, the Caribbean Islands, some Pacific Islands, several countries in South Africa, South America and Srilanka, and now spreading to other tropical, sub-tropical as well as warm temperate countries. Yadessa *et al*. (2010) reported that *R. solanacearum* race 3 biovar 2 (phylotype II) is the most destructive bacterial disease of tomato in Ethiopia. The distribution of disease has been shown in Fig. 1.6. In India, the disease has drawn attention of scientists of various research organizations under Indian Council of Agricultural Research, State Agricultural University and the Research and Development Units of private sector. The bacterial wilt resistant breeding lines of vegetables emanating from various Agricultual University and ICAR Instutions are tested in wilt-prone area of the country under Indian Institute of Vegetable Research at Varanasi, in which an extensive and multilocational network has been set up for coordinated resistance breeding and presently the national grid is recently operating at 8 centres, 18 subcentres 13 volantary centres of the country (AICRP-VC Annual Report 2011-12) and varieties are released for cultivation on zonal or national basis by All India Coordinated Vegetable Improvement Project. The disease has been reported from several states of India, from both the plains and plateau region of the west coast from Trivendrum in Kerala to Khera in Gujarat, from the Deccan and central plateau of Karnataka, Western Maharastra and Madhya Pradesh, in West Bengal, Orissa and the Chotanagpur plateau area of Jharkhand, Bihar, the eastern plains of Assam and the North Eastern Hills region of India (Kelman, 1954; Rao *et al*., 1975; 1976; Bahal and Shekhawat, 1993), Bombay, (Patel *et al*., 1952), Jharkhand (Sharma *et al*., 2005), Mysore (Coleman, 1909), Madras (Ranjan, 1924), Assam (Sen, 1930), Uttar Pradesh (Dey, 1947), Bihar (Dutta and Verma, 1969), West Bengal (Hutchinson, 1913; Dass and Chattopadhyay, 1955; Chattopadhyay and Mukhopadhyay,1968), Punjab (Mitra, 1937), Madhya Pradesh (Anonymous, 1969) and Uttarakhand (Singh *et al*., 2010). The distribution of bacterial wilt in solanaceous vegetables has been shown in Tables 1.7 & 1.8 and Fig. 1.6.

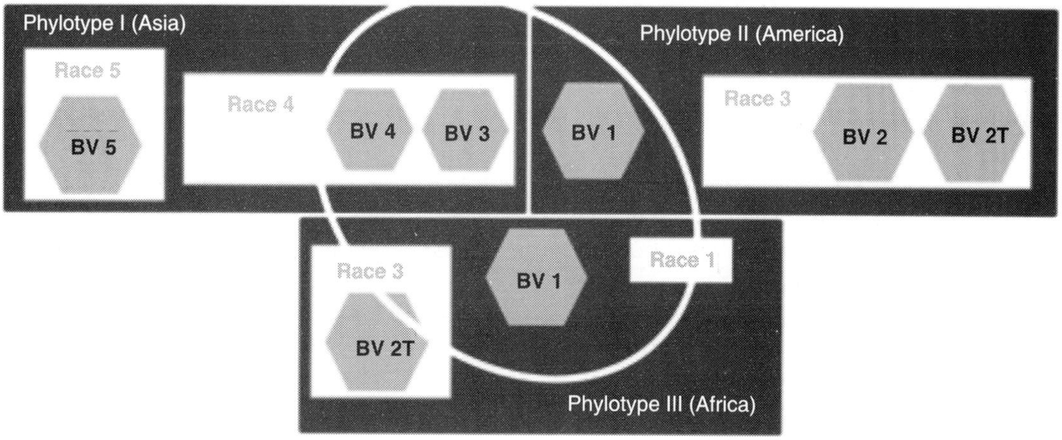

Fig. 1.6 Diversity within the *R. solanacearum* species complex. Correspondence among phylotypes, races and biovars of the pathogen. Phylotype IV is not represented because of its high heterogenicity

Table 1.8 Geographical distribution of bacterial wilt of solanaceous crops

Continent	*Country/State*	*References*
North America	Mexico	CABI, 2003
	Ontario (Canada)	Layme and Mckeen, 1967
	Florida	Cox, 1960; Sonoda, 1979; Kucharek, 1998 ; Pradhanang and Carlos, 2001
	Hawaii (USA)	Obrero, 1969
	North Carolina (USA)	Winstead and Kelman, 1952
	USA	Larson, 1944; Champoiseau and Momo, 2009b
	Wiscosin	Gallegly and Waker, 1949a,b; Hudelson, 2005
EPPO Region	Belgium	CABI, 2003
	France	CABI, 2003
	Germany	CABI, 2003
	Hungary	CABI, 2003
	Netherlands	CABI, 2003
	Lebanon	CABI, 2003
	Slovakia	CABI, 2003
	United Kingdom	CABI, 2003
	French Antilles	CABI, 2003
	Italy	Roubel, 1981, Mazzucchi *et al.* 2000
Asia	Bangladesh	CABI, 2003; Babu and Kelaniyangoda, 1993
	China	CABI, 2003
	Fujian	CABI, 2003
	Guangdong Guangxi	CABI, 2003

Contd...

Contd...

	Hebei,	CABI, 2003
	Israel	Volcam and Palti, 1960
	Jiangsu	CABI, 2003
	Taiwan	CABI, 2003
	Zhejiang	CABI, 2003
	India	CABI, 2003
	Indonesia (Java)	CABI, 2003
	Iran	CABI, 2003
	Japan (Kyushu)	CABI, 2003
	Nepal	CABI, 2003
	Pakistan	CABI, 2003
	Philippines	CABI, 2003
	Sri Lanka	CABI, 2003
	Republic of Korea	NIAST, 2002
	Malaysia	Possier *et al.* 1999
	Thailand	Horita and Tsuchiya, 2001; Chandrarikul, 1957
	Vietnam	Tung, 1985
	Fiji	Parham, 1934
	Georgia	Dukes *et al.*, 1965
	Japan	Arwinyanto *et al.*, 1994a
	Jiangxi(China)	Yuan *et al.*, 1995
	Malayasia	Lum and Wang, 1976
	Philippines	Reinking, 1918; Zehr, 1970
	Singapore	Yik *et al.*, 1994
	Saudi Arabia	Abdalla *et al.*, 1999
	Taiwan	Suzuki *et al.*, 1964; Michael *et al.*, 1996
	Pakistan	Begum *et al.*, 2012
	Nepal	Pradhanang and Elphinstone, 1996; 2000
Africa	Burundi	CABI, 2003
	Egypt	CABI, 2003
	Ethiopia	CABI, 2003
	Kenya	Harris, 1976; CABI, 2003
	Libya	CABI, 2003
	Réunion	CABI, 2003
	Rwanda	CABI, 2003
	South Africa	Anderson, 1947; CABI, 2003
	Zaire	CABI, 2003

Contd...

Contd...

	Zambia	CABI, 2003
	Zimbabwe	CABI, 2003
	Nigeria	Erinle, 1976
	Sierra Leone	Goodfrey, 1973
	Sub-Saharan Africa.	Lemaga *et al.*, 2001
	Argentina	CABI, 2003
South America	Goias	CABI, 2003
	Parana	CABI, 2003
	Pernambuco	CABI, 2003
	Rio Grande do Sul	CABI, 2003
	Santa Catarina	CABI, 2003
	Sao Paulo	CABI, 2003
	Chile	CABI, 2003
	Colombia	CABI, 2003
	Peru	CABI, 2003
	Uruguay	CABI, 2003
	Bolivia	CABI, 2003
	Brazil	Noda and Machado, 1993, Sikirou *et al.* 2009
	Venezuela	Anonymous, 1960
Central America and Caribbean	Cuba	French and Sequiera, 1970; French, 1998
	Costa Rica	Meseses *et al.*, 1990; CABI, 2003
	Dominican Republic	CABI, 2003
	El Salvador	CABI, 2003
	Guadeloupe	CABI, 2003
	Honduras	CABI, 2003
	Haiti	CABI, 2003
	Nicaragua	CABI, 2003
	Panama	French, 1998
	Trinidad	Thorold, 1949
	West Indies	Velez-fortune *et al.*, 1977
Oceania	Australia	CABI, 2003
	Guada Loupe	Kaan *et al.*, 1971
	Papua New Guinea	Shaw, 1984
	Queensland	Aberdeen, 1946
	Mauritius Renuion	Julien, 1973; Girard *et al.*, 1989
	New South Wales	CABI, 2003

Contd...

Contd...

	South Australia	CABI, 2003
	Victoria	CABI, 2003
	Western Samoa	Anonymous, 1976

Table 1.9 Distribution of bacterial wilt disease of solanaceous vegetable in India

State/Union Territory	*Place*	*References*
Andaman	Port Blair	Ansari, 1990; Sharma *et al.*, 1998
Andhra Pradesh	Hyderabad, Vijayrai (Eluru)	Kalloo *et al.* 1998
Arunachal Pradesh	Basar	Vishwardhar and Hazarika, 1988
Assam	Jorhat	Shekhawat 1976, Bahal and Shekhawat, 1993; Nath *et al.*, 1996; Bora *et al.*, 1996
Bihar	Sabour	Sinha, 1985; Sharma and Kumar, 1997
Delhi	New Delhi	Sinha, 1985
Gujarat	Junagarh	Kalloo *et al.*, 1998
Goa	Goa	Ramesh *et al.*, 2009
Himachal Pradesh	Katrain, Palampur, Kangra, Mandi (Solan)	Kumar *et al.*, 1993b; Kalloo *et al.*, 1998; Kulha *et al.*, 1994; Gupta *et al.*, 1998 CABI, 2003
Jammu and Kashmir	Jammu, bhast.	Singh *et al.*, 2010
Jharkhand	Ranchi	Ghufran and Chakravarti, 1960; Chand and Mishra, 1995; Sinha 1985; Sharma and Kumar, 1995-96; Singh *et al.*, 2010
Karnataka	Bangalore	Ram Kishun, 1987; Gadewar *et al.*, 1991; Tikoo *et al.*, 1988
Kerala	Vallanikkara	Paily, 1964; Devi and Raj, 1966
Madhya Pradesh	Jabalpur, Raigarh, Raipur	Kalloo *et al.*, 1998; CABI, 2003
Maharashtra	Dapoli, Dharwad	CABI, 2003
Manipur	Imphal	Verma and Gupta, 1979; CABI, 2003
Meghalaya	Barapani	Sahambi and Maiti, 1977; CABI, 2003
Nagaland	Jharnapani	Kumar and Sharma, 1979
Orissa	Bhubaneswar.Padua	Erocolani *et al.*, 1961; Mishra *et al.*, 1995; Singh *et al.*, 2010
Punjab	Ludhiana	Patel, 1972
Rajasthan	Durgapura	Kalloo *et al.*, 1978
Sikkim	Tadung	Sheoraj and Srivastava, 1979
Tripura	Agartalla	Venkatraman and Sen Gupta, 1979; CABI, 2003
Tamil Nadu	Coimbatore	Patel, 1972; CABI, 2003
Uttar Pradesh	Faizabad, Kalyanpur, Varanasi	Kalloo *et al.*, 1998; CABI, 2003
Uttarakhand	Nanital	Singh *et al.*, 2010
West Bengal	Kalyani	Chatterjee *et al.*, 1994, 1997; CABI, 2003; Singh *et al.*, 2010

Begum *et al.* (2012) reported that distribution of bacterial wilt caused by *R. solanacearum* in different vegetable growing areas of Pakistan (during April-September 2008-09) revealed that sweet pepper was highly susceptible with overall incidence percentage of 21.9% followed by hot pepper (16.6%), tomato (13.3%), potato (10.5%) and brinjal (5.5%). Aggregate incidence in the whole country was 13.8%. Most of the isolates were obtained from soil (17.2%) and plant samples (18.8%). The disease was mainly distributed in Islamabad Capital Territory and Punjab with becterial wilt incidence of 19.2% and 13.9% respectively. Most of the farmers were not certain about bacterial wilt symptomology and management. Mean values in becterial wilt affected areas were 9.9 soil pH, 29.5°C, 71.59% RH and 71.86 mm rainfall. The tests further revealed that biovar 2 and 3 were more prevalent in Pakistan with 80 % majority of strains belonging to biovar 3. Predominant biovar in potato was biovar 2 (mostly in Northern territory of the country). Studies further revealed that rest of strains obtained from tomato, sweet pepper, hot pepper and brinjal were mainly biovar 3. In addition, some biovar 3 strains were also found to be associated with sweet pepper and hot pepper seeds. Singh (1993) reported occurrence of a new biotype of *P. solanacearum* from India.

1.4.5.1 Tomato

The bacterial wilt has been reported on other crops like potato, eggplant, tomato chilli, etc. (Khan *et al.* 1980). It is attacked by 200 diseases causing organisms of diverse etiology (Jones *et al.*, 1991). The disease has been reported (Devi and Menon, 1980; Singh *et al.*, 2010; Gupta *et al.,* 1998) on tomato and other crops. Several workers have dealt with this disease in depth from time to time (Kelman, 1953; Anon, 1954; Stapp, 1965; Golfrey, 1973; Ashrfuzzaman and Islam, 1975, Sequeria and Kelman, 1976; Jaworshki *et al.*, 1988; Peixoto, 1997). The production of the crop is low due to bacterial wilt incidence (Bora, 1995) and therefore, the disease is considered to be a serious yield limiting factor particularly for the early grown tomato (Roy *et al.*, 1992) when temperature is favourable for the infection and spread of the pathogen.

1.4.5.2 Brinjal

Eggplant cultivation is severly impaired by three important wilt causing pathogens, *viz.*, *Pseudomonas solanacearum*, *Fusarium oxysporum* and *Meloidogyne incognita*, the causal agent of bacterial wilt, *Fusarium* wilt and *nemic* wilt, respectively and caused considerable damage of eggplant (Timm and Ameen, 1960; Talukder, 1974; Ahmed and Hossain, 1985; Mian, 1986; Ali *et al.* 1993). These are also the major limiting factors for eggplant production throughout the world (Hinata, 1986). Among the above diseases, bacterial wilt is one of the most devastating diseases and severely hampers its cultivation. Thirteen different diseases so far were recorded in Bangladesh (Khan *et al.*, 1998; Das *et al.*, 2000; Rashid, 2000). Among those diseases wilt of eggplant has been reported as one of the major constraints in its cultivation in Bangladesh (Rasid, 1976; Ali, 1993).

1.4.5.3 Chilli

Bacterial wilt is a widespread and devastating disease of many crops, ornamentals, and weeds in the tropics and subtropics all over the world (Hayward, 1991). In survey of 55 countries by Yoon *et al.* (1989) of the intertropical belt reported that 34% were as a major problem due to

R. solanacearum incidence in pepper cultivation with high to very high incidence, ranking first among the pepper disease. Bacterial wilt is a common disease in the Caribbean (Prior and Steva, 1990; Prior *et al.*, 1990), Latin America (Quesado-Soares and Lopes, 1995), West Africa, and Southeast Asia (Yoon *et al.*, 1989), where various biovars and races of the bacterium pepper are confronted. *R. solanacearum* was reported in temperate climates of Northern and Western Europe (Van der Wolf *et al.*, 1998) and serious problem in pepper in Florida when peppers grown after potato. Today, bacterial wilt is not a problem anywhere in southeastern United States (Momol *et al.*, 2008).

1.4.5.4 Potato

Potato brown rot (bacterial wilt) disease has been reported as an economic disease in hot and temperate regions of the world. Bacterial wilt has spread to three potato producing countries, *viz*, Australia, Southeastern United States and Latin America. The disease has been reported from all potato producing countries except Ecuador. It has been reported throughout central and southern Africa and is a serious production constraint in Uganda, Ethiopia, Kenya, Madagascar, Rwanda, Burundi, Nigeria and Cameroon. Although it is not a serious disease in Egypt, latent infection of tubers has resulted in a strong decline of potato exports to Europe. In South Asia, bacterial wilt occurs in the mid-hills of the Himalayas in India, Pakistan, Nepal and Bhutan; mid-hills and plains of India. In East and Southeast Asia, bacterial wilt is an important disease in Indonesia, the Philippines, southern Vietnam, Laos, Japan and southern China. In the early 1990s, becterial wilt became a serious threat to potato production in European countries including Belgium, England, France, the Netherlands, Spain, Italy and Portugal. It has also been reported in Russia. Messiha (2009) reported from the Netherlands bacterial wilt is a quarantine disease in the EU, and export of potatoes from Egypt is restricted to pest-free areas in the desert and has been reported as an economic disease in hot and temperate regions of the world. Potato brown rot, caused by race 3 biovar 2 (Phylovar II, sequevar 1) of *R. solanacearum*, is a serious endemic disease in the Nile Delta of Egypt is a quarantine disease with zero tolerance requirements hence quarantine measures have been followed (Commonision, 1998).

2
Diagnosis

Detection and identification of bacterial pathogen causing bacterial wilt is included in diagnosis. In diagnosis, it tries to prove that a bacterium isolated from a diseased plant, and identified, using different techniques, is really the cause of the disease observed. There are various steps involved in a diagnosis process, which are actually Koch's postulates. The steps are assessment of symptoms, isolation of pathogenic bacteria, growing pure culture of isolated bacteria, identification of pure culture, pathogenicity test, reisolation from inoculated plants, reidentification of reisolate and diagnosis report.

2.1 SYMTOMATOLOGY

2.1.1 Tomato

The wilting symptom of tomato can be induced by bacterial and fungal pathogen, root-knot nematode (Libman and Leach, 1962; Libman *et al.*, 1964; Nirmala Devi and Tikoo, 1992) and deficit or excess of soil moisture (Jone *et al.,* 1991). The most characteristics symptoms of tomato are very rapid witlings, especially where the plants are young and succulent. The flowering stage is the most critical stage where a plant shows sudden wilting (Gowda *et al.*, 1974; Sonada, 1975). Other primary indications of infection are stunting, downward curling of leaflets and petioles and, where pathogenesis is slow, excessive production of adventitious roots along the stem. Usually infected plants collapse quickly, but where this fails to happen there is a development of blackening of vascular system at the junction between stem and leaf. Further down the stem, the whole vascular system may be completely blackened and when cut oozes creamy bacterial slime. Buddenhagen and Kelman (1964) have summarized the syndrome of bacterial wilt in tomato. General symptoms of bacterial wilt have been shown in Fig. 2.1 and summarised in Table 2.1.

The first visible symptom is the flaccid appearance of the youngest leaves and in the field, this symptom does not happen until after flowering has begun (McCarter, 1991). Under favourable environmental conditions for the pathogen (soil temperatures of approximately 25°C; saturated humidity), epinasty and wilting of one side or of the whole plant follows within a few days leading to total plant collapse. Under less favourable conditions (soil temperature below 21°C), less wilting occurs, but large numbers of adventitious roots may develop on the stem. It is possible to observe water soaked streaks from the base of the stem which is evidence of necrosis in the vascular system. When the stem is cut crosswise, discoloured brown vascular tissues exude white or yellowish bacterial ooze.

The nature of barrier involved in limiting of bacterial spread in resistance cultivars can be seen by electron microscopy (Grimault *et al.,* 1994a,b,c) in the resistant cultivars, tyloses occluded the colonized vessels and the contiguous ones have limited bacterial spread. In the wilting of susceptible cultivars, no tyloses were observed in vessels and bacterial spread was not limited. Tyloses production in susceptible cultivars seemed delayed and less focused compared with resistant cultivars because numerous non-colonized vessels were found occluded by tyloses. Vascular colonization seemed generalized in the vascular bundles of susceptible cultivars as compared to resistant one. Other reactions involved in the resistance or susceptibility were observed, such as the formation of gum, cell wall breakdown and modification to the primary cell wall. The limitation of bacterial spread association with the resistance of tomato to bacterial wilt was mainly attributed to an induced, non-specific and physical barrier.

(a) Wilt of Tomato (b) Wilt of Brinjal (c) Wilt of Chilli (d) Wilt of Potato

Fig. 2.1 Symptoms of bacterial wilt in solanaceous crops under field conditions

At the early stages of disease, the first visible symptom is a subjective evidence of disease cr physical disturbance. It is an evident reaction by a plant to a pathogen, and is not necessarily visible. Different pathogens can induce similar symptoms of bacterial wilt. These are usually seen on the foliage of plants. These symptoms consist of wilting (Fig. 2.1a). At this stage, only one or half a leaflet may wilt, and plants may appear to recover at night, when the temperatures are cooler. As the disease develops under favorable conditions, the entire plant may wilt quickly and desiccate although dried leaves remain green, leading to general wilting and yellowing of foliage and eventually plant death. Another common symptom that can be associated with bacterial wilt in the field is stunting of plants. These symptoms may appear at any stage of plant growth, although in the field it is common for healthy-appearing plants to suddenly wilt when fruits are rapidly expanding. In well-established infections, cross-sections of stems may reveal brown discoloration of infected tissues. In young tomato stems, infected vascular, or conductive, bundles are responsible for long-distance transport of water and nutrients throughout the plant. Highly developed plants have two types of vascular tissues: the xylem and the phloem vascular bundles may become visible as long, narrow, dark brown streaks. In young, succulent plants of highly susceptible varieties, collapse of the stem may also be observed. A common sign of bacterial wilt of tomato observed at the surface of freshly-cut sections from severely infected stems is a sticky, milky-white exudates, which indicates the presence of dense masses of bacterial cells in infected vascular bundles, and particularly in the xylem, which is responsible for transportation of raw sap (water and nutrients) from roots to aerial parts of the plant.

2.1.2 Brinjal and Chilli

Sudden wilting of foliage of the youngest leaves, which may occur during night, is the initial symptom of the wilt disease (Fig. 2.1b, c). The disease progresses with permanent wilting of entire plant, with slight or no leaf yellowing. Infected young plants die immediately, while older plants may first show leaf drooping and discolouration only one side or part of the plant and eventually the whole plant wilts and dies. The vascular system in affected plants is discoloured from pale yellow to brown in colour. A simple diagnostic test is to place cross section of the infected tissue suspended in water that yields milky white exudates of the bacterial cells.

2.1.3 Potato

The early stage of infection in the field is recognized by wilting of the leaves towards the top of the plant at high temperatures during the day with recovery at night (Fig. 2.1d). In early stages of wilting leaves remain green, but later yellowing and brown necrosis develops. Epinasty also occurs in potato plants. Wilting of one shoot or whole plants becomes rapidly irreversible and results in the collapse and death of the plant. The vascular tissue of transversely cut stems from wilted plants usually appears brown and milky bacterial ooze exudes from the cut surface or can be expressed by squeezing. When a cut stem is placed vertically in water, threads of slime will stream from the vascular bundles. Potato tubers must be cut transversely close to the heel (stolon end) or longitudinally over the stolon end. Tubers may show bacterial ooze on the eyes and stolon end attachment. Soil sticks to the tubers at the eyes. The early stage of infection is recognized by a glassy yellow to light brown discolouration of the vascular ring from which a pale cream bacterial ooze emerges spontaneously after some minutes. Later, the vascular discolouration becomes a more distinct brown and necrosis can extend into the parenchymatous tissue. In advanced stages, infection breaks outwards from the heel end and the eyes from which bacterial slime may ooze causing soil particles to adhere. Reddish-brown slightly sunken lesions may appear on the skin due to collapse of vascular tissues internally. Symptoms may be confused with those of ring rot, caused by *Clavibacter michiganensis* subsp. *sepedonicus*. Wilting and other foliage symptoms may not occur under dry and cool conditions. Secondary development of fungal and bacterial soft rots is common in the advanced stages of the disease. Fucikovsky (1980) developed methods for diagnosing *P. solanacearum* of potato bacterial disease.

2.1.4 Other Hosts

Wilting symptoms are rarely observed in natural conditions on weed hosts *viz*; *Solanum dulcamara* and *S. nigrum* unless soil temperatures exceed 25°C or inoculum levels are extremely high (e.g. as for *S. nigrum* growing adjacent to diseased potato or tomato plants). When wilting does occur, the symptoms are as described for tomato. Non-wilting *S. dulcamara* plants growing with stems and roots in water may show internal light brown discolouration of vascular tissues on transverse section of the stem base or underwater stem parts. Bacteria may ooze from cut vascular tissues or form threads of slime if the cut stem is placed vertically in water, even in the absence of wilting symptoms.

Table 2.1 Wilt disease syndrome in tomato *(R. solanacearum.)* in relation to mechanisms of pathogenesis

Symptoms	*Cause*
External symptoms	
Foliage wilting	Interference with water movement by formation of extra cellular polysaccharides, bacterial cells and tyloses in the vessels and tracheids
Foliage yellowing	Breakdown of chlorophyll resulting from decreased supply of mineral nutrients and water, unknown effects to host and pathogen metabolites
Marginal leaf necrosis	Decreased supply of water and unknown factors
Leaf epinasty	Increased level of indole-acetic acid (IAA) and ethylene (formed by pathogen or host).
Adventitious roots	Increase in IAA, interference with downward movement in phloem resulting from effects of pathogen.
Stunting	Complex of above.
Internal Symptoms	
Vascular discolouration	Tyrosinase of pathogen
Tyloses Vessel Collapse Parenchyma proliferation	Increase in IAA levels.
Dissolution of pectic substances in middle lamella	Pectin methyl esterase, olygalacturonase
Degradation of cellulose on cell walls.	Cellulose.

Agrios (2005)

2.2 BACTERIAL OOZE

Ralstonia solanacearum is a limited xylem-invading pathogen and plants wilted by *R. solanacearum* have > 10^8 cfu/g of tissue. A common sign of bacterial wilt of tomato observed at the surface of freshly-cut sections from severely infected stems is a sticky, milky-white exudates, which indicates the presence of dense masses of bacterial cells in infected vascular bundles, and particularly in the xylem, which is responsible for transportation of raw sap (water and nutrients) from roots to aerial parts of the plant. Ooze also may accumulate on the cut surface (Fig. 2.2). Allen *et al.* (2001) reported that even if ooze does not form spontaneously a streaming test may be positive. Other wilt inducing pathogens do not produce comparable ooze. The ooze is usually an almost pure culture of *R. solanacearum,* which can be cultured on standard low ionic strength bacteriological media. This water streaming test (Ooze test) is of presumptive diagnostic value in the field.

2.3 ISOLATION OF *RALSTONIA SOLANACEARUM* FROM WILTED PLANTS

(a) Remove ooze or sections of discoloured tissue from the vascular ring in the potato tuber or from the vascular strands in stems of potato, tomato or other wilting host plants. Suspend in a small volume of sterile distilled water or 50 mM phosphate buffer and leave for 5 to 10 minutes.

(b) Prepare a series of decimal dilutions of the suspension.

Fig. 2.2 Bacterial ooze exudating out from the end

(c) Transfer 50-100 μl of the suspension and dilutions to a general medium (CPG) and/or to Kelman's tetrazolium medium and/or a validated selective medium (e.g. TTC, SMSA). Spread or streak with an appropriate dilution plating technique. If useful, prepare separate plates with a diluted cell suspension of *R. solanacearum* biovar 2 as a positive control.

(d) Incubate the plates for two to six days at 28°C.

2.4 MEDIA USED FOR ISOLATION

2.4.1 Non-selective Media

These media can also be used for isolation of *R. solanacearum* from fresh, symptomatic plants, due to the high density of the pathogen in the tissues. Non-selective media are generally used for growth of pure culture of *R. solanacearum especially in* retrieve cultures from frozen stocks or for successive plating of cells.

A. Casamino acid-Peptone-Glucose (CPG) medium (Kelman, 1954)

Casamino acid (casein hydrolysate)	1 g
Peptone	10 g
Glucose	5 g
Agar	17 g

Adjust pH to 6.5-7.0 if necessary. Autoclave at 121°C for 20 minutes.

On solid medium, colonies of *R. solanacearum* usually are visible after 48-72 hours of incubation at 28°C. Colonies of the normal (or virulent type) are white or cream-coloured, irregularly-round, fluidal, and opaque; and colonies of the mutant (or non-virulent) type are uniformly round, smaller, and butyrous (dry). This shift from virulent to non-virulent bacterial cells occurs during storage or under oxygen stress in liquid media.

B. Triphenyl tetrazolium chloride (TTC or TZC) medium (Kelman, 1954)

Prepare one litre of CPG medium as per details given above.

Adjust pH to 6.5-7.0 if necessary. Autoclave at 121°C for 20 minutes. After autoclaving cool the medium to 55°C and add 5 ml of a 1% stock solution of 2, 3, 5-tripheny tetrazolium chloride. The stock can be filter sterilized or autoclaved for 5 minutes at 121°C, and stored at 4°C or frozen.

2.4.2 Semi-selective Media

Semi-selective media can be used for isolation of the pathogen from non-symptomatic plants or from water and soil samples. Although widely used, they do not support growth of all *R. solanacearum* strains and/or do not suppress growth of all related or unrelated Gram-negative bacteria.

A. SM-1medium (Granada and Sequeira, 1983a)

Prepare 1 litre of TTC (or TZC) medium as per details given above.

Autoclave at 121°C for 20 minutes. After autoclaving cool the medium to 55°C and add 5 ml of a 1% stock solution of 2, 3, 5-tripheny tetrazolium chloride. The stock can be filter sterilized or autoclaved for 5 minutes at 121°C, and stored at 4°C or frozen. One litre of medium add Merthiolate tincture 5-50 μl. Also add Crystal violet 50 mg, Polymyxin β sulfate 100 mg, Tyrothricin 20 mg, Chloromycetin 5 mg, and Cycloheximide 50 mg.

[(Merthiolate tincture * contains 1 part Merthiolate per 1000 parts of 50% alcohol. Determine the best concentration to suppress local microflora, as suggested (Granada and Sequeira, 1981, 1983). Dissolve in 5 ml of 70% ethanol 30 minutes prior to use)].

B. Modified SMSA medium (Elphinstone *et al.*, 1996; French *et al.*, 1995) Prepare 1 litre of TTC (or TZC) medium, except substitute glycerol (5 ml per litre) for the glucose:

Casamino acid (casein hydrolysate)	1 g
Peptone	10 g
Glycerol	15 ml
Agar	17 g

Adjust pH to 6.5-7.0 if necessary. Autoclave at 121°C for 20 minutes. After autoclaving cool the medium to 55°C and add 5 ml of a 1% stock solution of 2, 3, 5-tripheny tetrazolium chloride. The stock can be filter sterilized or autoclaved for 5 minutes at 121°C, and stored at 4°C or frozen. One litre of medium add Crystal violet 5 mg, Polymyxin β sulfate 100 mg, Bacitracin 25 mg, Chloromycetin 5 mg, Penicillin 0.5 mg, When inhibition of fungal contaminants is desirable, Cycloheximide 100 mg may be added into the medium. Dissolve in 5 ml of 70% ethanol 30 minutes prior to use. The following chemicals are added in one litre of medium as Crystal violet *5 mg, Polymyxin β sulfate *100 mg, Bacitracin *25 mg, Chloromycetin *5 mg, Penicillin * 0.5 mg, When inhibition of fungal contaminants is desirable, Cycloheximide *100 mg may be added into the medium. * Dissolve in 5 ml of 70% ethanol 30 minutes prior to use.

Immunofluorescence colony staining (IFC), which combines bacterial colony-growth and serological methods, allows sensitive and quantitative detection of several plant pathogenic bacteria in complex environments with high microbial backgrounds like potato peel extracts (Van der Wolf *et al.*, 1998). Van der Wolf *et al.* (2000) reported that IFC was a sensitive method for quantitative detection of *R. solanacearum* bv 2 in soil. Elphinstone *et al.* (1996) found consistent positive results with IFC for the detection of *R. solanacearum* in potato tuber extracts only when 2×10^4 to 2.5×10^6 cfu/ ml were present. Imazaki and Nakaho (2010) used pyruvate amended modified SMSA medium to revive the viable but nonculturable (VBNC), which degrades hydrogen peroxide or serve an antitoxidant function in semiselective medium.

Sodium pyruvate @ 5 g/l most improved the sensitivity of *R. solanacearum* detection.

2.5 MORPHOLOGICAL, BIOCHEMICAL AND PHYSIOLOGICAL CHARACTERISTICS OF *R. SOLANACEARUM*

2.5.1 Potassium Hydroxide (KOH) Test

It is a preliminary test of Gram-staining to differentiate Gram-negative and Gram-positive bacteria within a short period. The destruction of the cell wall of Gram-negative bacteria and subsequent liberation of DNA, which is very viscid in water and produces the string of slime. Gram-positive wall is more resistant to KOH and it remains intact, and DNA is not released. 1-2 drops of 3% KOH is kept onto a clean microscopic glass slide. 24 h old culture of *R. solanacearum* from slant with help of a cooled sterile loop is mixed with KOH until an even suspension and lifts the loop from the slide. The *R. solanacearum* being Gram-negative bacteria will become gummy upon mixing with a loop and a string of slime is lifted.

2.5.2 Gram Staining

Gram reaction is essential for primary division of the bacteria. The Gram-staining differentiates Gram-negative and Gram-possitive bacteria on the basis of chemical composition of cell wall of the bacteria. Gram-staining involves the use of primary stain (crystal violet), the trapping reagent (iodine solution), decolouring agent (95% alcohol) and counter stain (safranin). Most plant pathogenic bacteria are the Gram-positive cells stain purple and the Gram-negative cells red.

Procedure:

- On a clean slide, dry a thin spread bacterial film in air, without heat. Then lightly flame the underside of the slide twice to fix the bacteria to the slide.
- Flood the smear with crystal violet solution for 1 min.
- Wash in tap water a few seconds. Drain off excess water, and lightly blot dry on a paper towel.
- Flood the smear with iodine solution for a minute.
- Wash in tap water a few seconds; blot dry.
- Decolourized with solvent, e.g., ethyl alcohol, until the solvent flows colourlessly from the slide (about 30 seconds) blot dry. (If decolourizer is used longer, the Gram-positive bacteria may lose colour).
- Rinse in tap water for about 2 seconds.
- Counter stain for about 10 seconds with safranin solution.
- Wash briefly in tap water. Blot dry and examine.

Ralstonia solancearum is stained red under microscope with 100 × objective lens and oil emulsion.

2.5.3 Arginine Hydrolase Test

This test is important for distinguishing *Pseudomonads*.

Medium

Peptone	1.0 g
NaCl	5.0 g
K_2HPO_4	0.3 g
Agar	3.0 g
Phenol red	0.01 g
L-arginine	1.0 g
Distilled water	1000 ml
pH	7.2

Dispense the medium in 5 ml quantities in tubes and autoclave. Stab inoculate the medium with 48 hr growth of the bacterium and cover the medium with sterile liquid paraffin to a depth of 1 cm. Incubate the tubes for 7 days at 25 – 30°C and observe daily. A change of the medium to red indicates the arginine – hydrolase activity. *Ralstonia solnacerum* shows negative arginine hydrolase test.

2.5.4 Nitrate Reduction

Medium

Peptone	10.0 g
Beef extracts	5.0 g
KNO_3 (nitrite free)	3.0 g
Distilled water	1000 ml

Dispense the medium about 10 ml into tubes, autoclave and cool. Then inoculate with the strain in question and plug each tube with 3% noble agar. Grow at 27°C up to 5 days is recorded as a positive test for denitrification. Add a few drops of sulphanilic acid (0.8% in 5N acetic acid and dimethyl-alpha-naphthylamine (0.5% in 5N acetic acid) to the nitrate broth culture. Nitrite is present if the mixture becomes distinct pink or red. No colour would mean that nitrate is present as such or has been reduced to ammonia and free nitrogen. To confirm either of these two possibilities add few zinc crystals to the above broth-reagent mixture and shake for a few minutes. Nitrates are present without reduction if the broth becomes pink or red. No colour in either of the above two tests would mean that nitrate is reduced to ammonia or free nitrogen. However, *Ralstonia* species cannot reduce nitrogen.

2.5.5 Oxidase Test (Fahy and Persley, 1983)

The oxidases are enzymes catalyzing the transfer of hydrogen directly to molecular oxygen, resulting in the formation of water.

$$2H_2 + O_2 = 2H_2O$$

For oxidase test, streak a 24-48 h slant growth of the test bacterium on a filter paper saturated with 1 % tetramethyl-para-phenylene-diamino-dihydrochloride. The reaction is positive if a red or purple colour appears within 10 seconds. The reaction is positive if the colour appears in 10-60 seconds (Kovacs 1956).

2.5.6 Fluorescent Pigments Production (Schaad *et al.,* 2001)

King *et al.*'s B medium (KB)

Proteose peptone	20.0 g
K_2HPO_4	1.5 g
$MgSO_4.7H_2O$	1.5 g
Glycerol	15 ml
Agar	15.0 g
Distilled water	1000 ml

The KB medium is used for detection of fluoresce in a fluorescent green or blue water soluble, chloroform insoluble Petri dish pigment (*i.e.*, pyoverdine siderophores). After 24-48 h growth at 27°C, colonies are examined for fluorescence with a long wavelength (366 nm) ultraviolet lamp. Impurities such as iron will repress pigment formation and quench the fluorescence of formed pigments. However, *R. solanacearum* did not produce fluorescent pigment on KB medium. Sometimes, it produces brown pigment during storage of the culture.

2.5.7 Hypersensitive Reaction on Tobacco Leaves

Most of the phytopathogenic bacteria produce hypersensitive reaction on tobacco leaves. Prepare a dilute suspension (10^7 cfu/ml) of the test bacterium and inoculate by Klements (1953) injection infiltration method. Record the observation for quick necrosis of the tissue within 24 h.

2.5.8 Detection of Poly-β-hydroxybutyrate Granules

(a) Prepare a smear of bacterial ooze from the infected tissue or from a 48-hour culture on YPGA or SPA medium on a microscope slide.

(b) Prepare positive control smears of a biovar 2 strain of *R. solanacearum* and, if considered useful, a negative control smear of a known PHB negative sp.

(c) Allow to air dry and pass the lower surface of each slide rapidly above a flame to fix the smears.

(d) Stain preparation with either Nile blue or Sudan Black and observe microscopically as described below.

2.5.9 Nile Blue Test

(a) Flood each slide with 1% aqueous solution of Nile blue A and incubate for 10 minutes at 55°C.

(b) Drain off the staining solution. Wash briefly in gently running tap water. Remove excess water with tissue paper.

(c) Flood the smear with 8% aqueous acetic acid and incubate for one minute at ambient temperature.

(d) Wash briefly in gently running tap water. Remove excess water with tissue paper.
(e) Re-moisten with a drop of water and apply a coverslip.
(f) Examine the stained smear with an epifluorescence microscope at 450 nm under oil immersion at a magnification of 600 to 1000 using an oil or water-immersion objective.
(g) Observe for bright orange fluorescence of PHB granules. Also observe under transmitted normal light to ensure that the granules are intracellular and that cell morphology is typical of *R. solanacearum*.

2.5.10 Sudan Black Test

(a) Flood each slide with 0.3% Sudan Black B solution in 70% ethanol and incubate for 10 minutes at ambient temperature.
(b) Drain off the staining solution and wash briefly in tap water, removing excess water with tissue paper.
(c) Dip the slides briefly in xylol and blot dry on tissue paper. *Caution: Xylol is harmful, take necessary safety precautions and work in a fume cupboard.*
(d) Flood the slides with 0.5% (w/v) aqueous safranin and leave for 10 seconds at ambient temperature. *Caution: Safranin is harmful, take necessary safety precautions and work in a fume cupboard.*
(e) Wash in gently running tap water, blot dry on tissue paper and apply a coverslip.
(f) Examine stained smears with a light microscope using transmitted light under oil immersion at a magnification of 1000 using an oil-immersion objective.
(g) Observe for blue-black staining of PHB granules in cells of *R. solanacearum* with pink-stained cell walls.

2.6 DIFFERENTIATION OF *RALSTONIA* SPECIES

Ralstonia contains only five recognized species and among them *R. insidiosa, R. mannitolilytica* and *R. pickettii* have been isolated from human clinical samples and latter two can be pathogenic (Ralston *et al.*, 1973; De Bacre *et al.*, 2001; Coenye *et al.*, 2003). *R. pickettii* has often been used as an out group in genetic studies of *R. solanacearum*, but R. *mannitolilytica* is more closely related to *R. solanacearum* (De Bacre *et al.*, 2001; Vandamme and Coenye, 2004; Vaneechoutte *et al.*, 2004). *R. solanacearum*, which causes lethal wilting disease (Table 2.2).

Table 2.2 Biochemical tests to distinguish *Ralstonia solanacearum* from some related non-fluorescent plant pathogenic bacteria, belonging to *Acidovorax, Burkholderia* and *Pseudomonas*

Test	*R.solanacearum*	*Burkholderia cepacia*	*Burkholderi gladioli*	*Burkholderia aryophylli*	*P. corrugate*	*Acidovorax avenae*
Diffusible pigment	+	+	+	+	–	–
Oxidase	–	+	V	+	+	+
Agrinine dihydrolase	–	–	–	+	–	–

Contd...

Contd...

Nitrate reduction	–	–	–	+	+	+
Growth at 41°C	–	V	V	+	–	+
Oxidation of:						
Galactose	+/V	+	+	–		
Glycerol	+/W	–	–	+W		
Mannose	+/V	–	+	–		
Utilization of:						
Cellobiose	V	+	+	V	–	
Trehalose	V	V	+	+		
D-Arabinose	–	+	+	+	V	–
D-Tartrate	+/V	-/W	+	–		
Mannitol	V	+	+	+		+
Sorbitol	V	+	+	+		+
L-Rhamnose	–	–	–	–	–	
Levulinate	V	+	-/W	–		
Sucrose	+	+	+	+		–
Glucose	+	+	+	+	+	+
Benzoate	V	–/V	+	–		
n-Propanol	+	+	–	+W	–	–
β-Alanine	V	–	–	–	–	+
Betaine	–	+	+	+		
L-Arginine	–	+	+	+		
L-lysine	–	+	–	V		
Heptanoate	–	+	–	–		
D-Fucose	–	+	+	+		
D-Raffinose	–	V	–	+		

+ = positive reaction; – = negative reaction; V = variable; W = weak

Table 2.3 Differentiation of two *Ralstonia* species from *Pseudomonas syzygii* and the blood disease bacterium of banana (Eden-Green, 1995)

Test	*R. syzygii*	*B. D. Bacterium*	*R. solanacearum*	*R. picketii*
Colonies on TTC medium	Tenacious, minute	Viscid, 5 mm	Fluidal, > 5 mm	ND
Motility	–	–	V	+
Growth at 37°C	–	+	+	+
Growth at 41°C	–	–	–	+

Contd...

Contd...

NaCl tolerance	< 1%	< 1.5 %	< 2.0 %	ND
Nitrate from nitrite	V	–	+	+
Gas from nitrate	–	–	V	+
Tobacco HR	–	–	+ (Systemic infection with stains from tobacco)	ND
Plant pathogenicity and host association	Clove	Banana	Solanaceae, Musaceae, etc.	Bacterium in humans; intracellular growth in *Acanthamoeba* spp.

ND = not determined; V = variable

2.6.1 Characterization of Biovar of *Ralstonia solanacearum*

In India, race 1 and race 2 and biovar 2, 3 and 4 (Hayward, 1976) are prevalent and race 2 and biovar I have not been recorded. In cool and humid hilly area race 3 and biovar 2 are prevalent whereas on the plain and plateau area of West Bengal race 1 and biovar 3 are prevalent. Biovar 4 is prevalent only at a few locations (Shekhawat, 1976; Sinha, 1985). Primarily the biovar 3, 4 and 5 are Asian type whereas biovar 1 and 2 are American type (Cook *et al*., 1989). Recently, the race 1 has been reported from Andaman (Ansari, 1990) and three races have been reported infecting different crops from various locations of Assam by Nath *et al.* (1996). Prior *et al.* (1990) has studied the aggressiveness of pathogen on tomato. The strain of *R. solanacearum* can be differentiated into 5 biovars (Hayward, 1964; He *et al.* 1983) according to their ability to oxidize several disaccharides (lactose, maltose and cellobiose) and hexose alcohols (mannitol, sorbitol and dulcitol). Singh, *et al.* (2010) surveyed in northern and eastern states of India such as Jammu & Kashmir, Himachal Pradesh, Uttarakhand, Jharkhand and West Bengal and isolated 68 isolates, out of which 58 isolates were categorized as biovar 3 and remaing 7 biovar 4. The biovar 3 and 4, causing wilt in solanacearous crops belong to race 1.

Symptom identification (Fig. 2.1) is the first step for early diagnosis of bacterial wilt of tomato. Accurate identification of *R. solanacearum* from either symptomatic or asymptomatic plants and from water or soil samples demands multiple microbiological and molecular methods. A battery of complementary tests that differ in their sensitivity and/or specificity should be used for field or laboratory analyses for unambiguous identification of bacteria to species and biovar. Screening tests can facilitate early detection and identification of bacteria in potentially infected plants or contaminated soil and water samples by *R. solanacearum*. They cannot be used to identify the race or biovar of the organism. For differentiation of biovars of *Ralstonia solanacearum* based on various tests including utilization of single alcohols and carbohydrates such as dextrose, mannitol sorbitol, dulcitol trehalose and oxidation of lactose, maltose, D(+) cellobiose and nitrite from nitrate and gas from nitrate are practiced.

Basal medium

$NH_4H_2PO_4$	1.0 g
KCL	0.2 g
$MgSO_4.7H_2O$	0.2 g
Peptone	1.0 g

Agar	3.0 g
Bromothymol Blue	80 mg
Distilled water	1000 ml
pH	7.0-7.1

Autoclave the medium at 121°C for 20-30 min and cool it to 50 to 60°C. Prepare 10% aqueous solution of different carbon sources such as dextrose, mannitol sorbitol, dulcitol trehalose and oxidation of lactose, maltose, D(+) cellobiose. Sterilize all the carbon sources with 0.22 μm membrane filter except dulcital which is sterilized by autoclave separately. Add 1% of the carbon sources into the basal medium. After thorough mixing, dispense 3 ml of the molten medium into the sterilized culture tubes and allow to solidify. Add 100 μl of 48 h old culture of *R. solanacearum* into the basal medium and incubate at 28-30°C and observe the tubes on 2, 4, 7 and 10 days. The change in colour of medium from an olivaceous green to yellow (acid pH, <6) indicates oxidation of the carbohydrate. Those biovars capable of oxidizing the disaccharides will take a few days more to give a clear positive result. The *R. solanacearum* has been divided into five biovars based on utilization and oxidation of carbon sources (Table 2.4 & Table 2.5). Hayward (1964) has classified the bacterium on the basis of bio-types (now biovars) or biochemical types. He used number 1, 2, 3, 4 and 5 biotypes for this subdivision, which appear to have little relation of the strain as pathogen. Biovar 2 further subdivided by using additional tests utilization of carbon sources and pectolytic activity. Bacterial wilt of potato in temperate and subtropical regions and at high altitudes in the tropics worldwide is caused by biovar 2 (race 3) strains with the phenotype D-ribose negative, trehalose negative and *meso*-inositol positive. This phenotype is RFLP group 26 in the classification of Cook *et al.* (1989). A dinstict phenotype of biovar 2 occurring in parts of Chile and Colombia, South America is D-ribose negative, trehalose positive and *meso*-inositol negative (Hayward, 1995; Hayward *et al.*, 1990). In RFLP group 27, most of these isolates do not produce nitrite from nitrate, a property of universal among all other biovar 2 strains. A third phenotype of biovar 2, which occurs mainly in Peru and Brazil, is D-ribose positive, trehalose positive and meso-inositol positive (Table 2.5). This type of biovar is also referred as N2 (Cook and Sequeira, 1994). Or biovar 2T to reflect its low land tropical origin (French *et al.*, 1995) corresponds with RFLP groups 29-31, 33, 36 and 39 (Cook and Sequeira, 1994). Prevalence of biovar 3 strains in different agro-climatic zones of India has been reported (James *et al.*, 2003; Singh *et al.*, 2010). High incidence of biovar 3 strains also exhibit that these have broad host range and are well adapted to various degrees of biophysical conditions and other biological factors such as interactions with various microfloras in soil. Biovar 2 (race 3) strains are widely distributed in Asia including India, Pakistan, Bangladesh, China and Philippines and Middle East countries (Lebonan and Iran) (CABI/EPPO, 1998). However, biovar 2 strains have more adaptation to cooler temperature. These strains have limited host range, affecting only potato as compared to biovar 3 strains (Lemessa and Zeller, 2007). Danial *et al.* (2006) reported that *R. solanacearum* strains race 3 biovar 2A were mainly responsible for outbreak of potato brown rot in Europe. Chakraborty *et al.* (1994) reported the occurrence of *P. seudomonas solanacearum* biovar 3 on eggplant, tomato and potato in West Bengal. Li, *et al.* (2006) have made genetic analysis of resistance to bacterial wilt and identification of an associated AFLP marker in eggplant.

Table 2.4 Differentiation of *R. solanacearum* biovars

Test	*Biovars*				
	1	*2*	*3*	*4*	*5*
Utilization of:					
Dextrose	+	+	+	+	+
Mannitol	–	–	+	+	+
Sorbital	–	–	+	+	–
Dulcitol	–	–	+	+	–
Trehalose	+	+	+	+	+
Oxidation of*:					
Lactose	–	+	+	–	+
Maltose	–	+	+	–	+
D (+) cellobiose	–	+	+	–	+
Nitrite from nitrate	+	+	+	+	+
Gas from nitrate	–	–	+	+	

+ + = 80% or more strains positive; – = 80% or more negative

*Disaccharides are oxidized to bionic acids, but are not utilized as source of carbon and energy.

Table 2.5 Test for differentiation of subphenotypes of biovar 2 of *R. solanacearum*

Test	*Biovar 2ARFLP group 26*	*Biovar 2ARFLP group 27*	*Biovar 2T or 2N*
Utilization of trehalose	–	+	+
Utilization of meso-inositol	+	–	+
Utilization of D-ribose	–	–	+
Pectolytic activity	Low	Low	High

Table 2.6 Global distribution of biovar and races of *R. solanacearum*

Biovar		*References*
1	Brazil Florida United States	Quezado-soares and Lopes, 1994 Norman and Yuen, 1998 Velupillai and Stall, 1984; Hayward, 1991
2	Ceylon (Sri Lanka) USA Nepal	Seneviratne and De, 1969 Hayward, 1976 Pradhanang *et al.*, 2000

Contd...

Contd...

3	Ceylon (Sri Lanka)	Seneviratne and De, 1969
	USA	Hayward, 1976
	Taiwan	Shaw, 1984; Hsu, 1991
	South-East Queensland	Peterson *et al.*, 1983
	Assam	Baruah and Deka, 1995
	West Bengal	Chattopadhaya *et al.*, 1994
	Bihar	Sinha, 1985
	Brazil	Quezado-soares and Lopes, 1994
	Nepal	Bhattrai *et al.*, 1998
	Chegalu (China)	Ling *et al.*, 1983; Shuai Zhengbin *et al.*, 1997
	Singapore	Yik *et al.*, 1994
	Kerala	Mathew and Peter, 2004
	Ranchi	Singh *et al.*, 2010
	Himachal Pradesh	Singh *et al.*, 2010
	Uttarakhand	Singh *et al.*, 2010
	Jammu and Kashmir	Singh *et al.*, 2010
4	Plains and plateau area of	Shekhawat, 1976; Sekahwat *et al.*, 1978
	West Bengal and Bihar	Sinha, 1985
	Japan	Kayamura *et al.*, 1985
	Papua New Guinea	Shaw, 1984
	Singapore	Yik *et al.*, 1994
	Saudi Arabia	Abadalla *et al.*, 1999
5	Singapore, Ranchi (Jharkhand)	Shuai Zhengbin *et al.*, 1997; Sharma and Kumar, 2005

Yadessa *et al.* (2010) reported that *R. solanacearum* race 3 biovar 2 (phylotype II) is the causal agent of bacterial wilt of tomato, the most destructive bacterial disease of this crop in Ethiopia.

2.6.2 Identification of races of *R. solanacearum*

There are differences in cultural and physiological properties among the isolates of *R. solanacearum* (Table 2.7). Race designation was determined by inoculating seedlings of 'Hick' Tobacco (*Nicotina tabacum*), '*Bonny Best'* tomato, 'Golden Summer' pepper (*Capsicum annuum*), and 'Mini Finger', eggplant (*Solanum melogena*) (Buddenhegen *et al.*, 1962). Five strains from the Quincy production area and sixth strains (K60) where isolated from tomato in North Carolina by Kelman and Person (1961). Three races of bacterium *R. solanacearum* have been recognized on the basis of host range studied (Buddenhagen and Kelman, 1964; Morton *et al.,* 1965). They reported that race 1 affecting solanaceae and other plants, race 2 from isolates affecting bananas and *Heliconias* and race 3 for isolates primarily affecting potato whereas Kishore *et al.* (1991) reported new record for race 1 in potato in high hills from India. Lozano and Sequeira (1970) reported that the race of *P. solanacearum* can be differentiated by a leaf infiltration technique. Duke *et al.* (1965) have identified race 3 from indigenous strain which were highly virulent to tomato, but only slightly virulent to flue cured tobacco in Georgia whereas 3 biotypes were reported from Ceylon by Seneviratne and De (1969) who reported that biological environmental factors play a part in differentiating the distribution of biotype. The global distribution of biovars of race 1 of tomato is shown in Table 2.6.

Table 2.7 Race determination in *R. solanacearum**

Host reaction	*Race*		
	1	*2*	*3*
Tomato/aubergine	Wilting	No reaction	Wilting
Tobacco cv. White Burley plants (stem inoculation)	Wilting	No reaction	No reaction
Tobacco cv. White Burley leaves (hypersensitivity test)	Necrosis (48 h) and wilting (7-8 days)	HR (12 – 24 h)	Chlorosis (2-8 days)
Musa accuminata	No reaction	Wilting	No reaction

*Race 4, pathogenic to ginger and a few other hosts and race 5, pathogenic to mulberry only, not included here (Janse, 1991)
Source: OEPP/ EPPO Bulletin 34: 173- 178.

2.6.3 Serological Characteristics of *R. solanacearum*

It is interesting to know whether there is a serological relationship between *R. solanacearum* and the other phytopathogenic or saprophytic bacteria which will be useful to know whether this species could be serologically related to the other phytopathologenic bacteria affecting the same host-plant. The technique used for serological studies on *R. solanacearum* is very similar to those used generally in the serology of other phytopathologenic bacteria.

Preparation and injection procedure of antigenic material: The antigenic material is usually prepared from whole bacterial cells. The whole cell may be living (Digat, 1968) or heat-killed at 100°C for 2 hours (Kojima and Buddenhegen, 1969). Sonicated cells were used once (Mortan *et al.*, 1965). These antigens are injected intravenously, but doses and frequency of injections were variable among the various authors.

Antiserum testing: The antisera can be tested by several techniques. The classical agglutination technique is the most frequently used. However, some different techniques, *viz.*, ring test (Perez, 1962), hemagglutination (Perez, 1967), and complement-fixing test have also been used.

2.6.4 Problem of Serological Specificity for Races and Strains

Serum specificity is sometimes questionable. The often-heard criticism of serology is that frequently the serological characters cannot be related to pathogenicity characters. However, a relationship could be shown between these two characters if the precise nature of pathogenicity characters was known. In other words, if it would be possible to extract and to purify protein molecules related to or controlling pathogenicity, it would be possible to use them as antigens. Antisera for the pathogenicity molecules could be obtained and a precise relationship between serology and pathogenicity could be finally established. In other words to become concrete, this hypothesis needs a systematic analysis of antigenic entities of the bacterium. Somatic antigen and the extracellular glycoprotein antigen are among these.

Preparation and injection procedure of antigen (Digat and Cambra, 1976)

Somatic antigen: The somatic antigens from 24 strains were extracted. Every extract was adjusted according to the method outlined in Table 2.8 to a concentration of 5 mg protein per millilitre (Biuret reaction).

Glycoprotein antigen: The glycoprotein antigens from different strains can be extracted according to the method outlined in Table 2.9. Each extract is adjusted to a concentration of 5 mg protein/ml. In some cases, the extract is concentrated by dialysis against Carbowax or by evaporation. For these two antigens, the usual injection procedure is very short: 0.5 ml intravenously on the 1st day and 0.8 ml on the 2nd day.

The antigens are very toxic for rabbits and mice. In these animals, the toxicity causes a hyper vascuarisation and frequently causes some serious necroses at the point of injection. The loss of hair for mouse and the paralysis of hind quarters for rabbits are common. If the antigen concentration is too high (*i.e.*, more than 10 mg/ml), the animal frequency dies within two days of the injection. If the antigen concentration is at the limit of toxicity, the animal can survive but its condition becomes quickly cachexic. It could be possible to reduce these problems by decreasing number, concentration and frequency of injection, or even by using intradermic and intravenous injections alternatively.

Antiserum testing: The kinetics of appearance of antibodies is followed every day by ear bleeding. Tests are performed with classical serological methods, *i.e.*, agglutination, Ouchterlony immunodiffusion and radial (Mancini) immunodiffusion, immunoelectrophoresis and immunofluorescence against the whole bacteria and bacterial extracts.

Antiserum absorption: Antisera are absorbed by whole cells subsequently from each isolate or simultaneously by a "pool" of isolates according to the usual technique. Several cycles of absorption and a careful filtration were necessary.

2.6.5 Imminodiffusion (ID) and Immunoelectophoresis (IE) Technique

Gel diffusion was made of agar 1.2%, sodium chloride 0.85%, sodium azide 0.02% in distilled water. Outchterlony ID, radial ID and are performed according to the usual methods. For radial ID, the melted gel diffusion was mixed with pre-warned antisera so that the final dilution of antisera was 1/40. Whole bacteria and glycoprotein extracts were used as antisera in the wells.

Immunofluorescence(IF): It is technique widely used in Europe for detection of bacterial pathogens in seed and preparative materials.

Its reaction performed according to the usual method (Coons and Kaplan, 1950). A Leitz Orthoplan microscope with ultra-violet epi-illumination system is used as observation system. It is used to screen potato pieces annually for the presence of R. solanacearum (Van der Wolf and Schoen, 2004).

2.6.6 Application of Serology for Studying Bacterial Wilt

Serology could not become useful in the following fields of research:

1. **Diagnosis and detection:** In isolating bacteria, *R. solanacearum* colonies can be confused with saprophytic colonies. Since it is very easy to examine cells from isolated colonies on the medium, agglutination tests can be made and allow a more efficacious screening

of the colonies. Otherwise, agglutination tests can be used directly with exudates from several infected plants. In this way, the bacterium can be directly detected in potato, tomato, eggplant, sweet pepper and banana. The most commonly used assays for bacteria detection and identification are agglutination, enzyme-linked immunosorbent assay (ELISA), immunofluorescence (Van der Wolf *et al.*, 2000), lateral flow strip tests or flow-through assays immunodiagnostic assays using *R. solanacerum* specific (Danks and Barker, 2000; Van der Wolf *et al.*, 2000). Antibodies (also known as immunoglobulins) are proteins that are used by the immune system to identify and neutralize foreign objects, such as bacteria. Due to their specificity, they are commonly used in biology for detection and identification of microorganisms.

A more suitable subtle detection technique is immunofluorescence, which provides a means for detecting the bacterium in the plant, water and soil (Berniac and Coleno, 1974). In this case, practical methods for a bacterial analysis of soil have to be improved since many serological cross-reactions with saprophytic soil bacteria could occur. The IF technique can be also used for detecting the bacterium in the host-plant and for studying the rate of spread of the pathogen.

2. **Taxonomy and Epidemiology:** The study of Digat and Cambra (1976) revealed that race and strain-specific agglutinating antisera could be obtained by absorbing them and using a convenient routine-dilution. Immunodiffusion technique is more practical for differentiating the subtle serological relationships among the strains.
3. **Disease resistance and host-pathogen relationship:** Bacterial cell immunolabelling provides a means of following the spread of bacterium from the roots to the top of the host-pathogen. Frequently, bacterial wilt resistant hybrids (tomato, potato) may be symptomless carrier of *R. solanacearum*. The very sensitive immunofluroescence technique provides a means for detecting small numbers of bacteria in the tolerant host-plant. The immunolabelling technique such as IF, immunoenzyme and immunoferritin techniques can contribute in clarifying the unknown cellular relationship, in particular by observing tissue sections.

Serology techniques are relatively low-cost and easily performed tests for routine *in situ* use. Immunoassays are being applied routinely for the detection of plant pathogens in plant material and soil and tests such as ELISA have demonstrated the sensitivity and specificity required to replace time-consuming and expansive assays like indicator-plant inoculation and dilution plating. Serological techniques also have the advantage that immunoassays are well established techniques for detection and identification of bacterial species.

Nucleic acid probes and PCR amplification offer highly specific and sensitive means to detect pathogens and their strains. Nucleic acid probes have the advantage over serological methods in that a test can be made that targets any specific surface components. Antigens present at one stage of development may differ in different media. Conversely, the genome of an organism is unlikely to change and give a false negative result depending on the developmental stage, growth conditions or extraction method employed. It is also unlikely that antibodies, even monoclonal, can be used in the same way as DNA probes to distinguish between very closely related strains. Although species-specific antibodies for some bacteria have been developed, relatively small genetic differences between isolates of the outside species would not usually be manifested as changes on the outside of the bacterial cell. DNA probes often provide the only reliable method of detecting these differences between closely related strains. However, for a rapid and reasonably sensitive assays of total bacterial numbers of a particular species, immunoassays may be the method of choice.

2.6.7 Immunological Assays

Protocol for soil assay:

1. Weigh 1 g of soil and suspend in 2 ml of buffer A.
2. Shake vigorecesly for 1 minute.
3. Allow settling for 5 min. and remove supernatant.
4. Coat the wall of immunoplate with this supernatant.
5. Incubate in a moist chamber at 37°C for one hour, or overnight at 4°C.

Protocol for plant assay:

1. Cut a 1cm section from the stem base (at or close to the intersection between stem and tap root), and weigh.
2. Macerate mechanically and resuspend in buffer B at the rate of 1g/ml.
3. Allow settling for 30 min. and remove supernatant.
4. Incubate in a moist chamber at 37°C for one hour, or overnight at 4°C.
5. Wash three times in buffer D by submerging the plate in the buffer and then flicking it out to empty it. Leave for 5 min., flick-out and then bang dry on sheet of tissue.
6. Dilute *R. solanacearum* antibodies 1:2000 in buffer add 100 µl to each wall.
7. Incubate and wash as in step 5 & 6.
8. Dilute anti-rabbit peroxidase conjugate 1:5000 in buffer C and add 100 µl to each cell wall.
9. Incubate and wash as in step 5 & 6.
10. Dissolve one TMB tablet in 10 ml of buffer and add 100 µl per well.
11. Leave at room temperature untill a colour change is obtained observed (5-20 min).
12. Stop the reaction by adding 25 µl to 3 M sulphuric acid to each wall.
13. Read the absorbance by 450 nm, or record the colour change visually .

Materials required:

(i) Immunno plates (ELISA Plates)
(ii) Polyclonal antibody of *R. solanacearum*
(iii) Anti-rabbit peroxides conjugate
(iv) 3,3,5,5 tetra methyl benzedrine (TMB) tablets

Stock preparation for immunological assay:

(i) 3M sulphuric acid
(ii) 10% Tween-20

(iii) Buffer A: Coating Buffer (pH 9.6), for soil assays

$NaCO_3$	0.318 g
$NaHCO_3$	0.586 g
Polivinyl pyrrolidone,(MW 40,000)	4.15 g
NaCl	2.00 g

(iv) Buffer B: Coating Buffer (pH 9.6) for plant assays

$NaCO_3$	0.318 g
$NaHCO_3$	0.586 g
$Na2SO_3$	0.1 g

Make up to 10 ml with distilled water

(v) Buffer C: Blocking buffer

Polivinyl pyrrolidone, (MW 40,000)	2.0 g
Full cream milk powder	0.5 g

Make up to 10 ml with buffer C

(vi) Buffer D: Washing Buffer

NaCl	16.0 g
KH_2PO_4	4.0 g
$Na_2HPO_4.12\ H_2O$	28.4 g

Make up to 2 litre with distilled water and add 10 ml 10% Tween-20.

(vii) Buffer E: Substrate Buffer

30% H_2O_2	4 µl
10% Sodium acetate (pH 5.8)	2 ml

Make up to 20 ml with distilled water

2.6.8 ELISA Technique

ELISA is well established methods for testing large number of samples in quality assurance programs worldwide. The procedure is mostly adapted to automation and procedures relatively reproducible results for large-scale testing. In ELISA techniques, both polyclonal and monoclonal antibodies are available for *R. solanacearum* and rapid detection kits are commercially available. Species specific or strain specific monoclonal antibodies produced for *R. solanacearum* and tested in more about 75 samples (Alveraz *et al.,* 1992; Chakraborti *et al.*, 1993; Robinson-Smith *et al.,* 1995; Elphinstone *et al.,* 1996; Griep *et al.,* 1998). The sensitivity of ELISA assays (10 – 10^6 cfu/ ml is adequate for identification of this bacteria from symptomatic plants and colonies on selective medium.

A model was proposed for determining resistance regardless of climatic condition prevalent during the experiment. This model was based on bacterial invasiveness term as the colonization index. Using a quantitative in print method it was confirmed that more resistant the genotype, the lower the bacterial colonization of the stem. The colonization index accounted both for wilted plant and for infected asymptomatic plant in which *R. solanacearum* population failed to produce wilt. The colonization index at mid stem was more useful in corn resistant under favourable condition. When environmental conditions were unfavourable to wilt, colonization index at collar level deserted resistant genotype clearly. This result framed the basis for a model for predicting the degree and stability of resistance (Prior *et al*., 1996). Rajeshwari *et al*. (1998) have developed a technique for detection of *R. solanacearum* in tomato. This serological assays so far the detection of pathogen were able to provide information as to the presence or absence of the pathogen in soil and plant, however, they could not discriminate between virulent and avirulent strains of the pathogen and were not specific to strain and races. In this technique, virulent bacterial cells (encapsulated with mucin) from tomato seeds were used as antigen and polyclonal antisera were developed in rabbit using a classified immunization protocol. Antisera thus developed were examined for the antibodies titre, sensitivity, specificity, rapidity and the efficacy of the antibody for identifying the potential for the application of this assay to the screening of the infected plant materials and seeds. The results indicate that anti-tomato *R. solonacearum* (i) has a good antibody titre of 1:10000; (ii) can detect a few as 100 bacterial cells/ml; (iii) *R. solonacearum* is tomato specific (it reacted with tomato *R. solonacearum* and not isolates from chilli (capsicum) or brinjal; (iv) is reactive to all isolates of *R. solonacearum* from tomato; (v) is not cross-reactive with non-pseudomonades (vi) is virulent strain specific as it recognizes the virulent exopolysaccharides component an anti-determinants and (vii) reactivity could be correlated well with the degree of infection in tomato seeds and plant materials. Thus ELISA devloped is sensitive, specific and rapid, therefore, suitable for detection of *R. solanacearum* isolates from tomato seeds during routine assays. Mc Garvey *et al*. (1999) have used this technique and reported that bacterial population in susceptible cultivars produced greater amounts of EPS 1 per plant than in resistant cultivars. Opina and Miller (2005) evaluated ELISA and immunostrip to detect *R. solanacearum* from 505 eggplant samples from 48 fields in the Philippines. They reported that both the ELISA and immunoStrip were the same for 91% of the samples, while remaining 9%, immunostrip results were positive, whereas ELISA were negative. Isolates of *R. solanacearum* belonging to biovar 1 and biovar 2A are successfully detected by double antibody sandwich- enzyme linked immunosorbent assay (DAS- ELISA) at low population levels after incubation of soil suspensions for 48 h at 30°C in a semi selective broth containing a potato tuber infusion. Detection thresholds of 20 and 200 cfu/ g of inoculated soil were obtained for biovar 1 and biovar 2A respectively. In naturally infested soils, average sensitivities of 97.6 and 100.9 cfu/g cells of soil were obtained for biovars 1 and 2A respectively (Priou *et al*., 2006). *R. solanacearum* was also detected from tomato irrigation ponds and aquatic weeds such as *Polygonum pensylvanicum* and *Hydrocotyle ranunculoides* in or surrounding the irrigation ponds to determine their role as alternative hosts or the source of the pathogen by using enrichment medium SMSA and ELISA.

Enrichment-ELISA Protocols

In this method, combining ELISA, which uses specific monoclonal antibodies, with prior enrichment of stem greatly, improves the sensitivity of detection of bacterial pathogens (Gorris *et al.,* 1994; Lopez *et al.,* 2001). It is effective because of low sensitivity of ELISA for bacterial detection using specific monoclonal antibodies (approximately 10^5-10^6 cfu/ml) and need to improve this sensitivity to detect latent infections of quarantine bacteria (Janse, 1988,

2006; Gorris *et al.,* 1994; Louws *et al.*, 1999). The use of optimized enrichment for each plant pathogenic bacterium allows its specific multiplication in the sample before detection. Specific protocol is required for *R. solanacearum*. The medium temperature, duration and incubation conditions of the enrichment are crucial for optimizing it and improve the detection sensitivity (Gorris *et al.*, 1994; Lopez *et al.*, 1997). To improve the detection of *R. solanacearum* biovar 2 from water samples at low temperature, enrichment for 72 h in modified Wilbrink liquid medium was set up that can be combined with DASI-ELISA using monoclonal antibodies. It was able to detect viable but nonculturable *R. solanacearum* cells (1-10 cfu/ml) (Caruso *et al.*, 2005). *R. solanacearum* biovar 2A was detected from stems of symptomsless plants before harvest of the potato crops by using double antibody sandwich (DAS)-ELISA and indirect ELISA on nitrocellulose membrane after enrichment of the plant extracts in a semi-specific broth (Priou *et al.,* 2010). A simple method is required for rapid detection of plant pathogenic bacteria for testing large number of samples by non-experienced technicians. For this purpose, tissue print ELISA and lateral flow devices have been designed for bacteria. Although detection specificity is very high when using the appropriate monoclonal antibodies. The sensitivity is low for detecting bacteria and they are more appropriate for analyzing plants with symptoms. The lateral flow devices kits are based on the existing technology similar to a pregnancy test kit and such kits are only available for a few bacteria. A lateral flow device kit developed by Central Science Laboratory, U. K., permits detection of *R. solanacearum* in 3 min, in single step (Danks and Baker, 2000). Rapid ImmunoStrip® is available from Agdia Inc.

2.6.9 Flow Cytometry

Immunodiagnostic detection has been enhanced by the development of flow cytometry. It can be used for identification and quantification of cells or other particles as they pass individually through a sensor in a liquid stream. Cells are identified by fluorescent dyes conjugated to specific antibodies and multiple cellular parameters are determined simultaneously based on the cells, fluorescence and its ability to scatter light. The cells may be sorted electronically, permitting purification and/or culture of subpopulations of selected cells for further confirmatory tests (Alvarez and Adams, 1999; Alvarez, 2001). Sample is prepared as cell suspensions are filtered to remove large particles then stained with fluorochrome labelled antibodies. Fluorescent markers for viability (vital stains, such as propidium and hexidium iodide for red fluorescent staining of dead cells and corboxy fluorescein diacetate and calcein AM for green fluorescent staining of viable cells) can be used to differentiate live from dead cells (Van der Wolf and Schoen, 2004). This technique has been applied for detection of *R. solanacearum* and to determine viability of bacteria in seed potatoes (Van der Wolf *et al.*, 2004).

Lateral Flow Device

The principles used for rapid lateral flow devices are primarily those of ELISA, but various types of filters are used as the solid support for the initial binding reaction (Danks and Baker, 2000; Elphinstone, 2005). A lateral flow device test kit developed by Central Science Laboratory, U. K., permits detection of *R. solanacearum* in 3 min in a single step (Danks and Baker, 2000).

Immunostrips

Several rapid screening tests,such as immunostrips (Agdia), are available commercially for rapid and field detection of *R. solanacearum*.The strain of *R. solanacearum* race 1, 2, 3 and 4 were found positive using immunostrip. The assay consistently detected *R. solanacearum* populations ranging from 10^8 to 10^4 cfu /ml. The detection limit varied from *R. solanacearum* strain in plant debris of artificially infected ginger. It has been reported that the immune strip assay always tested positive when population densities were approximately 3.5 cfu /ml in effluent water and potting medium. The immunostrip assay always tested negative when the population densities of *R. solanacearum* were < log 3.0 cfu/ ml (Paret *et al.*, 2008).

Table 2.8 "O"somatic antigen extraction with trichloroacitic acid (modified from Staub, A.M. in Methods in Carbohydrate Chemistry 5; 92-93)

1. *P. solanacearum* 48 h culture on LPG Agar in Roux bottle are harvested in sterile water (10 ml per Roux bottle) and washed twice in sterile water in a weighed centrifuge tube. The packed cells are weighed and suspended in 5 times their weight of 0°C water
2. Equal volume of 0.5 N trichloroacitic acid is added. The mixture is cooled to 4°C for 3 h and then centrifuged.
3. Supernatant neutralized to pH 6.5 and precipitated with 2 volumes of ethanol at – 4°C

   ↓ ↘ Pellet eliminated

4. Precipitate is sedimented overnight at –4°C and centrifuged.

   ↓ ↘ Supernatant eliminated

5. Pellet is dissolved in 0.1 original volume of sterile water and dialyzed for 4 days against distilled water, and then it is centrifuged at 27000XG

   ↓ ↘

6. Supernatant is called "O" Somatic antigen — Pellet eliminated (bacterial debris)

Table 2.9 Glycoprotein antigen extraction with ammonium sulfate

1. *R. solanacearum* 48 h culture on LPG Agar in Roux bottles are harvested in sterile water (10 ml per Roux bottle) shaken vigorously for 30 minutes, centrifuged at 3000 g for 15 minutes and then at 6000 g for 15 minutes.

   ↙ Supernatant is filtered, neutralized at pH7 and precipated with equal volume of saturated ammonium sulphate solution at +4°C.

   ↘ Pellet eliminated

Contd...

Contd...

2. Supernatant is filtered, neutralized at pH7 and Precipitated with equal Volume of saturated Ammonium sulphate solution at +4°C.

↓

Sediment eliminate

3. Precipitate is sediment overnight at +4°C and centrifuged at 20,000 × G for 15 minutes

↓

4. Pellet is dissolved in original volume of sterile water and dialyzed for 4 days against distilled water. Cruide glycoprotein fraction is named as "Glycoprotein antigen".

Infectivity titration: This method is used for comparison of virulence of *R. solanacearum* strain. Infectivity titration can be used in two ways: to define resistance in host pathogen: and to quantify the virulence of plant pathogen (Lum and Kelman, 1981; Bora and Addy, 1982; Ercolani, 1984). There is an extensive litreature on application of this technique to bacterial disease of plants. Anyone contemplating work in this area should first read the authoritative review of Ercolani (1984) who states that infectivitivity titration is an easy and efficient procedure for measuring virulence of the pathogen and susceptibility of the host in bacterial infection of plants. Both host and pathogen effects can be of either a quantal or a quantitative type. Quantal effects occur as an all-or none response, which by definition can only take either of two mutually incompatible forms, such as alive or dead, healthy or diseased, turbid or wilted, whereas quantitative effects occur as response that can take any of a series of values, e.g., number of local lesions and time to response. Infectivity titration with quantitative response has greater potential for achieving practical results than titration with quantal responses.

Procedure

(i) Prepare serial tenfold dilutions in phosphate buffer so that final concentration of bacteria in colony forming units (cfu's) per ml is ca.10^6, 10^5, 10^4 and 10^3.

(ii) For each of the dilutions inoculate 5 plants. Twenty microlitres aliquots are inoculated by stabbing gently into the axil of the third leaf from the apex of the tomato plants, using micropipette tips, which are left, in place to allow uptake of inoculums. Thus the plants received doses of 20000, 2000, 200 and 20 viable cells, respectively

(iii) A control series of plants is inoculated with 20 microlitres of sterile phosphate buffer (0.05 M, pH 7.0).

(iv) Examine the plants daily for 210 days for the appearance of wilt symptoms using the disease rating scale following Winstead and Kelman (1952).

(v) The infectivity titration of 3 strains of *R. solanacearum*, race 3, biovar D (R. Sol-1 $^{r+c+e+}$, R. Sol-2 $^{r+c+e+}$ and R. Sol-3 $^{r+c+e+}$, rifamycin, chloramphenicol and streptomycin resistant) was studied using tomato plants evaluated 9 and 14 days after inoculation. Two strain R. Sol-1 $^{r+c+e+}$, R. Sol-3 $^{r+c+e+}$, showed lower virulence than wild type(R. Sol-O $^{r-c-e-}$) which did not differ from *R. solanacearum* 2 $^{r+c+e+}$. The evaluation at 14 days showed larger differences between the averages. The infectivity titration can defect strain with similar virulence to the wild type and those which lose virulence during screening to antibiotic resistance (Yosses *et al.*, 1998).

2.6.10 Gas Chromatography

Toxin is detected by this method (Baruah and Daka, 1995). This method is for identification of bacteria and statistical analyses of FAME. Profile data were conducted using the MIDI Microbio Identification system, Network, DEL, VERSION 3.6 (Sasser, 1990).

2.7 MOLECULAR CHARACTERISTICS

2.7.1 DNA Marker

DNA genetic markers, such as restriction fragment length polymorphisms (RFLPs) and random amplified polymorphic DNAs (RFLPs), promise to increase understanding of bacterial wilt resistance dramatically. DNA markers make it possible to monitor inheritance throughout an entire plant genome at high resolution and dissect the contributions of specific genetic loci to complex characters. Aarons *et al.* (1993) have used DNA genetic marker mapping of genes for bacterial wilt resistance in tomato.

Another important application of DNA markers is in breeding for complex, polygenic characters. Often, it is difficult to retain all the genes involved in a complex resistance character throughout the course of a breeding programme. Retaining resistance loci can be specifically challenging if creating an artificial epidemic is difficult or if the disease is associated with a specific environmental condition or plant developmental stage. However, DNA marker tightly linked to the underlying resistance loci can be used as a basis for selection, thereby ensuring that the genes are retained during the course of the breeding programme. The possibility of antigenic contamination (especially by cell wall material or bacterial debris) must be kept in mind.

Harai *et al.* (1998) has used DNA marker for bacterial wilt to evaluate resistance in tomato breeding. However, Bartolla *et al.* (1999) used plasmid DNA and pathogen clone (*R. solanacearum*) GM 11000 for expressing the marker gene was actively multiplied inside the plant vessels. Moreover, experiment in which *R. solanacearum* strain harboring different marker gene were co-inoculated into plant demonstrated that infecting bacterial strain exchange genetic information in the plant by a mechanism of transformation. These assays are based on the use of antibiotics in various test formats to detect and identify any molecules or cells (including bacteria). A biochemical growth test is used for biovar determination of *R. solanacearum*. This test is based on the differential ability of strains of the pathogen to differentially produce acid from several carbohydrate sources, including disaccharides and sugar alcohols. At the sub-species level, identification of strains of *R. solanacearum* can be assessed with several nucleic-acid based methods such as DNA probe hybridization uses the ability of two complementary single-stranded nucleic acids to combine into a single molecule.

2.7.2 DNA Probe Hybridization

Nucleotide probe of known sequence is used to bind complementary strand of undetermined organism for identification. The polymerase chain reaction is a technique that consists of amplifying a DNA molecule exponentially.

2.7.3 Nucleic Acid Based Detection of *Ralstonia solanacearum*

Polymerase Chain Reaction (PCR)

In this method a specific part of the nucleic acid of the target bacterium is artificially multiplied before detection takes place. This multiplication is reached by repeated cycles of denaturation (95°C), annealing and extension of nucleic acids (72°C). Theoretically, the sensitivity of PCR is very high, *i.e.*, one copy of target DNA in a sample can be detected. However, in practice usually a lower sensitivity 10^3 cells/ml is reached due to inhibitory substances in the plant extract. Although PCR can reach high sensitivity and specificity, its introduction for routine detection has been hampered by a lack of robustness. PCR is the most widely used molecular technique for detection of bacteria. Their main advantages are specificity and rapidity. Specificity is directly related both to the design of the primers or probes and to the amplification or hybridization protocols. The time required until the final result is usually less than 24 h whereas that required for conventional microbiological detection of bacteria was of the order of several days in the case of a negative result and 5–10 days in the case of a positive result due to the tests necessary to confirm it. Furthermore, the possibility of designing a multiplex PCR saves time and reagent costs compared with monospecific PCR, which requires several reactions for the same number of tests. Colorimetric detection of PCR products, on membranes or in microtitre plates, has been employed successfully, increasing sensitivity and facilitating interpretation of results for the use of the technique in routine analyses. Detection of bacteria in a given sample by PCR is not only dependent on the performance on the PCR assay itself, but also on the efficiency of the procedure employed to extract the nucleic acids from the plant material. The sensitivity of detection can be reduced by inhibitors that may be present in the extract of 237 nucleic acids. To check for substances that may interfere with the amplification process, internal controls can be designed for each pair of primers, or real-time PCR can be employed. PCR efficiency is controlled by many parameters, such as polymerase type, buffer composition and stability, purity and concentration of dNTPs, cycling parameters as well as the characteristics of the starting template. The specific primers have been designed based on either the amplification of specific genes from the chromosome or plasmids, or on different approaches such as sequences selected from RAPD differential bands obtained by subtractive hybridization (Luows *et al.*, 1999). Some specific primers have been developed for detection of *R. solanacearum* from soil, seed and planting material and water as well as for characterization are given in Table 2.10. Stefani *et al.* (2005) used PCR for detection of latent infection of *R. solanacearum* biovar 2, race 3 in tomato crops. Grover *et al.* (2009) developed a reliable and sensitive protocol for detection of *R. solanacearum* using multiple displacement amplification PCR (MDA-PCR). It could detect as low as 1 cfu/ml of bacteria within 8 h including DNA isolation. However, *R. solanacearum can* be detected from seeds and planting materials by using various PCR techniques as given in Table 2.11.

2.7.3.1 *Validated PCR Protocols and Reagents*

NB: Preliminary testing should permit reproducible detection of 10^3 to 10^4 cells of *R. solanacearum* per ml of sample extract. Preliminary testing should also show no false positive results with a panel of selected bacterial strains.

1. PCR Protocol (Seal *et al.* 1993)

Oligonucleotide primers

Forward primer OLI-1 5′-GGG GGT AGC TTG CTA CCT GCC-3′

Reverse primer Y-2 5′-CCC ACT GCT GCC TCC CGT AGG AGT-3′

Expected amplicon size from *R. solanacearum* template DNA = 288 bp

PCR Reaction Mix: The following regents are used for study.

Reagent	*Quantity per reaction*	*Final concentration*
Sterile UPW	17,65 µl	
10X PCR buffer (15 mM MgCl2)	2,5 µl	1X (1,5 mM $MgCl_2$)
dNTP mix (20 mM)	0,25 µl	0.2 mM
Primer OLI-1 (20 µM)	1,25 µl	1 µM
Primer Y-2 (20 µM)	1,25 µl	1 µM
Taq polymerase (5U/µl)	0,1 µl	0,5 U
Sample volume	2,0 µl	–
Total volume	25 µl	–

PCR reaction conditions

Run the following programme:

Step I: 96°C for 2 min (denaturation of template DNA) } 1 cycle

II: 94°C 20 s (denaturation of template DNA)

III: 68°C for 20 s (annealing of primers) } 35 cycles

IV: 72°C for 30 s (extension of copy)

V: 72°C for 10 min (final extension) } 1 cycle

VI: hold at 4°C forever

This programme of the duration steps of cycles (II), (III) and (IV) may be changed as per thermal cycler used for PCR amplification.

Restriction Enzyme Analysis of Amplicon

PCR products amplified from *R. solanacearum* DNA produce a distinctive restriction fragment length polymorphism with enzyme *Ava* II after incubation at 37°C.

2.7.3.2 PCR Protocol (Pastrik and Maiss, 2000)

Oligonucleotide Primers

Forward primer Ps-1 5′- AGT CGA ACG GCA GCG GGG G -3′

Reverse primer Ps-2 5′- GGG GAT TTC ACA TCG GTC TTG CA -3′

Expected amplicon size from *R. solanacearum* template DNA = 553 bp.

PCR reaction mix: The following regents are used for study.

Reagent	*Quantity per reaction*	*Final concentration*
Sterile UPW	16,025 µl	
10X PCR buffer (15 mM $MgCl_2$)	2.5 µl	1X (1,5 mM $MgCl_2$)
BSA (fraction V) (10 %)	0.25 µl	0.1 %
dNTP mix (20 mM)	0.125 µl	0.1 mM
Primer Ps-1 (10 µM)	0.5 µl	0.2 µM
Primer Ps-2 (10 µM)	0.5 µl	0.2 µM
Taq polymerase (5U/µl)	0.1 µl	0.5 U
Sample volume	5.0 µl	
Total volume	25.0 µl	

N.B. Originally optimised for MJ Research PTC 200 thermocycler with Gibco Taq Polymerase. Perkin Elmer AmpliTaq and buffer can also be used at the same concentrations.

PCR reaction conditions

Run the following programme:

Step I: 95°C for 5 min (denaturation of template DNA) } 1 cycle
II: 95°C for 30 s (denaturation of template DNA)
III: 68°C for 30 s (annealing of primers) } 35 cycles
IV: 72°C for 45 s (extension of copy)
V: 72°C for 5 min (final extension) } 1 cycle
VI: hold at 4°C forever

This programme of the duration steps of cycles (II), (III) and (IV) may be changed as per thermal cycler used for PCR amplification.

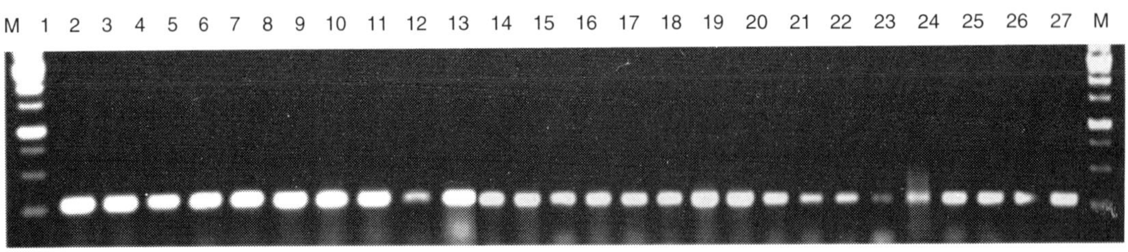

Fig. 2.3 Amplification of fragments of the 16s rRNA gene from *R. solanacearum* isolates separated on an agarose gel showing the 288 bp. Lane 1 & 2: isolates from J& K, Lane 3 - 8: isolates from H. P., Lane 9-19 from Uttarakhand, Lane 20 – 24 from Jharkhand and Lane 25-27 from West Bengal (Singh *et al.,* 2010)

Pastrik and Maiss (2000) have detected *R. solanacearum* in potato tubers by polymerase chain reaction

Restriction Enzyme Analysis of Amplicons

PCR products amplified from *R. solanacearum* DNA produce a distinctive restriction fragment length polymorphism with enzyme *Taq* I after incubation at 65°C for 30 minutes. The restriction fragments obtained from *R. solanacearum* specific fragment are 457 bp and 96 bp in size.

2.7.3.3 Multiplex PCR Protocol with Internal PCR Control (Pastrik *et al.* 2002)

Oligonucleotide primers:

Forward primer RS-1-F 5′- ACT AAC GAA GCA GAG ATG CAT TA -3′

Reverse primer RS-1-R 5′- CCC AGT CAC GGC AGA GAC T -3′

Forward primer NS-5-F 5′- AAC TTA AAG GAA TTG ACG GAA G -3′

Reverse primer NS-6-R 5′- GCA TCA CAG ACC TGT TAT TGC CTC -3′

Expected amplicon size from *R. solanacearum* template DNA = 718 bp (RS-primer set)

Expected amplicon size from the 18S rRNA internal PCR control = 310 bp (NS-primer set).

PCR Reaction Mixture: The following regents are used for study.

Reagent	*Quantity per reaction*	*Final concentration*
Sterile UPW	12.625 µl	–
10X PCR buffer (15 mM $MgCl_2$)*	2.5 µl	1X (1,5 mM $MgCl_2$)
BSA (fraction V) (10 %)	0.25 µl	0.1 %
dNTP mix (20 mM)	0.125 µl	0.1 mM
Primer RS-1-F (10 µM)	2.0 µl	0.8 µM
Primer RS-1-R (10 µM)	2.0 µl	0.8 µM
Primer NS-5-F (10 µM)**	0.15 µl	0.06 µM
Primer NS-6-R (10 µM)**	0.15 µl	0.06 µM
Taq polymerase (5U/µl)	0.2 µl	1.0 U
Sample volume	5.0 µl	–
Total volume	25.0 µl	–

*Methods were validated using *Taq* polymerase from Perkin Elmer (AmpliTaq) and Gibco BRL.

**Concentration of primers NS-5-F and NS-6-R were optimized for potato heel end core extraction using the homogenization method and DNA purification according to Pastrik (2000). Re-optimisation of reagent concentrations will be required if extraction by shaking or other DNA isolation methods are used.

PCR Reaction Conditions

Run the following programme:

Step		
	I: 95°C for 5 min (denaturation of template DNA)	1 cycle
	II: 95°C for 30 s (denaturation of template DNA)	35 cycles
	III: 58°C for 30 s (annealing of primers)	
	IV: 72°C for 45 s (extension of copy)	
	V: 72°C for 5 min (final extension)	1 cycle
	VI: hold at 4°C forever	

This programme of the duration steps of cycles (II), (III) and (IV) may be changed as per thermal cycler used for PCR amplification.

Restriction Enzyme Analysis of Amplicons

PCR products amplified from *R. solanacearum* DNA produce a distinctive restriction fragment length polymorphism with enzyme *Bsm* I or an Isoschizomere (e.g., Mva 1269 I) after incubation at 65°C for 30 minutes.

2.7.3.4 *Ralstonia solanacearum Biovar-specific PCR Protocol* (Pastrik *et al.* 2002)

Oligonucleotide Primers

Forward primer Rs-1-F 5′- ACT AAC GAA GCA GAG ATG CAT TA -3′

Reverse primer Rs-1-R 5′- CCC AGT CAC GGC AGA GAC T -3′

Reverse primer Rs-3-R 5′- TTC ACG GCA AGA TCG CTC -3′

Expected amplicon size from *R. solanacearum* template DNA:

with Rs-1-F/Rs-1-R = 718 bp

with Rs-1-F/Rs-3-R = 716 bp.

PCR Reaction Mixture

(a) Biovar 1/2-specific PCR: The following regents are used for study.

Reagent	*Quantity per reaction*	*Final concentration*
Sterile UPW	12.925 µl	–
10X PCR buffer (15 mM $MgCl_2$)*	2.5 µl	1X (1,5 mM $MgCl_2$)
BSA (fraction V) (10 %)	0.25 µl	0.1 %
dNTP mix (20 mM)	0.125 µl	0.1 mM
Primer Rs-1-F (10 µM)	2.0 µl	0.8 µM
Primer Rs-1-R (10 µM)	2.0 µl	0.8 µM
Taq polymerase (5U/µl)	0.2 µl	1.0 U
Sample volume	5.0 µl	–
Total volume	25.0 µl	–

(b) Biovar 3/4/5-specific PCR: The following regents are used for study.

Reagent	*Quantity per reaction*	*Final concentration*
Sterile UPW	14.925 µl	–
10X PCR buffer (15 mM $MgCl_2$)*	2.5 µl	1X (1.5 mM $MgCl_2$)
BSA (fraction V) (10 %)	0.25 µl	0.1 %
dNTP mix (20 mM)	0.125 µl	0.1 mM
Primer Rs-1-F (10 µM)	1.0 µl	0.4 µM
Primer Rs-3-R (10 µM)	1.0 µl	0.4 µM

Contd...

Contd...

Taq polymerase (5U/μl)	0.2 μl	1.0 U
Sample volume	5.0 μl	–
Total volume	25.0 μl	–

PCR Reaction Conditions

Run the following programme for both biovar 1/2- and biovar 3/4/5-specific reactions:

Step I: 95°C for 5 min (denaturation of template DNA) } 1 cycle

II: 95°C for 30 s (denaturation of template DNA)

III: 58°C for 30 s (annealing of primers) 35 cycles

IV: 72°C for 45 s (extension of copy)

V: 72°C for 5 min (final extension) } 1 cycle

VI: hold at 4°C forever

This programme of the duration steps of cycles (II), (III) and (IV) may be changed as per thermal cycler used for PCR amplification.

Restriction Enzyme Analysis of Amplicons

PCR products amplified from *R. solanacearum* DNA using primers Rs-1-F and Rs-1-R produce a distinctive restriction fragment length polymorphism with enzyme *Bsm* I or an Isoschizomere (e.g., Mva 1269 I) after incubation at 65°C for 30 minutes. PCR products amplified from *R. solanacearum* DNA using primers Rs-1-F and Rs-3-R have no restriction sites.

Preparation of the Loading Buffer

Bromphenol blue (10%-**stock solution**)

Bromphenol blue	5 g
Distilled water (bidest)	50 ml
Loading buffer	
Glycerol (86%)	3.5 ml
Bromphenol blue (10% **stock solution**)	300 μl
Distilled Water (bidest)	6.2 ml

10x Tris Acetate EDTA (TAE) buffer, pH 8.0

Tris buffer	48.40 g
Glacial acetic acid	11.42 ml
EDTA (disodium salt)	3.72 g
Distilled water	1.00 L

Dilute to 1x before use.

Table 2.10 List of primer sets for identification of *R. solanacearum* through PCR.

Pathogen	*Probe*		*References*
R. solanacearum	Y2: 5'- CCC ACT GCT GCC TCC CGT AGG AGT-3' OLII: 5'- GGG GGT AGC TTG CTA CCT GCC-3'	288bp	Seal *et al.*, 1993
R. *solanacearum* and related spp.	759: 5'- GTC GCC GTC AAC TCA CTT TCC-3' 760: 5'- GTC GCC GTC AGC AAT GCG GAA TCG-3'	281bp	Ito *et al.*, 1998 Opino *et al.*, 1997
R. solanacearum	PS96-H: 5'- TCA CCG AAG CCG AAT CCG CGT CCA TCA C-3' PS96-I: AAG GTG TCG TCC AGC TCG AAC CCG CC-3'	148bp	Hartung *et al., 1998* Seal *et al.*, 1992
R. solanacearum	pehA#3: 5'- CAG CAG AAC CCG CGC CTG ATC CAG-3' pehA#6: 5'- ATC GGA CTT GAT GCG CAG GCC GTT-3'	504bp	Gillings *et al.*, 1993
R. solanacearum race 1 and 3 biovar 1, 2, 3 and 4	RALSF:5'-GCTCAAGGCATTCGTGTGGC-3' RALSR:5'-GTTCATAGATCCAGGCCATC-3'	932bp	Kang *et al.*, 2007
R. solanacearum race 1 biovar 3 & 4	Hrp_rs2F:5'-AGAGGTCGACGCGTACAGT-3' Hrp_rs2R:5'-CATGAGCAAGGACGTCA-3'	323bp	Singh *et al.*, 2014b

2.7.3.5 *Fluorescence in situ Hybridization (FISH)*

Fluorescence *in situ* hybridization is dependent on the hybridization of DNA probes to species-specific regions of bacterial ribosome. It can detect single cells but in practice, the detection level is near 10^3 cells/ml of plant extract and even though has been employed in some recent works (Ercolini *et al.,* 2006). There is a high affinity and selectivity of DNA probes because FISH takes place under very stringent hybridization conditions, where a difference of one nucleotide in a 15-20 oligonucleotide probe targeted to 23S rRNA is sufficient for discrimination. *R. solanacearum* race 3 biovar 2 was detected from potato peels by using this method (Wulling *et al.*, 1998; Van der Wolf and Schoen, 2004) and it has been included in official diagnostic protocols in the EU (Directive 2006/63/CE for *R. solanacearum*) and recommended in the EPPO protocol for the same pathogen (EPPO, 2004). The procedure for detecting bacteria from samples is given below.

2.7.3.6 *Validated Reagents for FISH Test*

Oligo-probes

R. solanacearum-specific probe OLI-1-CY3: 5′-GGC AGG TAG CAA GCT ACC CCC-3′

Non-specific eubacterial probe EUB-338-FITC: 5′-GCT GCC TCC CGT AGG AGT-3′

Fixative solution

(i) Heat 9 ml molecular grade water (e.g., ultra pure water (UPW)) to about 60°C and add 0.4 g paraformaldehyde. Paraformaledhyde dissolves after adding 5 drops of 1N NaOH and stirring with a magnetic stirrer.

Table 2.11 Detection of *R. solanacearum* from seed and planting materials by using PCR based techniques.

PCR Genomic assay	*Host*	*Detected from*	*Race/biovar of Bacteria*	*Reference*
SCAR primers				
Rep-PCR	Ginger, mioga and cucuma	–	*R. solanacearum* race 4	Horita *et al*., 2004
RAPD	Brinjal, chilli and tomato	–	*R. solanacearum* race 3	James *et al*., 2003
16S r RNA	Tomato	–	*R. solanacearum* race 1 biovar 3 and 4	Singh *et al*., 2010.
PCR (RNA operon)				Pastrik *et al*., 2002
Fli C	–	Soil	*R. solanacearum, R. pickettii*	Schonfeld *et al*., 2003
	Geranium	Plant part	*R. solanacearum* race 3 biovar 2	Swanson *et al*., 2007
Nested- PCR	Banana, tomato, chili, eggplant and tobacco	Soil	*R. solanacearum*	Khakvar *et al*., 2008
hrp gene, PCRhrp gene, Nested- PCR	Tomato, potato tobacco, eggplant, pepper *Pelargonium asperum*	Plant and tuber	*R. solanacearum R. solanacearum* biovar 1, N2, 3, 4 & 5	Poussier *et al*., 1999; Poussier *et al*., 2000
Bio-PCR	Tomato, melon	Soil, Plant water	*R . solanacearum* race 1 biovar 3	Lin *et al*., 2009; Singh *et al*., 2014b
Co-operational PCR 16 S r RNA Co-operational PCR	*Pelargonium* spp	Cutting	*R. solanacearum* *R. solanacearum* race 3 biovar 2	Bertolini *et al*., 2003 a; 2003b Macro- Noales *et al*., 2008
PCR (insertion sequence)	Eight different host		*R. solanacearum* race 1	Lee *et al*., 2001
16 r RNA, PCR16 r RNA, PCR16 r RNA, Multiplex PCR	– Potato	– Potato tuber	*R. solanacearum* subdivision 2a *R. solanacearum, R. solanacearum* biovar1, 2, N2, 3, 4 & 5	Boudazin *et al*., 1999; Pastrik and Maiss, 2000; Seal *et al*., 1999
PCR -RFLP	–	–	*R. solanacearum*	Poussier *et al*., 1999; 2000
PCR-DGGE	–	–	*R. solanacearum*	Van Overbeek *et al*., 2002

Contd...

Contd...

Multiplex–PCR (16S-23S r RNA ITS)	Potato	Tuber	*R. solanacearum*	Pastrik *et al.*, 2002; Bertolini *et al.*, 2003a)
Real–time PCR	Potato	Tuber and plant	*R. solanacearum*	Stead *et al.*, 2003; Weller *et al.*, 1999; Weller *et al.*, 2000; Ozakman and Achaad, 2003 Swanson *et al.*, 2007
Real Time Bio-PCR	Potato Ginger	Tuber and plant	*R. solanacearum* race 3 biovar 2 *R. solanacearum race 4*	
Real Time Bio-PCR	Geranium	Rhizomes plant	*R. solanacearum* race 3 biovar 2	
RNA based approaches (NASBA)			*R. solanacearum*	Betsink *et al.*, 2002
16S r (NASBA) RNA	Potato	Tuber, irrigation water	*R. solanacearum race 3 biovar2*	Van der Wolf *et al.*, 2004
IC-PCR	Tomato, pepper, weeds (*Physalis minima*, *Amaranthus spinosus* and *Euphorbia hirta)*	Plant and soil	*R. solanacearum race 1*	Dittapongpitch and Surat, 2003
Putative *Ipx* C gene	–	–	*R. solanacearm*	Villa *et al.*, 2003
Cytochromec1 signal peptide gene	Tomato	Soil and plants	*R. solanacearum* race 1 & 3 biovars 1, 2, 3 &4	Kang *et al.*, 2007

(ii) Adjust pH to 7.0 by addition of 1ml of 0.1 M phosphate buffer (PB; pH 7.0) and 5 drops of 1N HCl. Check pH with indicator strips and adjust if necessary with HCl or NaOH. *(Do not use a pH meter in solutions with paraformaledhyde.)*

(iii) Filter the solution through a 0.22 µm membrane filter and keep dust-free at 4°C until further use.

(The fixative contains paraformaldehyde which is toxic. Wear gloves and do not inhale, it is advisable to work in a fume cupboard.)

3X Hybmix

NaCl	2.7 M
Tris-HCl	60 mM (pH 7,4)
EDTA (filter sterilised and autoclaved)	15 mM

Dilute to 1x as required.

Hybridization solution

1x Hybmix	
Sodium dodecyl sulphate (SDS)	0.01 %
Formamide	30 %,
probe EUB 338	5.0 ng/µl
probe OLI-1 or OLI-2	5.0 ng/µl

Prepare quantities of hybridization solution according to the calculations in Table 2.6. For each slide (containing 2 different samples in duplicate) 90 µl hybridization solution is required.

Formamide is very toxic so wear gloves and take necessary safety precautions.

Table 2.12 Suggested quantities for preparation of hybridization mix

Number of slides	*1*	*4*	*6*	*8*	*10*
Sterile UPW	23.1	92.4	138.6	184.8	231.0
3x hybmix	30.0	120.0	180.0	240.0	300.0
Formamide	27.0	108.0	162.0	216.0	270.0
1 % SDS	0.9	3.6	5.4	7.2	9.0
Probe EUB 338 (100 ng/µl)	4.5	18.0	27.0	36.0	45.0
Probe OLI-1 or OLI-2 (100 ng/µl)	4.5	18.0	27.0	36.0	45.0
Total volume (µl)	90.0	360.0	540.0	720.0	900.0

Store all solutions containing light sensitive oligo-probes in the dark at – 20°C. Protect from direct sunlight or electric light during use.

0.1M Phosphate buffer, pH 7.0

Na_2HPO_4	8.52 g
KH_2PO_4	5.44 g
Distilled water	1.00 l

Dissolve ingredients, check pH and sterilize by autoclaving at 121°C for 15 minutes.

Co-operational PCR

It is a new PCR, which has high sensitivity for the amplification bacterial targets from plant material. The Co-PCR (co-operational amplification) technique can be performed easily in a simple reaction based on the simultaneous action of four or three primers. The reaction process consists of the simultaneous reverse transcription of two different fragments from the same target, one internal to the other, the production of four amplicons by the combination of the two pairs of primers, one pair external to the other, and the co-operational action of amplicons for the production of the largest fragment. Co-PCR usually produces the largest amplicon, in contrast to nested- PCR, which requires two sequential reactions and obtains the smallest fragment. Metal block and capillary air thermal cyclers have been employed for the detection of a bacterium, but by using only three primers (Caruso *et al.,* 2003), which shows the possibilities of this new approach. The low amount of reagents (ten times less than in conventional PCR) probably increases susceptibility to inhibitors. However, this step was not necessary when analyzing the presence of *R. solanacearum* in water (Caruso *et al.,* 2003). Co-PCR requires only one reaction, minimizing manipulation and reducing risk of contamination.

Real-time (Quantitative) PCR

Quantitative real time PCR has become possible by the development of detectors that can measure fluorescence that is emitted during the PCR cycle. This method is based on the 5'-3' exonuclease activity of the Taq DNA polymerase, which results in cleavage of fluorescent dye-labeled with different probes (TaqMan®) during PCR. The exponential nature of PCR in theory allows the amount of starting material to be calculated from the amount of product at any point in the reaction. In practice, however, reaction conditions can interfere with exponential amplification and affect product concentration. Early attempts at quantization involved stopping the PCR reaction at various points to generate standard curves, which resulted in a laborious, low-throughput process.

TaqMan probe

In this systems, an oligonucleotide probe sequence of approximately 25-30 nucleotides in length is labeled at the 5' end with a fluorochrome, usually 6-carboxyfluorescein (6-FAM) and a quencher fluorochrome, usually carboxytetramethyl-rhodamine (TAMRA), at the 3' end. The probe is degraded by the 5'-3' exonuclease activity of the Taq polymerase as it extends the primer during each PCR amplification cycle and the fluorescent chromophore released. The amount of fluorescence is monitored during amplification cycle and is propotional to the amount of PCR product generated. Real-time reaction monitoring with specific instruments and fluorescent probes combine amplification, detection and quantification in a single step. TaqMan probes consist of single-stranded oligonucleotides that are complementary to one of the target strands. A fluorescent dye adorns the 5' end and a quencher is bound to the 3' end. Fluorescence occurs when the polymerase replicates a template on which a TaqMan probe is bound and the 5' exonuclease activity cleaves the probe (Varma-Basil *et al.,* 2004). They have been proposed for detection of *R. solanacearum* (Weller *et al.,* 2000a; Ozakman and Schaad, 2003). Huang *et al.* (2009) detected *R. solanacearum* strain G 1000 from soil using specific primer pair R. sol 1-R. sol 2 and the TaqMan probe Rs-pro. The detection limit was 100 fg/µl in conventional PCR and 1.2 fg/μl in real-time PCR by combining it with modified protocol to extract DNA from soil.

Fluorescent Resonance Energy Transfer (FRET)

The probes require labeling of two oligonucleotide probe sequences that are adjacent to each other within the PCR target fragment. Probe 1 contains fluorescein label at its 3' end, whereas probe 2 is labeled at its 5' end with another label (Light Cycler red 640). It is imperative that the two probes are designed so that when they hybridize to the amplified PCR product, they are aligned head to tail to bring the two fluorescent dyes in close proximity to each other. The fluorescein dye attached to the Cycler ™ or RAPID (ruggedized advanced pathogen identification device) unit and it emits a green fluorescent light at a slightly longer wavelength. When the second probe is in close proximity, the energy emitted by the first probe excites the light Cycler Red 640 dye attached to the second probe and red fluorescent light at longer wavelength is now emitted that can be detected at 640 nm. Fluorescence is measured during the annealing step of each of the amplification cycles when both probes hybridize to the PCR amplicons (Didenko, 2001; Cockerill and Smith, 2002).

Molecular Beacons

These are fluorescent oligonucleotide probes that are designed to include stem loop folding. They are complementary nucleotide sequences that are complementary to the target amplicons. A fluorescent chromophore is attached at the 5' end of the probe and a quencher molecule is attached at the 3' end. A stem structure is formed by annealing of the complementary arm sequences that are added on both the sides of the probe sequences. When a stem structure is formed, the fluorophore transfer energy to the quencher and no fluorescence is emitted. However, when the probe hybridizes to the target amplicons during PCR amplification, the fluorophore and quencher become separated from each other and fluorescence can be detected (Didenko, 2001; Cockerill and Smith, 2002). Other probes are based on stem-loop hairpin structures, molecular beacons being the simplest hairpin probes. Scorpion probes covalently couple the stem-loop structure to a PCR primer and, because hybridization of probe sequence to amplicon is intramolecular, Scorpion probes are more efficient than binary systems such as molecular beacons.

Several plant pathogenic bacteria have been detected in plants including *Clavibacter michiganennsis* sub sp. *spedonicus* (Schaad *et al.*, 1999), *R. solanacearum* (Weller *et al.*, 2000b) in potato tubers, *Agrobacterium* strains from plants (Weller and Stead, 2002) and *Xyllella fastidiosa* in asymptomatic grape vines (Schaad *et al.*, 2001) by using several real-time PCR (Fig. 2.4). Weller *et al.* (2000a) were able to detect 100 cfu/ml of pure culture of *R. solanacearumm* using TaqMan® and the 7700 sequence detection system using either a genus-specific 16S probe or a biovar 2 specific probe. However, when spiked potato tuber extracts were assayed, the threshold for detection dropped significantly to 10^5 and 10^7 cfu/ml with the species specific and biovar 2 specific probes respectively (Weller *et al.*, 2000a) in tuber extracts diluted by 1:10 and in undiluted extract, the threshold was 10^7 cfu/ml because of presence of PCR inhibitors. It was also found severe reaction inhibitors in potato tubers in a real time TaqMan® assay for *C. michiganensis* subsp. *sepedonicus* (Schaad *et al.*, 1999).

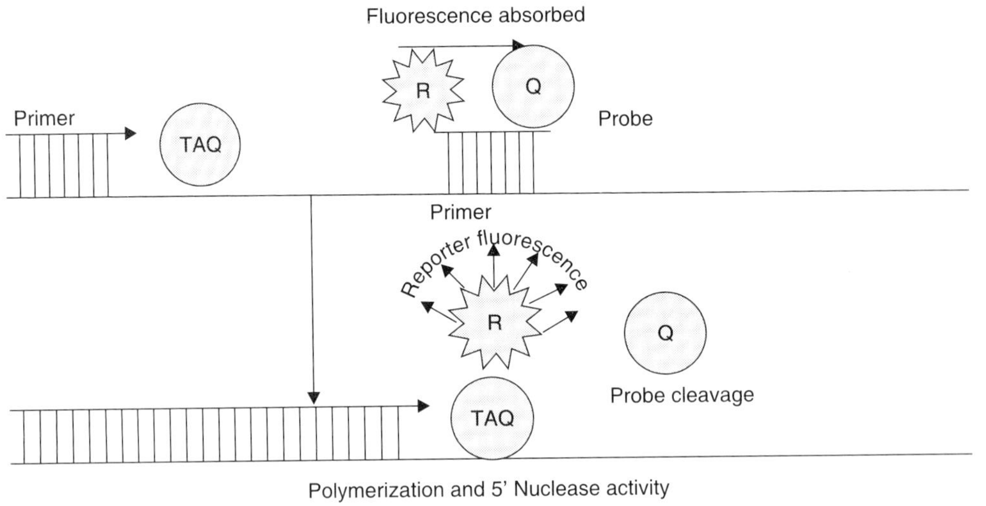

Fig. 2.4 TaqMan TM, chemistry in real-time PCR

Bio-PCR

In this PCR, combine the viable enrichment of mostly growth media with an enzymatic amplification, which is known as BIO-PCR. The target bacterium is enriched in liquid or solid media and detected at extremely low levels in seeds and other propagative materials. The BIO-PCR assay includes the following simple steps as (i) extracting a sample (ii) plating a sample onto agar media (half are retained for visual recovery or adding sample to liquid medium (iii) incubating for 15-72h, depending on the growth of target bacterium; (iv) washing plates to remove bacteria or centrifuge liquid medium and using 1 or 10 μl for direct PCR.

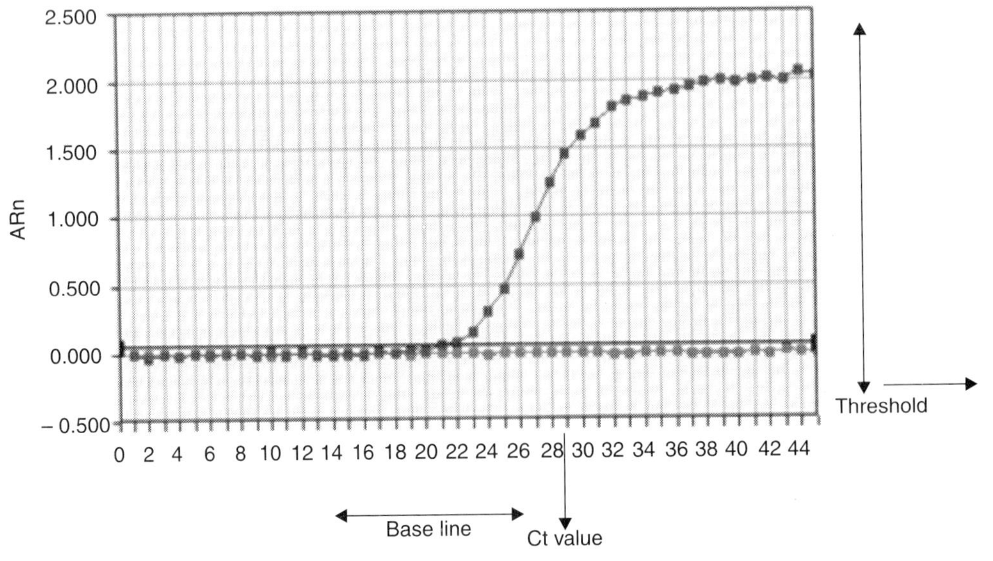

Fig. 2.5 Amplification plot from positive and negative real-time PCR reactions.

In general, bacterial assays can be very fast since extraction of DNA is not needed; a simple 2–3 min hot start is adequate to break the cells. Also, direct PCR reduces the chances of cross-contamination. However, even when used as a direct assay, real-time PCR has several disadvantages compared with isolation techniques when used to detect very low levels of bacteria as often occurs in seeds. False negatives can result from using small volumes (1–10 µl) and presence of PCR inhibitors is a problem (Rasmussen and Wulff, 1991; Prosen *et al.,* 1993; Weller *et al*., 2000a). False positives can result from presence of dead cells or free DNA. To avoid these inherent disadvantages and to utilize the best of both techniques, PCR can be used in combination with pathogen isolation on agar media. Samples can be archived at –20°C. BIO-PCR has several advantages over traditional PCR, including increased sensitivity, elimination of PCR inhibition, and detection of viable cells only. The keys to developing a successful real-time BIO-PCR protocol are (i) choose an appropriate semiselective liquid or solid medium and (ii) determine precisely the time for enrichment. For solid media, the time for pinpoint-size colonies to appear is the critical step for the success of BIO-PCR. Too long an incubation time permits saprophytes to grow and interferes with PCR. Ideally, pinpoint-size colonies of the target organism should be visible before the saprophyte colonies reach 1 mm diameter or greater. If such a medium is not available for the target organism, a new one should be designed. For most plant pathogens, 24–48 h is sufficient time to form pinpoint colonies. Faster growing species such as *Erwinia* and *Ralstonia* only require an enrichment time of 10–15 h. A single pinpoint-size colony normally contains over 1000 cells/ml, which is the threshold for PCR. A medium that does not produce mucoid growth is best. A major disadvantage of BIO-PCR is the chance for cross-contamination during plating or washing the samples. Also, 15–48 h is needed to enrich for growth of the target organism. If time is of greatest concern, BIO-PCR is not recommended. However, if sensitivity is most important, BIO-PCR should be considered.

For even greater sensitivity, BIO-PCR can be combined with membranes. This works extremely well when sampling large volumes of water. Using surfactant-free membranes made especially for use in BIO-PCR by Sartorius North America (Edgewood, N.Y.) and a 10-20 ml syringe (Schaad *et al.,* 1996), samples are filtered and the membrane is placed right side up onto semi-selective agar media and incubated for 1–3 days, depending upon the growth rate of the target organism. After incubation, the filter is removed and placed in a microfuge tube with 50 µl water and vortexes, as described above for identification of a pure culture. By filtering 100 ml of water, it would not be possible to detect one cell because of the dilution factor alone; however, with BIO-PCR, a single cell trapped on the filter will result in thousands of targets after a 15-to 36-h enrichment. The sample is then used for direct PCR. Real time-PCR with using enrichment media protocol has been developed for *R. solanacearum* (Ito *et al.,* 1998) and *R. solanacearum* bv-2 in potato tubers (Weller *et al.,* 1999, 2000a).

Singh *et al.* (2014b) developed BIO-PCR protocol using nucleotide sequences of virulence, hypersensitive response and pathogenicity (hrp) gene and they reported sensitivity of the primer was 200 cfu/ml and improved further detection level by using non-specific enrichment medium casamino acids-pepton-glucose broth followed by PCR. Out of 130 samples of asymptomatic tomato plants, irrigation water, and soil collected from bacterial wilt infested fields in different agro-climatic regions of India, *R. solanacearum* was detected in 86.9, 88.5, and 90.9 per cents samples using BIO-PCR, respectively (Fig. 2.6 and Fig. 2.7).

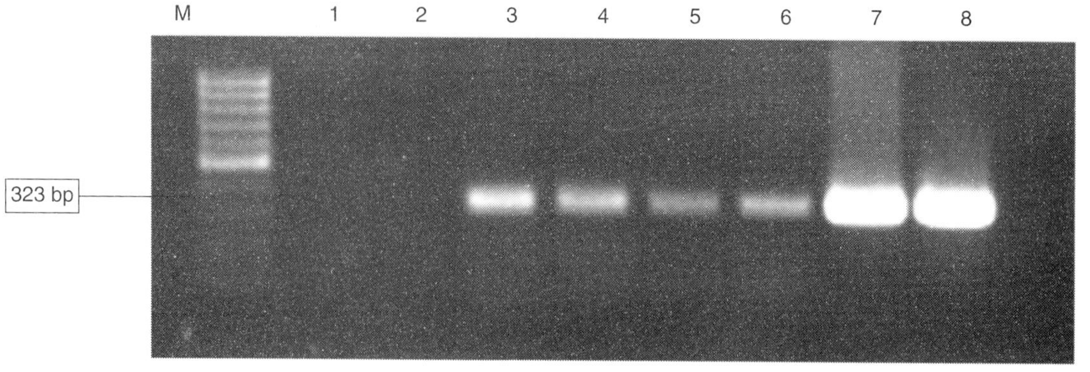

Fig. 2.6 Detection of *R. solanacearum* from irrigated water in farmer's tomato field by *hrp* gene based a set of primer hrp_allF and hrp_all R amplified at 323 bp. Lane M: 100 bp DNA ladder, lanes 1&2: soil samples, 3&4: water samples (BIO-PCR), 5&6: Direct irrigation water samples, 7: oozing from wilt infected tomato plant, 8: +ve control amplified.

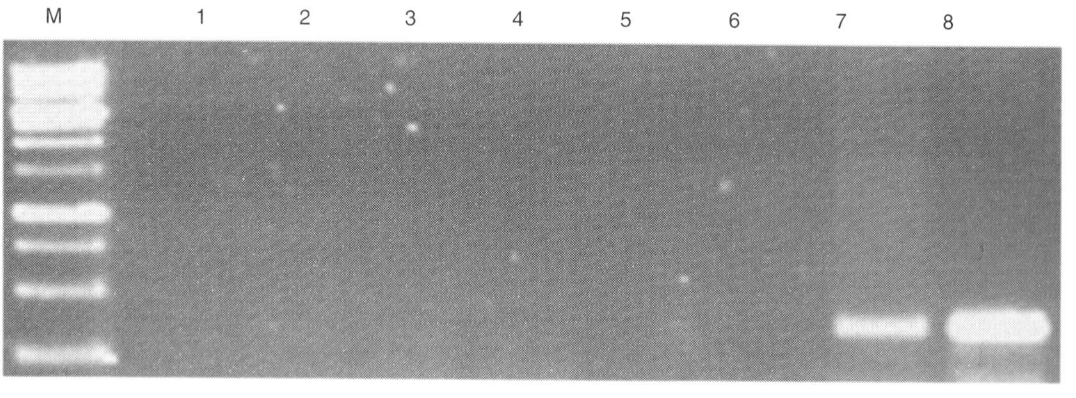

Fig. 2.7 Amplification of fragments of the *hrp* gene from *R. solanacearum* separated on an agarose gel showing the 323 bp. Lanes M: 1 Kb DNA ladder, 1-6: soil DNA, 5-6: direct water, 7: Bio culture from soil, 8: positive control

Multiplex PCR

Multiplex PCR allows the simultaneous and sensitive detection of different DNA targets in a single reaction. On the other hand, PCR detection protocols can be designed to verify the presence of more than one pathogen in plant material by looking for common specific sequences in two or more of them, or to detect related bacteria on multiple hosts. Multiplex PCR is useful in plant pathology because different bacteria and/or RNA viruses frequently infect a single host and consequently sensitive detection is needed for the propagation of pathogen-free plant material. There are several examples in plant pathology of simultaneous detection of several targets and the amplification by multiplex PCR.

Nested-PCR

Sensitivity and specificity problems associated with conventional PCR and RT-PCR can be reduced by using nested PCR-based methods, based on two consecutive rounds of amplification

(Simmonds *et al.,* 1990; Porter-Jordan *et al.,* 1990). Usually, the products of the first amplification are transferred to another tube before the nested PCR is carried out using one or two internal primers (hemi-nested or nested amplification respectively). Sensitivity is increased by two orders of magnitude reaching about 10^2 bacterial cells/ml of extract (Pradhanang *et al.,* 2000). However, the two rounds of amplification in different tubes also increase the risk of contamination, especially when the method is used on routine in a large scale. One limitation of the nested PCR approach concerns the need to accurately establish the ratio between external and internal primers and the use of limiting amounts of external primers to avoid interference during the second amplification.

This newly developed method combines the advantages of multiplex RT-PCR with the sensitivity and reliability of nested RT-PCR carried out in a single closed tube. It enables the simultaneous detection bacterial DNA targets in a single analysis, performed in plants. It also saves time and reagent costs because it can be performed in a single reaction, although accurate design of compatible primers is needed. The compartmentalization of a single Eppendorf tube with a pipette tip allowed multiplex-PCR and nested PCR to be combined effectively. During the first amplification reaction there is no interference of the external with internal primers because they are physically separated from the initial reaction cocktail. Once the multiplex RT-PCR ends, the internal primers are mixed with the products of the first reaction before proceeding to the nested multiplex. It has been shown that primers OLI-1, OLI-2, Y2 and JE2 are not completely specific for species of *R. solanacearum* but nested PCR using with these primers provide a very specific tool for identification of *R. solanacearum* strains in soil and water samples. The primers with their positions, which have been used for conventional and nested PCR is described by Khakvar *et al.* (2008) as given Fig. 2.8.

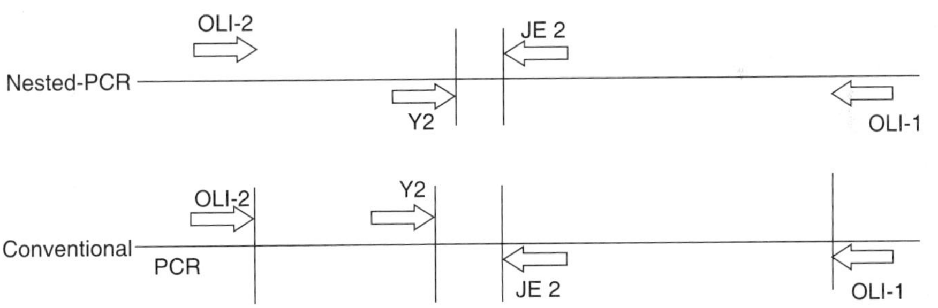

Fig. 2.8 The primers with their positions that have been used for conventional and nested–PCR.

Loop-mediated Isothermal Amplification (LAMP)

Loop-mediated isothermal amplification (LAMP) is another type of isothermal amplification that it is being increasingly used in the diagnostic field offering sensitivity and economic costs (Notomi *et al.*, 2000). The method requires a set of four specifically designed primers that recognize six distinct sequences of the target and a DNA polymerase with strand displacement activity. The amplification products are stem-loop DNA structures with several inverted repeats of the target and cauliflower-like structures with multiple loops, yielding >500 g/ml. The LAMP reaction was enhanced by the addition of loop primers (Nagamine *et al.,* 2001, 2002a,b), reducing time and increasing sensitivity. The amplification takes place at 60-65°C for 60 min. Although it was initially developed for DNA it can be adapted to amplify RNA (RT-LAMP) (Fukuta *et al.*, 2003). Kubota *et al.* (2008) developed four sets of primers to replicate the gene coding for the flagellar subunit, Fli C and conditions for detection of *R. solanacearum* were optimized to complete in 60 min at 65°C. Magnesium pyrophosphate resulting from the amplification reaction could be detected optically as an increase in the solution turbidity, and the DNA products spread in a reproducible ladder – like banding pattern after electrophoresis in an agarose gel. This method is rapid but and sensitive and detection limit of this LAMP assay is between 10^4-10^6 cfu/ml.

Microarray Technology

Microarray technology used for detection of bacterial plant pathogens is very limited. Microarrays are generally composed of thousands of specific probes spotted onto a solid surface either nylon or glass and each probe is complementary to a specific DNA sequence like genes, ITS, ribosomal DNA. Hybridization with the labeled complementary sequence provides a signal that can be detected and analyzed. Although there is great potential for microarray technology in the diagnosis of plant diseases, the practical development of this application is still in progress. For example, following the methodology utilized for genetic analysis (Brown and Botstein, 1999) large numbers of DNA probes used in two-dimensional arrays have allowed thousands of hybridization reactions to be analyzed at the same time (Hadidi *et al.,* 2004). Until now, the microarray technology focuses its use in multiplex format of similar or very different pathogens, taking advantage of the number of probes that can be employed in one chip (Bonants *et al.,* 2002; Schoen *et al.*, 2002, 2003; Fessehaie *et al.*, 2003; Franke-Whittle *et al.*, 2005; Bonants *et al.,* 2005; Boonham *et al.,* 2007; van Doorn *et al.,* 2007; Pasquini *et al.*, 2008). With the availability of genomic sequences of pathogens and the rapid development of microarray technology, as well as a renewed emphasis on detection and characterization of quarantine pathogens, there is a rush in the European Union to set up this technology and apply it to detection. The probes can be prepared in at least three basic formats: a) PCR fragments arrayed on nylon membranes, hybridized against cDNA samples radioactively labeled, called microarrays (Richmond *et al.,* 1999); b) PCR products spotted onto glass slides and DNA labeled with fluorescent dyes (Richmond *et al.,* 1999; Zimmer *et al.,* 2000; Wei *et al.,* 2001); and c) oligonucleotides of different lengths (from 18 to 70 bp) arrayed and hybridized with the same type of labelled DNA material (Lockhart *et al.,* 1996; Loy *et al.,* 2002, 2005; Fessehaie *et al.,* 2003; Peplies *et al.,* 2003). For bacterial detection, the material spotted until now is almost universally oligonucleotides targeting the 16S- 23S rDNA genes (Crocetti *et al.* 2000; Loy *et al.* 2002; Fessehaie *et al.,* 2003; Peplies *et al.,* 2003; Loy *et al.,* 2005; Franke-Whittle *et al.*, 2005). The microarrays are analysed either by scanning or by a direct imaging system. Another type of microarray under development is called the nanochip (Sosnowski *et al.* 1997; Nanogen, Inc., San Diego, CA 92121, USA) based on an electronically addressable electrode array that provides direct electric field control over the transport of charged molecules to selected micro locations and concentration over an immobilized substrate. A particular feature of this system is that biotinylated immobilized molecules can be either oligo capture probes or amplified PCR samples. Hybridization is detected and analyzed by fluorescent oligo probes. By regulating the electric-field strength, hybridization stringency can be adjusted for homologous interactions. Nano chips have shown high specificity and accuracy to diagnose bacterial pathogens affecting potato, due to their ability to discriminate single nucleotide changes (Ruiz-García *et al.* 2004). The potential of microarray technology in the detection and diagnosis of plant diseases is very high, due to the multiplex capabilities of the system. Moreover, it can be coupled with other systems, *i.e.,* to perform nucleic-acid extraction on the chip (Liu *et al.,* 2007), achieve PCR reactions and their detection on the same device (van Doorn *et al.,* 2007) or even mix all the systems in one (Lee *et al.* 2006), providing the possibility of automation that can be of great importance and utility. This possibility, with the coupling with previous steps of the analyses (extraction, PCR, detection) promises a wider use in future protocols (Bonants *et al.,* 2005; Lee *et al.,* 2006; Boonham *et al.,* 2007; van Doorn *et al.,* 2007; Liu *et al.,* 2007). Furthermore, new developments, like the labeling of total bacterial RNA (François *et al.,* 2003), the direct detection of DNA or RNA without previous PCR amplification (Call *et al.,* 2003), or multiplex detection based on padlock probe technology (pUMA) (Bonants *et al.,* 2005), may make this technique simpler.

Guidot *et al.* (2009) used microarray representative of the phylotype I strain GMI1000, to which were added 660 70-mer oligonucleotide representatives of new genomic islands detected in the phylotype IIB1 strain IPO1609. The brown rot strain-specific genes thus identified were organized in nine clusters covering 2 to 29 genes within the IPO1609 genome and 6 genes isolated along the genome. They reported that these specific genes, 29 were parts of mobile genetic elements. Based on the findings they reported that multiple probes are required to consistently detect all IIB1 strains and they recommended that the use of probes, which are not part of genetic mobile elements.

Identification of Phylotyping of *R. solanacearum*

130 isolates of *R. solanacearum* isolated from different crops of solanaceous crops (tomato, capsicum, chilli, brinjal and potato) from states of Jharkhand, Orissa, Himachal Pradesh and Uttarakhand and Jammu & Kashmir (Fig. 2.9). Phylotype affiliation of each isolate was determined as described (Frgan and Prior, 2005; Prior and Fegan 2005). Multiplex–PCR was carried in 25 µl volume of master mix, containing 5X PCR buffer (5 µl), 25mM $MgCl_2$ (1.5 µl), 10mM dNTPs (0.5 µl), (primers: Nmult:21:1F : 1 µl, Nmult:21:2F : Nmult:21:2R 1 µl, primer Nmult:22:InF : 1 µl, Nmult:22InR : 1 µl, Nmult:23:AF : 1 µl, OLI1: 1 µl, Y2: 1 µl), Taq polymerase: 0.25 µl and Water : 18.25 µl. The following cycling program was used in the Master cycler Gradient: Initial deturation: 96°C for 5 min, 30 cycles of 94°C for 15 s, annealing 59°C for 30 s, extension 72°C for 30 s followed by final extension 72°C for 10 min. A 10 µl aliquot of each primer amplified product was subjected to electrophoresis on 1.5 % agarose gel, stain with ethidium bromide and visualized as phylotype 1, which produce 2 band, 144 bp & 288 bp. 288 bp band amplified by 16s rRNA primer using in master mix, that sequence conserved in all *R. solanacerum* bacteria Nmult:21:1F: 5'-CGTTGATGAGGCGCGCAATTT-3'; Nmult:21:2F: 5'-AAGTTATGGA CGGTGGAAGTC-3'; Nmult:22:InF: 5'- ATTGCCAAGACGAGAGAAGTA-3'; Nmult:23: AF: 5'- ATTACGAGAGCAATCGAAAGATT-3'), with a unique and conserved reverse primer 13 (Nmult:22:RR:5'-TCGCTTGACCCTATAACGAGTA-3'),

Recently, a new classification scheme has been described for strains of *R. solanacearum*, based on comparison of DNA sequences is commonly used for classification studies of strains

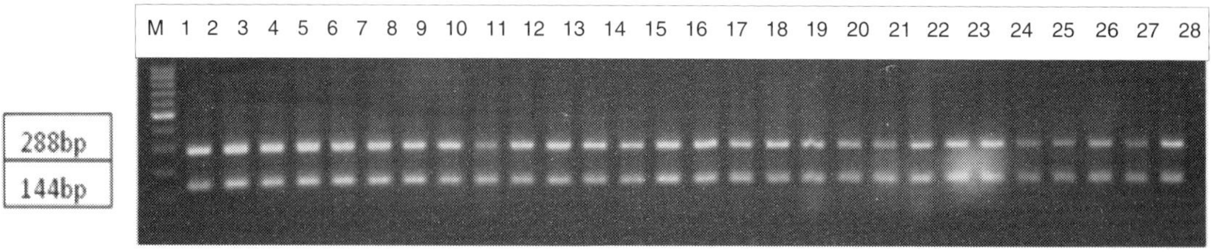

Fig. 2.9 130 isolates of *Ralstonia solanacearum* isolated from different crops of solanaceous crops (tomato, capsicum, chilli, brinjal and potato) from states of Jharkhand, Orissa, West Bengal, Himachal Pradesh, Uttarakhand and Jammu & Kashmir. All isolates of *R. solanacearum* belong to phylotype I, which is mostly dominated in Asian countries.

of microorganism. It is basically assumed that the higher the homology is between two strains, the more closely related the strains are in terms of evolution. These types of studies are known as phylogenetic studies. A sequevar or sequevance variant is defined as a group of strains with a highly conserved sequence within the area sequenced sequevars. A phylotype is defined as a group of strains that are closely related based on phylogenetic analysis of sequence data. Phylotypes were identified within the species that broadly reflects the ancestral relationships and geographical origins of the strains *In R. solanacearum* four phylotypes *viz*., Phylotype I, Phylotype II, Phylotype III and Phylotype IV have been reported by Fegan and Prior (2005). Each phylotype is composed of a number of sequevars (Fig. 2.10).

Amplified Fragment Length Polymorphism (AFLP) - PCR

The AFLP fingerprinting method is usually applied for discrimination at low taxonomic level, *i.e.*, species and strain level. Basically, genomic DNA from a sample is digested to completion, typically with two different restriction enzymes in order to produce a large number of fragments. Specific nucleotide adapters of 25-30 bp are ligated to the restricted DNA fragments. Oligonucleotide primers that anneal to these adapters are used, however, to impart additionally selectivity, the primers vary at their 3' end, such that they will amplify only a subset of the restricted DNA fragments. These 3'- end primer extensions must match the target sequence for amplification to occur. The amplified DNAs are are separated on a denaturing polyacrylamide gel and the amplified fragments are visualized by autoradiography. This method is highly discriminative and very useful in population studies, but expensive and labour intensive (Rademaker *et al.*, 2000). AFLP methodology has already been used to study the diversity of race 3 isolates of *R. solanacearum* (Van der Wolf *et al.*, 1998) but has never been used to analyse a worldwide collection of *R. solanacearum* strains. Although AFLP analysis cannot be performed on mixed cultures, such studies of pathogen diversity using pure cultures of representative strains will eventually permit rapid identification of unknown isolates and potentially be useful in diagnostics. Pousser *et al.* (2000) used AFLP method to analyse 96 *R. solanacearum* strains, one BDB strain and one *Ralstonia pickettii* strain collected from worldwide as described by Janssen *et al.* (1996) and Vos *et al.* (1995) with slight modifications. They used *Msp*I and *Sac*I to digest DNA instead of *Eco*RI and *Mse*I. Genomic DNA (200 ng per sample) was digested for 2 h at 37°C in 50 µl

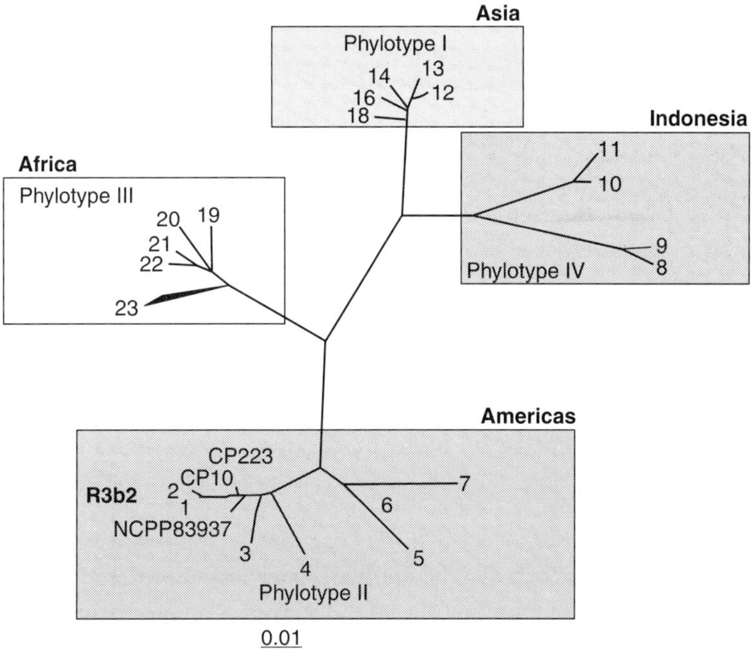

Fig. 2.10. Classification and geographic origins of *R. solanacearum* strains based on sequence analysis of endoglucanase is a type of cellulose, which is class of enzymes produced by fungi, bacteria and protozoans that catalyze the cellulolysis of endoglucanase gene sequences. Numbers indicate sequevars (1 to 23) *(Source: Fegan and Prior, 2005)*

(final volume) containing 5 U *Msp*I, 5 U *Sac*I, 0.125 μl BSA (10 μg μl^{-1}) and 2.5 μl 10x`One Phor All ' buffer. Next, 50 pmol double-stranded *Msp*I-adapter (5'-GACGATGAGTCCTGAA-3', 5'-CGTTCAGGACTCATC-3') (50 pmol μl^{-1}), 5 pmol double- stranded *Sac*I-adapter (5'-CTCGTAGACTGCGTACAAGCT- 3', 5'-TGTACGCAGTCTAC-3') (5 pmol μl^{-1}) 1 μl ATP (10 mM), 1 U T4 DNA ligase (5 Uμl^{-1}) and 2.5 μl 10x 'One Phor All' buffer were added to the digested DNA and the ligation reactions were performed for 3 h at 20°C. Digested (D) and ligated (L) DNA were diluted (D) eightfold and the resulting DLD DNA was then stored at 4°C until used. Selective amplifications were done with two primers (*Msp*I-primer and *Sac*I-primer) complementary to the adapter sequences, and the *Msp*I and *Sac*I restriction sites respectively, with additional selective nucleotides at their 3'- ends (cytosine for the *Sac*I-primer and cytosine plus guanine for the *Msp*I-primer). The *Sac*I-primer was labelled with γ^{-33P} [10 μCi (370 kBq) per DNA amplification and T4 polynucleotide kinase. DNA amplifications were carried out in a 50 μl reaction mixture. DLD DNA (5 ll) was added to 45 μl mixture containing 5 μl 10x PCR buffer, 1.5 μl labeled *Sac*I-primer (50 μg μl^{-1}), 2 μl unlabelled *Msp*I-primer (30 μg μl^{-1}), 1.5 μl $MgCl_2$ (50 mM), 8 ll of each dNTP (1.25 mM) and 0.6 ll *Taq* polymerase (5 U μl^{-1}). Amplifications were performed with a thermocycler by using the following protocol: 30 cycles consisting of denaturation at 94°C for 30 s, annealing at 56°C for 1 min and extension at 72°C for 1 min. Amplified fragments were separated by electrophoresis on 5% polyacrylamide gels and fingerprint patterns were visualized as described by Vos *et al.* (1995). They found that 159 DNA fragments which were reproducibly detected, ranging from 50 to 350 bp. Different DNA fingerprints were generated from strains belonging to the same biovar (for example 23, 12 and 13 from biovars 1, 2 and 3, respectively), but also originating from the same geographical area (for instance 6 from biovar 2 strains from Reunion Island). Identical DNA patterns could be produced by strains originating from different countries (for example IV 19 was found in strains from 7 countries). Depending on the strain, DNA patterns contained 50-80 different DNA bands. Comparison of AFLP fingerprints revealed that 95% of the fragments were polymorphic. Three fragments appeared to be specific to *R. solanacearum* species. Many of the DNA fragments were useful for distinguishing subgroups within *R. solanacearum* species.

Detection and Identification of *R. solanacearum* in Water

Principle

The validated detection scheme, described in this section, is applicable for pathogen detection in samples of surface water and can also be applied for testing samples of potato processing or sewage effluents. However, it is important to note that the expected sensitivity of detection will vary with the substrate. Sensitivity of the isolation test is affected by populations of competing saprophytic bacteria which are generally much higher in potato processing and sewage effluents than in surface water. Whereas the scheme below is expected to detect as few as 10^3 cells per litre in surface water the sensitivity of detection in potato processing or sewage effluents is likely to be significantly lower. For this reason, it is recommended to test effluents after any purification treatments (e.g., sedimentation or filtration) during which saprophytic bacterial populations are reduced. The limitations in sensitivity of the test scheme should be considered when assessing the reliability of any negative results obtained. Whereas this scheme has been successfully used in survey work to determine presence or absence of the pathogen in surface water, its limitations should be realised when used in similar surveys of potato processing or sewage effluents.

Sample Preparation

In sample preparation following note is important for better result.

1. Detection of *R. solanacearum* in surface water is most reliable during late spring, summer and autumn seasons when water temperatures exceed 15°C.
2. Repeated sampling at different times in the above mentioned period at designated sampling points will increase the reliability of detection by reducing the effects of climatic variation.
3. Take into account the effects of heavy rainfall and the geography of the watercourse to avoid extensive dilution effects that may obscure presence of the pathogen.
4. Take surface water samples in the vicinity of host plants if these hosts are present.

At selected sampling points, collect water samples by filling disposable sterile tubes or bottles at a depth if possible below 30 cm and within 2 m from the bank. For processing and sewage effluents, collect samples from the point of effluent discharge. Sample sizes up to 500 ml per sampling point are recommended. If smaller samples are preferred, it is advisable to take samples on at least three occasions per sampling point, each sample consisting of two replicated sub-samples of at least 30 ml. For intensive survey work, select at least three sampling points per 3 km of watercourse and ensure that tributaries entering the watercourse are also sampled.

Transport samples in cool dark conditions (4 to 10°C) and test within 24 hours.

If required, the bacterial fraction may be concentrated using one of the following methods:

(a) Centrifuge 30 to 50 ml sub-samples at 10,000 g for 10 minutes (or 7000 g for 15 minutes) preferably at 4 to 10°C, discard the supernatant and resuspend the pellet in 1 ml pellet buffer (10 mM phosphate buffer, pH 7.2).

(b) Membrane filtration (minimum pore size 0.45 μm) followed by washing the filter in 5 to 10 ml pellet buffer and retention of the washings. This method is suitable for larger volumes of water containing low numbers of saprophytes. Concentration is usually not advisable for samples of potato processing or sewage effluent since increased populations of competing saprophytic bacteria will inhibit detection of *R. solanacearum*.

Detection and Identification of *R. solanacearum* in Soil

Principles

The validated detection scheme, described in this section, is applicable for pathogen detection in soil samples but can also be used to test samples of solid potato processing waste or sewage sludge. However, it should be noted that these methods are insufficiently sensitive to guarantee detection of low and/or irregularly dispersed populations of *R. solanacearum* that may occur in naturally infested samples of these substrates. The limitations in sensitivity of this test scheme should be considered when assessing the reliability of any negative results obtained and also when used in surveys to determine presence or absence of the pathogen in soils or sludges. The most reliable test for presence of the pathogen in a field soil is to plant a susceptible host and monitor it for infection, but even with this method low levels of contamination will escape detection.

Sample Preparation

Sampling of field soil should follow standard principles used for nematode sampling. Collect 0.5 to 1 kg of soil per sample from 60 sites per 0.3 ha from a depth of 10 to 20 cm (or in a grid of 7 × 7 metres) If the pathogen is suspected to be present, increase the number of collection points to 120 per 0.3 ha. Maintain samples at 12 to 15°C prior to testing. Sample potato processing and sewage sludges by collecting a total of 1 kg from sites are representing the total volume of sludge to be tested. Mix each sample well before testing. Disperse sub-samples of 10 to 25 g of soil or sludge by rotary shaking (250 rpm) in 60 to 150 ml extraction buffer for up to two hours. If required, addition of 0,02% sterile Tween-20 and 10 to 20 g sterile gravel may assist dispersion. Maintain the suspension at 4°C during testing.

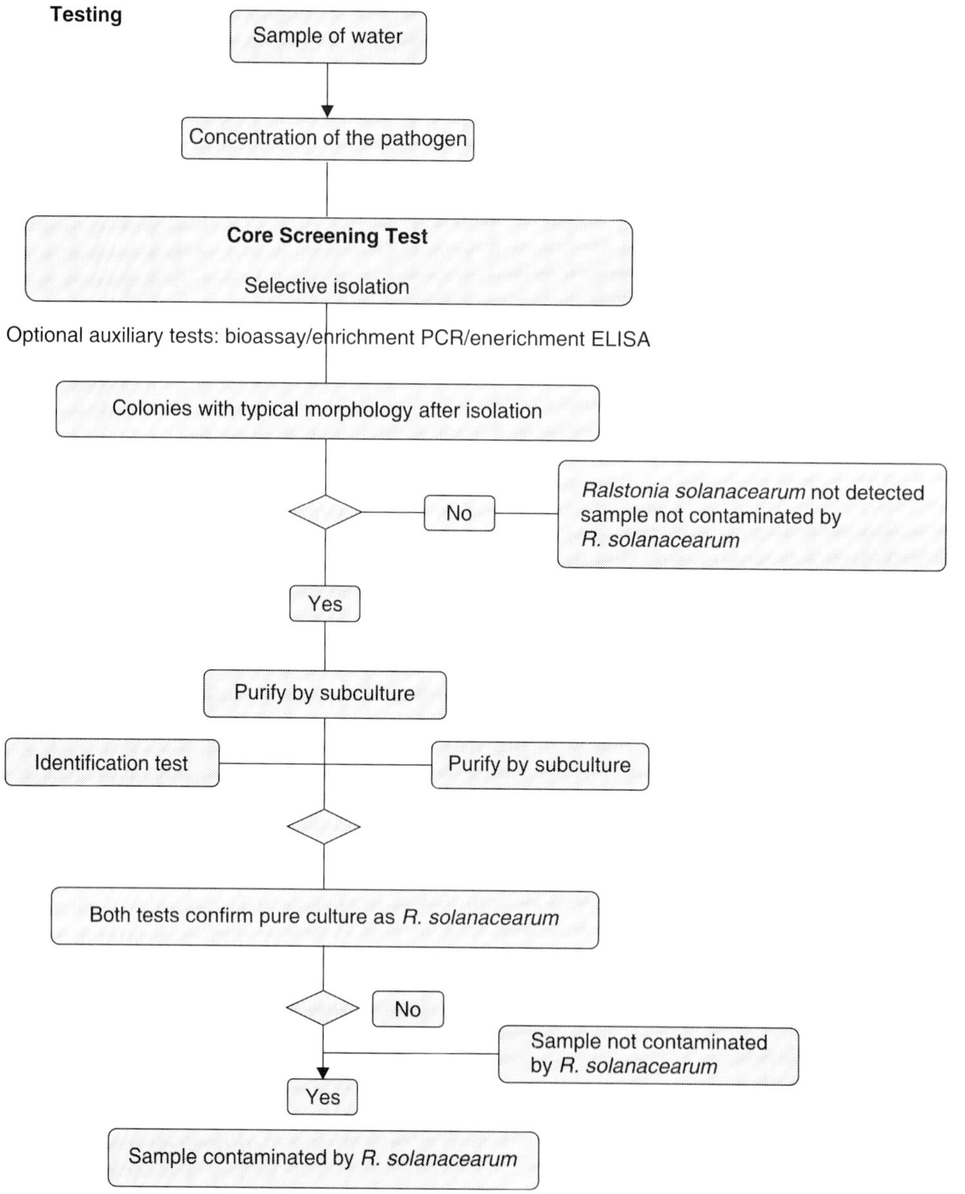

Fig. 2.11 Flow chart for detection and identification of *R. solanacearum* in water

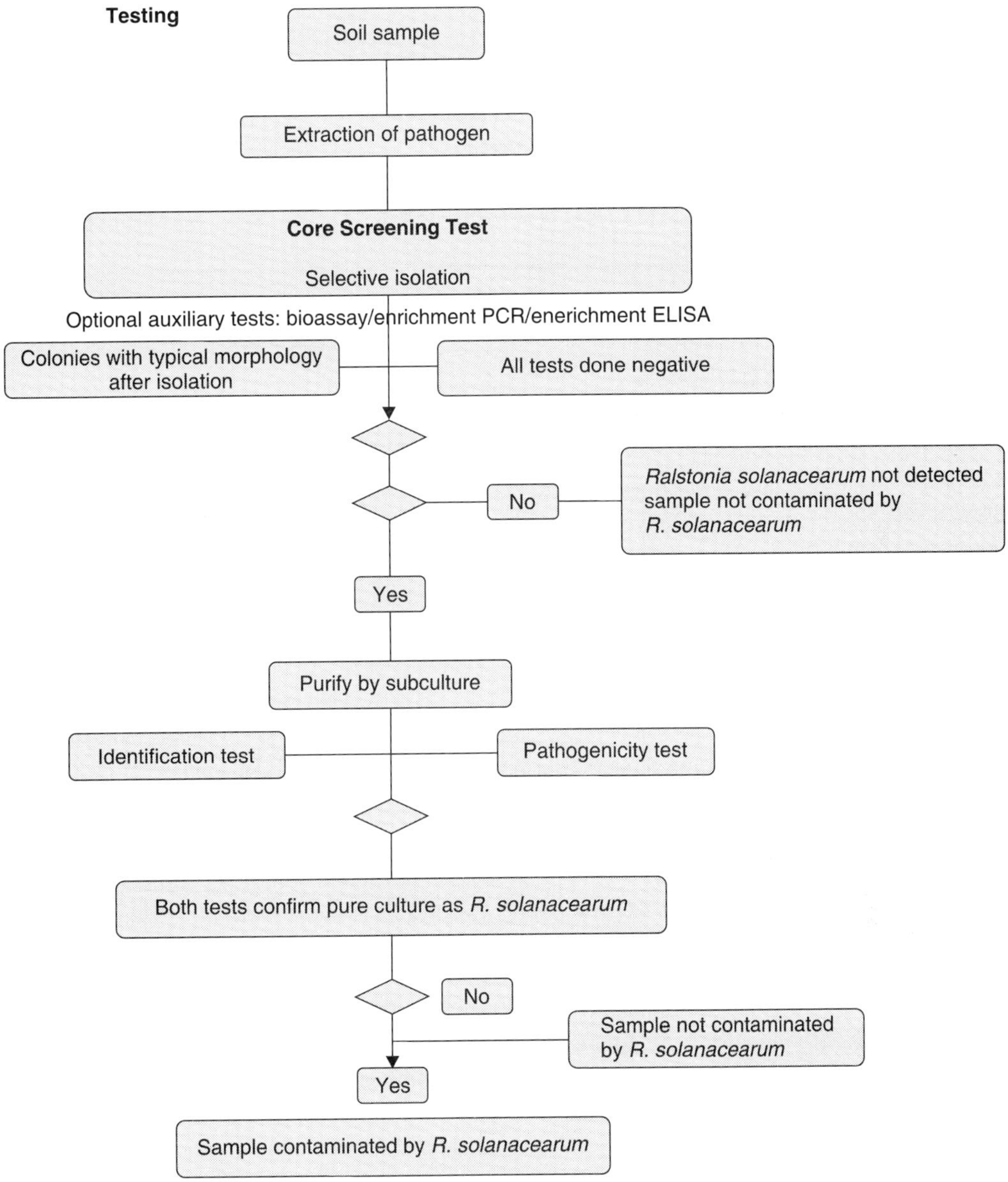

Fig. 2.12 Flow chart for detection and identification of *R. solanacearum* in soil

3
Maintenance and Preservation of Culture

It is very difficult to maintain and preserve the culture of *Ralstonia solanacearum* alive, to keep them in a genetically stable form and save labour. The selection of method depends on the objectives of preservation. The method is to be used depends on whether the culture is to be preserved for a few days until a positive identification is made or for future long term reference. Preservation methods have a common objective of reducing the organism's metabolic rate as low as possible, while still maintaining viability. A high recovery or survival rate with minimum of damage or change to the surviving bacteria is also highly desirable. The various methods such as sterile tap water, slant cultures covered with sterile paraffin oil, cryostorage and lyophylization are used to maintain and preserve *R. solanacearum* culture.

3.1 STERILE TAP WATER

Ralstonia solanacearum can be stored for several years in distilled water/deionized water or tap water boiled to eliminate chlorine without significant loss of virulence or change in phenotype. Cultures should be streaked first on the TZC medium and examine contamination of other saprophytic bacteria and fungi carefully. The well-isolated fluidal colonies should be restreaked on to the CPG containing plates. It has been reported that some strains are sensitive to the formazan pigment produced for TZC. Single distinct colony from a 48-72 h old culture of *R. solanacearum* is selected and 1 large or 2 small loops of the bacterial culture are to 5 -8 ml of sterile water in screw cap test tubes. Suspensions should be turbid and contain 10^7 to 10^8cfu/ml (Kelman and Person, 1961). the suspension should be stored about at 20°C with minimum 15°C and maximum 28°C. The stored culture should be examined at six month interval by streaking on the medium. The bacteria can be preserved for 8 -10 years in water and maintain viability of the culture. Sunaina *et al.* (1998) have developed a technique for multiplication of *R. solanacearum* in pure water. In this technique the initial bacterial concentration remains less than 10^6 cfu/ml followed by incubation period of first four days till it reaches a level of 10^{6-7} cfu /ml. This continuation of growth is not desirable for long-term storage of bacteria in its virulent form, as growth during storage in water can invite the problem of contamination (Kelman and Person, 1961). When initial bacterial population at the time of storage at 20°C in water, *i.e.*, equal to or more than 10^6 cfu/ml, the bacterial population will remain constant for a long period. As there will be no further multiplication of bacteria in this case, so the appearance of a virulent type of *R. solanacearum* will be rare and the bacteria in this condition can be stored for a long period at constant concentration without losing virulence.

3.2 SLANT CULTURES COVERED WITH STERILE PARAFFIN OIL

It is a simple method of preserving culture of *Ralstonia solanacearum* on agar slopes covered by sterile mineral oil. Medicinal grade liquid paraffin oil is used, which protects the controlled drying. The interval between subculture may be extended for several months to years. Long-term preservation is often further improved, if the oil layered cultures are stored at 4°C. Although this technique of preservation is simple but for purpose of transport and handling is not as convenient as the drying methods such as freeze drying or freezing in liquid nitrogen. Another disadvantage with this method is that there is a slow diffusion of oxygen through the oil, which allows growth to continue at a slow rate and therefore genetic stability may be poor. A loop of 48-72 h old bacterial culture should be transferred on a fresh CPG or PDA slants and incubate it at 30°C for 48 h. After the growth of bacteria on the slant, it should be covered with sterile paraffin oil up to at least 10 mm above the top of the slope and stored at 20°C (Kelman and Jensen, 1951). It is better to grow culture on short slopes to reduce the cost of oil.

These two methods of storage, if properly done, are adequate for storing *R. solanacearum* for several years. Cultures have maintained viability for 9-10 years in water and under mineral oil. Routine checks should be made at 6 months to 1 yr. intervals on stored cultures by streaking them out on TZC agar to determine the stability of the virulent type. Mutants may develop due to changes in temperature, pH, etc., so a culture can be lost if this happens. If a culture reaches as much as a 50% mutant level, it is wise to prepare a new stock of the virulent form.

3.3 CRYOGENIC STORAGE

In cryogenic storage, bacteria can survive long term storage in the frozen state by markedly reducing their metabolic rate. *R. solanacearum* has been stored in freezers at less than –70ºC (–20 to –70°C) temperature, which is the most convenient method of long-term storage with minimal phenotypic changes. The lower the temperature the less is the loss of viability. The bacteria and temperature higher than – 70°C should not be used for long storage but may be satisfactory for up to 1 year. The bacteria survive freezing in the growth medium, the addition of cryoprotectants such as 5 -10% glycerol or dimethyl sulphoxide affords some protection from the stress of freezing. For storing *R. solanacearum,* 0.6 ml of 20% (v/v) glycerol in water is added into each screw-capped freezer vials, capped loosely, and autoclaved for 20 min. The tubes are allowed to cool to room temperature, caped tightly, and stored at 4ºC until needed. To stock a strain, 0.5 ml of CPG broth is added to each of two vials. *R. solanacearum* is streaked on CPG or TZC medium and incubated at 30°C for 48 h. Then the cells from one of the heavily streaked two-day old is transfered to each vial; thoroughly mixed, sealed tightly, and placed in the ultralow freezer. One vial serves as the working stock and the second vial as a backup stock in a separate ultralow freezer to reduce loses, if one of the freezers fails. A vial from the ultralow freezer is removed quickly, while the contents remain frozen, a sterile wooden applicator or hypodermic needle is used to scrape off a small quantity of the sample from the surface of the frozen stock and streaked onto a TZC plate to recover strains (Sly, 1984; Denny and Hayward, 2001).

3.4 LYOPHILLIZATION (FREEZE DRYING)

Lyophillization has been widely used as the preferred method for long-term preservation of for many years. This technique is well suited to culture collections which are in infrequent demand. A as large number of ampoules may be produced and are easily distributed. In lyophillization, there are combined two most successful long-term preservation methods, *i.e.*, freezing and drying. The overall process involves the removal of water vapour by vacuum sublimation from the frozen state. This method overcomes the problems related to drying from the liquid state and the dried ampoules may be stored at room temperature in the dark, although long-term survival is improved with storage in refrigerators. *R. solanacearum* tolerates lyophillization very well, which was used earlier by Sequeira to maintain cultures. For freezing suspension medium is prepared by separately autoclaved 14% (w/v) Bacto peptone and 14 % (w/v) sucrose in water and then combined equal volumes of the sterile solutions. Cells of 48 h old growth of *R. solanacearum* on CPG plate are removed and a very dense (about 10^{10} cfu/ml) suspension is made in a small quantity of the suspension medium. Bacterial suspension is added in each ampoule about 10% of the total volume of the ampoule. A Pasteur pipette is used to dispense the bacterial suspension into the bottom of sterile lyophillization ampoules. The samples are frozen in a dry ice ethanol bath and attached the ampoules to a freeze drying machine. When lyophillization is complete, the flame seals the neck of each ampoule under vacuum. The ampoules should be stored at room temperature and protected from light. By using this technique of preservation, some lyophilized cultures are viable for 30 years (Denny and Hayward, 2001).

4
Mechanism of Pathogenesis and Genomics

4.1 GENOME

The genomes of three *R. solanacearum* strains with very different host ranges have been sequenced at Genoscope (http://www.genoscope.cns.fr, France). The first was that of strain GMI 1000 (Salanoubat *et al.,* 2002), isolated from a tomato plant with symptoms (Boucher *et al.*, 1985), and classified as phylotype I, race 1, biovar 3, with a wide host range (Genoscope, 2010). Its genome has a size of 5.8 Mb with a high G+C content (average value of 67%) and a coding potential for approximately 5,120 proteins. It is organized in two independently replicating circular replicons: a 3.7 Mb chromosome and a 2.1 Mb megaplasmid (Genin and Boucher, 2004).The chromosome harbors genes for basic cellular functions, while the megaplasmid encodes many essential genes for pathogenicity and genes that may be related to the overall fitness of the bacterium (Genin and Boucher, 2004). In Genoscope, the genomes of two other strains of *R. solanacearum*, Molk 2 and IPO 1609, have been sequenced on regions of special interest.Although both strains belong to phylotype II, Molk 2 is a strain of race 2 isolated from a banana plant and strictly limited to this host, whereas IPO 1609 is a race 3 strain isolated from potato (van Elsas *et al.*, 2001). The genomes of Molk 2 and IPO 1609 are comparable to that of GMI 1000 (about 6 Mb, GC estimated at 67%). However, when approximately 100 kb from the Molk 2 genome were sequenced at random and compared with GMI 1000 genome, up to 30% of the genome of Molk 2 was absent in the genome of GMI 1000 (Genin and Boucher, 2004).Comparison of the three genomes will allow analysis of pathogenicity determinants related to host specificity (Genin and Boucher, 2004). A draft genome of *R. solanacearum* strain UW551, a race 3 biovar 2 strain isolated from geranium in the United States, was obtained and annotated (Gabriel *et al.* 2006). GC content was approximately 64.5%, and about 62% of the open reading frames (ORFs) had a functional assignment (Gabriel *et al.,* 2006). The sequences were compared with the genome of strain GMI 1000 (Salanoubat *et al.,* 2002). Most genes encoding pathogenicity determinants appeared to be common to both strains, while one 22-kb region present in GMI 1000 apparently by horizontal gene transfer and absent from UW551, likely encodes enzymes essential for utilization of the three sugar alcohols that distinguish biovars 3 and 4 from biovars 1 and 2 (Gabriel *et al.,* 2006). In spite of the existence of a core genome in the four strains, some genomic instability has been reported for *R. solanacearum*, since the genomes harbor many transposable elements that may take active part in acquisition, loss and alteration of genetic material, and so contributing to generate genomic variation (Genin and Boucher., 2004).This flexibility would be related to the great diversity within the *R. solanacearum* species complex.

Ho *et al.* (2009) reported that Phylotype I/race 1 strains of *R. solanacearum* are known to be variable in their virulence. To explore the candidate genes associated with virulence, differential hybridization on the genomic microarray of GMI1000 was conducted between two groups of strains with distinct virulence levels.The genome of GMI1000 (Phylotype I/race 1/ biovar 3; isolated from tomato) has been completely sequenced and annotated and a genomic microarray has been developed. Four strains with high virulence (Pss190, 1308, 365, 749) and another four strains with low virulence (Pss216, 965, 991, and 79) were hybridized on the array individually. After genome-wide comparison of two groups, 44 candidate genes were identified, including genes coding for Type III secretion system effector, hrpB regulator, transmembrane transport protein, and mobile genetic elements. A large portion of the candidate genes belong to the alternative codon usage region (ACUR). To understand the importance of these candidate genes on virulence, distribution of the candidate genes was determined over 45 *R. solanacearum* strains in a population isolated from tomato in Taiwan. Virulence of the 45 strains in this population has been assayed on three tomato varieties, 'Hawaii 7996', L180, and L390. A dot-blotting method was used by spotting the genomic DNA of the 45 strains on a nylon membrane and using specific markers designed from each candidate gene and amplified from GMI1000 as probes. Presence (+) or absence (–) of each marker in individual strains was recorded. A mean final percentage of wilted plants from the three tomato varieties in the two groups were compared. Almost all markers tested are associated with virulence on susceptible variety L390, but only two markers were shown to be associated the virulence on the resistant variety 'Hawaii 7996'. These results indicate the GMI1000 genome might not contain the genes required for the high virulence phenotype of Hawaii 7996.

Remenant *et al.* (2011) reported that the genomes of the blood disease bacterium and *R. syzygii* are closely related to *R. solanacearum* strains from Indonesia, even though these highly host-specific and insect-transmitted strains are phenotypically different from broad host range *R. solanacearum* strains. Our comparative genomic analysis demonstrated that these specialized organisms, which belong to *R. solanacearum* phylotype IV, are part of the same genomic species. This suggests that strains in the *R. solanacearum* species complex should be divided into three genomic species.

Remenant *et al.* (2011) reported comparative analysis of genome sequences and gene content and confirmed the differentiation of *R. solanacearum* species complex strains into four phylotypes. Genetic distances between strains, in conjunction with CGH analysis of a larger set of strains, revealed differences great enough to consider reclassification of the *R. solanacearum* species complex into three species. The data are still too fragmentary to link genomic classification and phenotypes, but these new genome sequences identify a pan-genome more representative of the diversity in the *R. solanancearum* species complex (Fig. 4.1 and Fig. 4.2).

Ralstonia solanacearum, which causes bacterial wilt of diverse plants, produces copious extracellular polysaccharide (EPS), a major virulence factor. The function of EPS in wilt disease is uncertain. Leading hypotheses are that EPS physically obstructs plant water transport, or that EPS cloaks the bacterium from host plant recognition and subsequent defense. Tomato plants infected with *R. solanacearum* race 3 biovar 2 strain UW551 and tropical strain GMI1000 up regulated genes in both the ethylene (ET) and salicylic acid (SA) defense signal transduction pathways. The horizontally wilt-resistant tomato line Hawaii 7996 activated expression of these

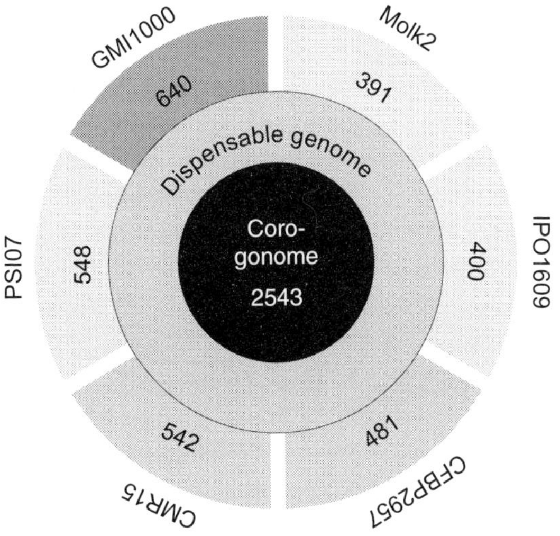

Fig. 4.1 Number of genes in the *R. solanacearum* pan-genome (*Remenant et al.*, 2010)

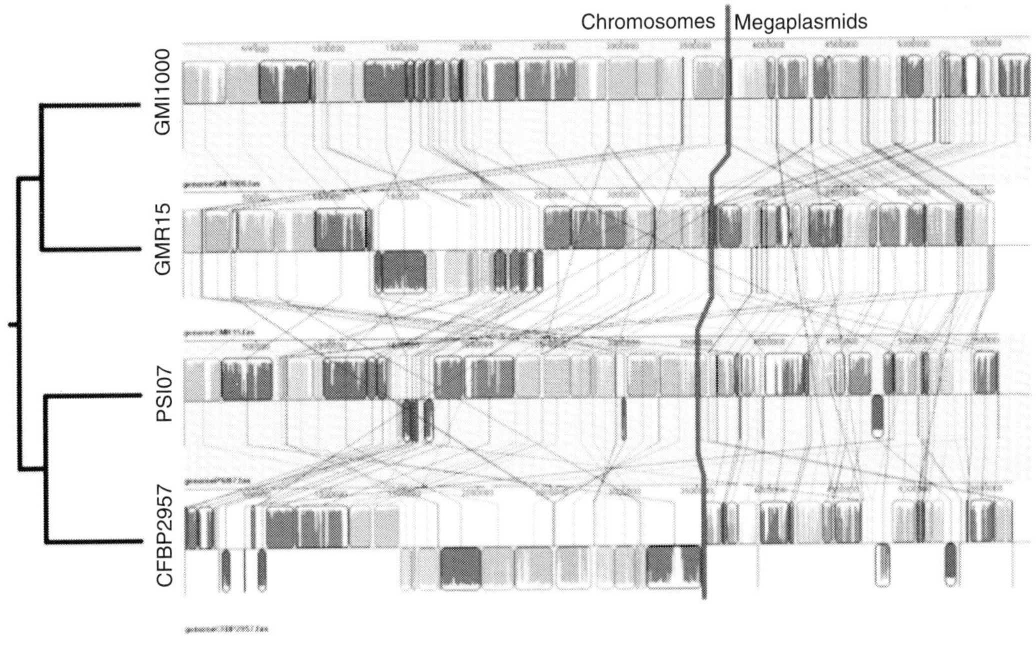

Fig. 4.2 Multiple genome alignment for strains GMI1000, CMR15, PSI07 and CFBP2957. Fine colored lines represent rearrangements or inversions relative to the GMI1000 genome. Chromosomes are separated from megaplasmids by the thick red line (Remenant *et al.*, 2010)

defense genes faster and to a greater degree in response to *R. solanacearum* infection than did susceptible cultivar Bonny Best. However, EPS played different roles in resistant and susceptible host responses to *R. solanacearum*. In susceptible plants the wild-type and eps2 mutant strains induced generally similar defense responses. But in resistant Hawaii 7996 tomato plants, the wild-type pathogens induced significantly greater defense responses than the eps2 mutants, suggesting that the resistant host recognizes *R. solanacearum* EPS. Consistent with this idea, purified EPS triggered significant SA pathway defense gene expression in resistant, but not in susceptible tomato plants. In addition, the eps2 mutant triggered noticeably less production of defense-associated reactive oxygen species in resistant tomato stems and leaves, despite attaining similar cell densities *in planta*. Collectively, these data suggest that bacterial wilt-resistant plants can specifically recognize EPS from *R. solanacearum* (Milling *et al.,* 2011).

Salanoubat *et al.* (2001) have presented the complete genome sequence and its analysis of strain GMI1000. The 5.8-megabase (Mb) genome is organized into two replicons: a 3.7-Mb chromosome and a 2.1-Mb megaplasmid. Both replicons have a mosaic structure providing evidence for the acquisition of genes through horizontal gene transfer. Regions containing genetically mobile elements associated with the percentage of G+C bias may have an important function in genome evolution. The genome encodes many proteins potentially associated with a role in pathogenicity. In particular, many putative attachment factors were identified. The complete repertoire of type III secreted effector proteins can be studied. Over 40 candidates were identified. Comparison with other genomes suggests that bacterial plant pathogens and animal pathogens harbor distinct arrays of specialized type III-dependent effectors.

Presently, the whole genome sequence six strains of *Ralstonia solanaceaum* has been completed (Remenant *et al.*, 2010). However, a general feature of the *R. solanacearum* strain G1100 described in Table 4.1 (Salanoubat *et al.*, 2002).

Table 4.1 General features of the *R. solanacearum* strain GMI 1000 genome

Genome feature	*Chromosome*	*Megaplasmid*	*Genome*
Length (bp)	3,716,413	2,094,507	5,810,922
G + C ratio	67.04%	66.86%	66.97%
Protein coding regions	87.8%	86.5%	87.3%
tRNAs	55	3	58
Ribosomal RNA operons	3	1	4
Protein-coding genes	3,448	1,681	5,129
Average length of protein coding genes (bp)	946	1,077	989
Genes with functional assignment	1,609	652	2,261
Orphan genes*	12.6%	18.7%	14.6%
Regulatory genes*	7.2%	9.6%	8.0%
Insertion sequences phage sequences†	3.4%	2.5%	3.1%

*Percentage of total protein-coding genes;
†Percentage of replicon size

Both replicons have a mosaic structure providing evidence for the acquisition of genes through horizontal gene transfer. Regions containing genetically mobile elements associated with the pertcentage of G + C bias may have an important function in genome evolution. The genome enocodes many proteins potentially associated with a role in pathogenicity. In particular,

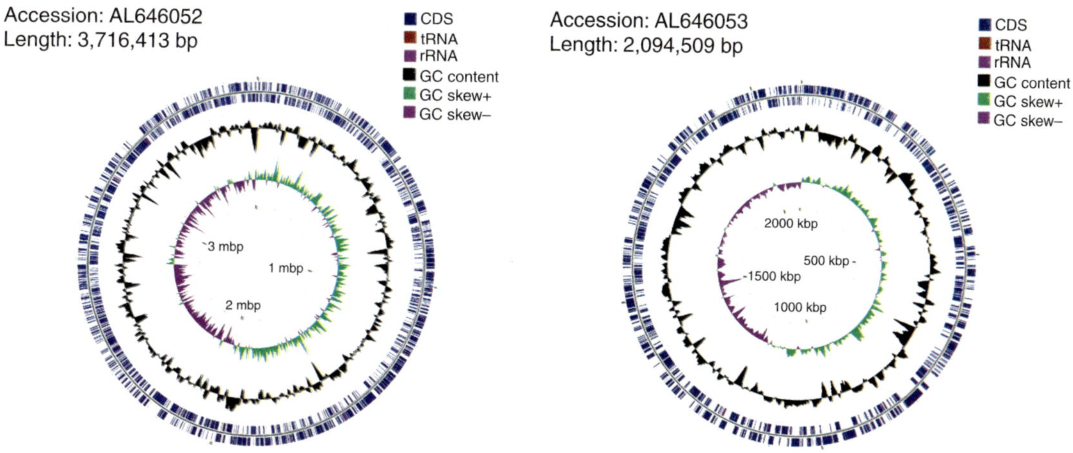

Ralstonia solanacearum GMI1000 complete genome

Ralstonia solanacearum GMI1000 plasmid pGMI1000MP complete sequence

Fig. 4.3 Genome sequence of *Ralstonia solanacearum* (Salanoubat *et al.,* 2002)

many putative attachment factors were identified. The complete repertoire of type III secreted effector proteins an be studied. Over 40 candidates were identified. Comparison with other genomes suggests that bacterial plant pathogens and animal pathogens harbor distinct arrays of specialized type III-dependent effectors.

Genin and Denny (2012) reported that *R. solanacearum* is a major phytopathogen that attacks many crops and other plants over a broad geographical range. The extensive genetic diversity of strains responsible for the various bacterial wilt diseases has in recent years led to the concept of an *R. solanacearum* species complex. Genome sequencing of more than 10 strains representative of the main phylogenetic groups has broadened our knowledge of the evolution and speciation of this pathogen and led to the identification of novel virulence-associated functions. Comparative genomic analyses are now opening the way for refined functional studies. The many molecular determinants involved in pathogenicity and host-range specificity are described, and we also summarize current understanding of their roles in pathogenesis and how their expression is tightly controlled by an intricate virulence regulatory network. Guidot *et al.* (2007) have investigated the gene distribution among strains of the highly polymorphic plant pathogenic β-proteobacterium *R. solanacearum*, paying particular attention to the status of known or candidate pathogenicity genes. Guidot *et al.* (2009) reported a list of 70 genes and 15 intergenes specific to the potato brown rot strains of *R. solanacearum* from phylotype IIB1. This list was identified by comparative genomic hybridization on microarray and subsequent polymerase chain reaction validation with 14 IIB1 strains against 45 non-IIB1 strains that covered the known genetic diversity in *R. solanacearum*. The microarray used consisted of the previously described microarray representative of the phylotype I strain GMI1000, to which were added 660 70-mer oligonucleotides representatives of new genomic islands detected in the phylotype IIB1 strain IPO1609. The brown rot strain-specific genes thus identified were organized in nine clusters covering 2 to 29 genes within the IPO1609 genome and 6 genes isolated along the genome. Of these specific genes, 29 were parts of mobile genetic elements. Considering the known instability of the *R. solanacearum* genome, we believe that multiple probes are required to consistently detect all IIB1 strains and we recommend the use of probes which are not part of genetic mobile elements.

4.2 PATHOGENICITY DETERMINANTS

R. solanacearum possesses diverse genes involved in colonization and wilting of host plants, such as those coding for lytic enzymes and EPS, hypersensitive reaction and pathogenicity (*hrp*) genes, structural genes encoding effector proteins injected by a type III secretion system (T3SS) from the bacterium into the plant cell, genes coding for factors implicated in cell adherence, and others (Schell, 2000; Genin and Boucher, 2004; Denny *et al*., 1990). Schell *et al.* (1988) reported about role of polygalacturonase encoded by pglA and its involvement in phytopathogenicity of *P. solanacearum.*

4.3 QUORUM SENSING

A novel strategy for controlling pathogenesis of bacteria can be the engineered interference with their cell-to-cell communication. Besides multicellular organisms, cell-to-cell communication is of vital importance for single-celled organisms, too, bacteria included. Quorum sensing (QS) is a mechanism to control gene expression in response to cell density (Fuqua *et al.,* 1994; Bassler, 1999), *i.e.,* when a population has reached a minimal population size, the 'quorum'. QS is mediated by a group of diffusible signal molecules, also called 'quormones' because they function similarly to insect pheromones.The most common quormone group in Gram-negative bacteria consists of acyl homoserine lactones (AHLs), though for example in *R. solanacearum* another type of quormone (3-hydroxypalmitic acid methyl ester) is also known (Flavier *et al.,* 1997a,b).

4.4 MAIN PATHOGENICITY GENES

Ralstonia solanacearum pathogenicity is distinctly regulated in early or late stages of infection in response to environmental conditions, such as the presence of host plant cells and bacterial population densities (Schell, 2000; Hikichi *et al.,* 2007). Adsuar (1961) developed a root dipping techniques for the identification of wilt resistance in tomato cultivars.

4.4.1 Early Stages of Infection

Ralstonia solanacearum senses the presence of a plant cell by an outer membrane receptor, PrhA, which transduces the signal through a complex regulatory cascade progressively integrated by PrhR, PrhI, PrhJ, HrpG and HrpB regulators (Cunnac *et al*., 2004a,b; Brito *et al.,* 2002; Aldon *et al.,* 2000; Hikichi *et al.,* 2007; Brown, 2009). The nature of the plant cell wall inducer is not known although it is likely to involve a carbohydrate from the pectic/cellulosic portion (Aldon *et al.* 2000). HrpG co-regulates the expression of two independent pathways (Valls *et al*., 2006). The first one controls the expression of genes likely involved in adaptation to life *in planta* (Valls *et al*., 2006) (Fig. 4.4). HrpG has thus been proposed to have a role in the molecular transition between the environment and the host by shifting the expression of a large set of genes in addition to those concerning the T3SS (Valls *et al.*, 2006). The second pathway is dependent on HrpB (Valls *et al.*, 2006; Cunnac *et al.*, 2004a; Brown, 2009). Among the components of the transducing cascade, HrpB is a major regulator for *R. solanacearum* pathogenicity since it activates expression of the *hrp* genes, which encode the T3SS (Schell, 2000; Brown, 2009) (Fig. 4.4). Translocators PopF1 and PopF2 are regulated by HrpB, and also effector proteins that will be injected out of the bacterial cell into the plant cell (Cunnac *et al.*, 2004a; Hikichi *et al.,* 2007; Brown, 2009). In consequence, transcriptional activation of *hrp* genes by HrpB allows the pathogen to invade the host and proliferate in intercellular spaces (Hikichi *et al.* 2007). During this stage of infection, T2SS and T3SS seem to co-regulate pathogenicity factors and, for instance, T2SS would influence secretion of PopB, while the *pehC* gene is positively regulated by HrpB (Occhailini *et al.,* 2005; Hikichi *et al.,* 2007). Furthermore, HrpB action extends beyond the T3SS. This regulator activates an operon of six genes responsible for the synthesis of a HrpB-dependent factor (HDF), a tryptophan derivative presumably involved in acyl homoserine lactone (AHL) receptor-mediated activity (Delaspre *et al.* 2007). AHLs are autoinducers taking part in bacterial quorum sensing systems, regulatory networks that activate gene expression at high cell densities and mediate long-distance intercellular communication (Flavier *et al.,* 1997a,b). The HDF produced by *R. solanacearum* might be interfering with quorum sensing systems of other bacteria present during the infection (Delaspre *et al.,* 2007). Occhialini *et al.* (2005) reported that genome-wide analysis of the *hrp* gene acts as a regulatory switch controlling multiple virulence pathways for expression in *R. solanacearum.*

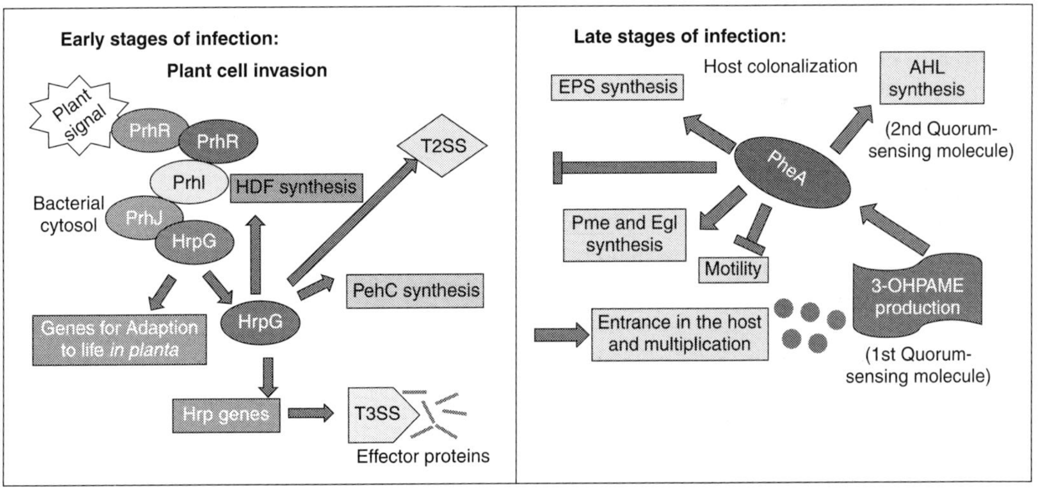

Fig. 4.4 Control of the main pathogenicity genes in early and late stages of host infection by *R. solanacearum*. Adapted from (Hikichi *et al.,* 2007; Brown, 2009)

4.4.1.1 *Control in Late Stages of Infection*

During colonization T2SS contributes to *R. solanacearum* invasion into xylem vessels, and so to systemic infection of the host (Hikichi *et al.,* 2007). At this stage, production of extracellular pathogenicity determinants is transcriptionally controlled by an extensive and complex regulatory network of distinct, interacting signal transduction pathways (Schell, 2000). The main transcriptional regulator is PhcA (Phc, phenotype conversion) (Brumbley *et al.,* 1993), which simultaneously activates diverse virulence genes such as those of EPS biosynthesis and production of Pme and Egl exoproteins, and represses others such as *hrp* genes and those related to production of polygalacturonases, and motility (Schell 2000; Genin and Boucher, 2004; Brown, 2009). PhcA is involved in the five-gene cell density-sensing system (Schell, 2000), an essential part of the complex regulatory network, which controls activity of PhcA protein in response to the levels of an endogenous signal molecule, the 3-hydroxypalmitic acid methyl ester (3-OH PAME) (Schell, 2000a,b). Thus, at low bacterial cell densities (early stages of host colonization), 3-OH PAME is not produced, PhcA regulator remains inactive and *R. solanacearum* cells are motile and highly pectolytic. When bacterial cell densities are high, as in the xylem vessels (late stages of host colonization), 3-OH PAME accumulates extracellularly, promoting the activation of PhcA production, which results in activation of EPS and hydrolytic enzyme synthesis, and repression of motility (Schell, 2000a; Genin and Boucher, 2004) (Fig. 4.4). 3-OH PAME is then an intercellular signal autoregulating virulence gene expression in *R. solanacearum*, acting in the quorum sensing system of this pathogen (Flavier *et al.*, 1997b). The PhcA regulator also positively controls production of a second quorum sensing molecule in *R. solanacearum*, an AHL, which takes part in an additional regulatory system mediated by the SolI-SolR regulators to improve virulence gene expression (Schell, 2000a; Flavier *et al.*, 1997a). Moreover, *R. solanacearum* megaplasmid has a pair of ORFs showing homology to SolI-SolR, what suggests the existence of a third quorum sensing system in the pathogen (Genin and Boucher, 2004). Mutation in *phcA* results in the phenotypic conversion of *R. solanacearum* colonies into the afluidal type (Brumbley and Denny, 1990; Brumbley *et al.,* 1993; Denny *et al.*, 1994). The conversion involves the loss of one or more components of the EPS and a strong reduction in endoglucanase activity but, an increase in that of endopolygalacturonase (Brumbley and Denny, 1990). Furthermore, cell motility has been shown to be increased in PC-type variants (Brumbley and Denny, 1990; Kelman and Hruschka, 1973). Although the variants are impaired in wilting the host (Kelman, 1954), they are infective *in planta* (Denny and Baek, 1991) causing symptoms such as stunting, stem necrosis and adventitious root formation (Denny and Baek, 1991). Several PC-type variants carrying localized mutations within *phcA* were able to revert to the wild type *in planta*, and *in vitro* when in the presence of root exudates of susceptible hosts (Poussier *et al.,* 2003).

4.4.1.2 *Hydrolytic Enzymes*

Phytopathogenic bacteria have often developed enzymes to hydrolyze plant cell wall components to obtain nutrients and energy, which are further involved in early stages of the infective process, favoring the entry and advance of the pathogenic agent in host tissues (Boucher *et al.*, 2001). *R. solanacearum* produces several plant cell wall-degrading enzymes, secreted via the type two secretion systems (T2SS) (Huang and Allen, 1997; Tans-Kersten *et al.*, 1998)]. These include one β-1, 4-cellobiohydrolase (CbhA) and some pectinases whose activities have been identified respectively as one β-1, 4-endoglucanase (Egl) (Schell, 1987; Roberts *et al.*, 1988), one endopolygalacturonase (PehA), two exopolygalacturonases (PehB and PehC) (Huang and Allen, 1997; Schell *et al.*, 1988), and one pectin methyl esterase (Pme) (Schell,1987). *R. solanacearum* Egl is a 43-kDa protein (Schell, 1987) that has proved to be involved in pathogenicity (Roberts *et al.*, 1988). Inactivation of *egl*, *pehA* or *pehB* genes revealed that each contributes to *R. solanacearum* virulence, and a deficient mutant lacking the six enzymes wilted host plants more slowly than the wild-type (Liu *et al.*, 2005). Since pectin catabolism does not significantly contribute to bacterial fitness inside the plant (Gonzalez and Allen, 2003), it seems that cellulase and pectinolytic activities are preferably required for host colonization than for bacterial nutrition (Valls *et al.*, 2006). Thus, *R. solanacearum* hydrolytic enzymes are thought to be involved in pathogenicity *in planta* (Hikichi *et al.,* 2007).

4.4.1.3 Exopolysaccharide

Several phytopathogenic bacterial species produce high amounts of EPSs either in pure culture or during *in planta* multiplication. Although usually related to pathogenicity, it is often difficult to know if the EPSs take active part in symptom production or if they indirectly favor infection (Schell, 2000a,b; Boucher *et al.*, 2001). In *R. solanacearum*, it has been reported that all virulent wild-type strains (mucoid colonies) produce EPS (Husain and Kelment, 1958; Buddenhagen and Kelman, 1964; Kelman, 1954; Boucher *et al.*, 1992; Poussier *et al.,* 2003), while EPS-deficient mutants (non-mucoid colonies) are avirulent. *R. solanacearum* EPS appears to be highly heterogeneous, since it has a varying composition among strains (Drigues *et al.*, 1985). The main virulence factor is an acidic, high molecular mass extracellular polysaccharide (EPS I), a long (>106 Da) polymer with a trimeric repeat unit of N-acetyl galactosamine, 2-N-acetyl-2-deoxy-L-galacturonic acid, and 2-N-acetyl-4-N-(3-hydroxybutanoyl)-2-4-6-trideoxy-D-glucose (Schell, 2000; Orgambide *et al.* 1991). EPS I is more than 90% of the total *R. solanacearum* EPS produced, and approximately 85% appears as a released, cell-free slime, whereas 15% has a cell surface-bound capsular form (Schell, 2000a,b). EPS I-deficient mutants, it was found that EPS I caused wilting in infected plants (Denny and Baek, 1991; Saile *et al.*, 1997). *In planta*, EPS would probably act by occluding xylem vessels, interfering directly with normal fluid movement of the plant, or by breaking the vessels due to hydrostatic overpressure (Schell, 2000a). On the other hand, EPS I might also favor stem colonization by the pathogen, since EPS I-deficient mutants were shown to multiply more slowly, and colonized poorly the stem of infected plants (Saile *et al.*, 1997; Araud-Razou *et al.*, 1998). In that sense, EPS I would be contributing to minimizing or avoiding the recognition of bacterial surface structures such as pili and/or lipopolysaccharide by plant defence mechanisms (Araud-Razou *et al.* 1998; Young and Sequeira, 1986). As EPS-deficient mutants can infect and multiply to some extent *in planta* without inducing wilting symptoms, EPS might take part mainly in late stages of the process, modulating disease severity rather than the infective ability of the bacterium. In *R. solanacearum*, EPS is thought to be the main factor accounting for the virulence of the pathogen (Schell, 2000b; Hikichi *et al.,* 2007) Fig. (4.4).

The extracellular polysaccharide (EPS) is an important factor required by *P. solanacearum* str. AWI to wilt of tomato (Denny and Beak, 1991) the disease development was strongly suppressed when the roots of tomato seedling were treated with lipopolysaccharides (EPS) obtained or extracellular polysaccharides (EPS) obtained from *P. glumae*, before inoculation with

P. solanacearum. Disease occurrence was also significantly suppressed by application of EPS solution. It is suggested that EPS could be involved in the mechanism of disease resistance. Tanigawa *et al.* (1991) suggested the appearance wilt symptom is linked to accumulation polysaccharides chain component and not with the peptide. EPS is consistently required for wild type of virulence of *P. solanacearum (*Kao *et al.,* 1992*)*. The tomatine is induced in tomato plant by avirulent strain of *R. solanacearum* (Arwiyanto *et al.,* 1994a,b). They found that plant inoculated with avirulent strain produced a growth inhibitory substance at a higher level than those of the uninoculated control in the root system, but not in the stem.

Exopolysaccharides are high-molecular-weight polymers that are composed of sugar residues and are secreted by a microorganism into the surrounding environment. Microorganisms synthesize a wide spectrum of multifunctional polysaccharides including intracellular polysaccharides, structural polysaccharides and extracellular polysaccharides or exopolysaccharides (EPS). Exopolysaccharides generally consist of monosaccharides and some non-carbohydrate substituents (such as acetate, pyruvate, succinate, and phosphate). Owing to the wide diversity in composition, exopolysaccharides have found multifarious applications in various food and pharmaceutical industries. Many microbial EPSs provide properties that are almost identical to the gums currently in use. With innovative approaches, efforts are underway to supersede the traditionally used plant and algal gums by their microbial counterparts. Moreover, considerable progress has been made in discovering and developing new microbial EPS that possess novel industrial significance (Suresh and Mody, 2009).

Many plant pathogenic or symbiotic bacteria, including several species of *Clavibacter, Erwinia, Pseudomonas, Ralstonia, Rhizobium*, and *Xanthomonas*, produce large amounts of extracellular polysaccharide (EPS) during growth and pathogenesis (Denny, 1995). EPSs have multiple functions and appear to provide a selective advantage for bacteria during their epiphytic or saprophytic life period in that they protect bacteria from desiccation, concentrate nutrients, and enhance attachment to surfaces. During pathogenesis, EPSs regulate and minimize interaction with plant cells thereby reducing the effect of host defense responses (Király *et al.*, 1997) and contact with toxic substances while promoting colonization. In addition, EPSs may play a primary role in the development of disease symptoms, e.g., wilting caused by plugging of xylem vessels. At least three gene clusters, which are organized more or less similarly in various phytopathogenic bacteria, are important for EPS biosynthesis. In *R. solanacearum*, an 18-kb operon (eps) with at least nine genes is responsible for the acidic component of EPSs (Huang and Schell, 1995). Another gene cluster (ops) that contains at least seven structural genes seems to be necessary for nucleotide sugar components of both EPSs and lipopolysaccharides (Kao and Sequeira, 1991). Kao *et al.* (1994) have identified a regulator gene whose over-expression resulted in decreased EPS production and reduced virulence in *R. solanacearum*. Though expression of the eps operon in *R. solanacearum* appears to be controlled by a complex regulatory network (Huang *et al.,* 1995) that is environmentally regulated, it may be interesting to express either the regulator gene or another mutated regulatory system in the apoplast of transgenic plants to test whether they affect EPS production and virulence of *R. solanacearum*. An attractive way to understand the role of EPS in more depth and eventually increase disease resistance by reduced EPS production in plant pathogenic bacteria could be through the use of polysaccharide depolymerase enzymes. These observations indicate that correct expression of the gene in plants may be useful for testing this approach to control bacterial diseases. In addition, these bacteriophages apparently use additional ways to facilitate the infection of their host, which includes a lysozyme and a holin that may form a pore to support cell lysis by the lysozyme (Kim and Geider, 2000). It should be mentioned that numerous bacteriophages of bacterial plant pathogens have been described in the past (Okabe and Goto, 1963) including phages of *R. solanacearum* (Hayward, 1964) and its banana-attacking strain (Buddenhagen, 1960). At least one of these phages has been found to produce a bacteriolytic protein against *R. solanacearum* (Ozawa *et al.,* 2001). Alvarez *et al.* (2010) reported that *R. solanacearum in* the presence of host plant cells, a regulatory cascade activates secretion of cell wall-degrading enzymes and the cluster of hypersensitive reaction and pathogenicity genes encoding components of a type III secretion system. Once in the plant tissues, high densities of the pathogen increase expression of virulence genes and production of exopolysaccharide, the main pathogenicity determinant. These genes are controlled by a density-dependent regulatory network taking part in a quorum sensing system. After destroying the plant, the bacterium can persist in the environment through diverse survival forms until contact with a new host. Lin *et al.* (2010) reported that mutation in *R. solanacearum* pathogenicity determinant Rsc2860 (Frur) causes altered extracellular polysaccharides biogenesis.

4.4.1.4 Toxin

The tomatine content reached 450 µg/g 9 days after inoculation. In an *in vitro* experiment 100 µg/disc of authentic tomatine and C150 µg/disc of extracted tomatine were sufficient for growth inhibition of *R. solanacearum.* Both extracted and authentic tomatine exhibited bacterostatic activity against the bacterium. Treatment with heat killed cells or culture filtrate of strain 10-OP type failed to increase tomatine content in the root tissue. Baruah and Daka (1995) reported that the toxin is responsible for causing wilt disease in tomato due to *P. solanacearum* race 1 biovar III. Characterization of toxin revealed that presence of glucose, mannose and polysaccharides. Cell lines of 5 wilt susceptible tomato varities, growing in modified MS medium, were treated with purified toxin (EC50). Cell showing toxin intensitively were isolated and regenerated to plants. One regenerant showed a high level of resistant to *P. solanacearum*, while 2 other had moderate resistance. Field test of the regenerants for 3 years showed segregation for the character. Resistant plant had a significant increase in reshitin production following toxin and pathogen treatments. The toxin induced physical damage of the membrane system in susceptible cells.The mode of action and structure of bacterial toxins and of extracellular polysaccharides.

Frey *et al.* (1994) reported that *Hrp* mutants of *Pseudomonas solanacearum* as potential biocontrol agents of tomato bacterial wilt. In *R. solanacearum*, the *hrp* genes control induction of both, disease development and the hypersensitive reaction (HR) (Boucher *et al.*, 2001, Boucher *et al.*, 1992). Therefore, *hrp* mutants are unable to induce symptoms in susceptible host plants and a HR in resistant plants or non-hosts (Boucher *et al.*, 1992). HR is a plant defense mechanism preventing the spread of pathogen infection to other parts of the plant. It is associated with plant resistance and characterized by a rapid and programmed plant cell death localized in the region surrounding an infection (Nimchuk *et al.*, 2003). It can be visualized as necrotic areas in the plant tissues affected. *R. solanacearum hrp* mutants do not seem to be involved in the infection process, since most of them could be isolated from the stems of infected plants but, they showed an impaired ability to multiply *in planta* (Trigalet and Demery, 1986), not observed when cultured on minimum media (Boucher *et al.*, 1985), which may indicate a possible role of *hrp* genes in diverting certain plant metabolites from the plant to the bacteria (Boucher *et al.*, 1992). The *hrp* genes are clustered on the megaplasmid (Genin and Boucher, 2004), and encode components of a T3SS (van Gijsegem *et al.*, 1993) and effector proteins. In all *hrp* clusters, conserved genes (*hrc* genes) might be forming the core of the T3SS (Bogdanove *et al.,* 1996). T3SSs have an important role in pathogenesis, since they are thought to secrete effector proteins translocated inside host cells, and accessory proteins supporting the translocation called translocators (Genin and Boucher, 2004; He *et al.*, 2004; Meyer *et al.,* 2006). Effect or proteins would act in the invasive stages of the infection by either inhibiting plant defenses or inducing nutrient release from the host cell (Boucher *et al.*, 1992; Boucher *et al.*, 2001; Meyer *et al.,* 2006) In some cases effectors can elicit HR due to recognition by specific plant resistance genes. These effectors are then referred to as avirulence (Avr) proteins, since they would be hindering pathogenicity on the host having a corresponding resistance gene (Keen, 1990). T3SS includes extracellular appendages as the Hrp pili in plant pathogens, believed to function either in the attachment to plant cells and/or as conduits for protein translocation, since they might penetrate the plant cell wall. *In vitro*, Hrp pili-deficient mutants were impaired in secretion of effectors and accessory proteins (He *et al.*, 2004). *R. solanacearum* produces Hrp-dependent pili, in addition to the polar fimbriae which were independent on the expression of the *hrp* genes; both types of pili are located at the same pole of the bacterium (van Gijsegem *et al.*, 2000) *R. solanacearum* Hrp pili are mainly composed of the HrpY protein, essential for T3 protein secretion but not for attachment to plant cells (van Gijsegem *et al.*, 2000). Two proteins secreted via the T3SS, PopF1 and PopF2, were identified as translocators, with PopF1 playing a more important role in virulence and HR elicitation than PopF2 (Meyer *et al.,* 2006.) Among the effector proteins, *R. solanacearum* T3SS secretes PopA, PopB, PopC and PopP1 under control of the transcriptional regulator HrpB. PopA produces a HR-like response when infiltrated into plant tissue at high concentration (Arlat *et al.*, 1994), and may allow nutrient acquisition *in planta* and/or the delivery of other effector proteins into plant cells (Lee *et al.*, 2001). PopB has a nuclear localization signal which enables it to be transported to the plant cell nucleus (Gueneron *et al.*, 2000). PopC contains sequences analogous to those of some plant resistance gene products (Gueneron *et al.*, 2000). PopP1 acts as an *avr* determinant towards resistant plants (Lavie *et al.*, 2002). PopA, PopB, PopC or PopP1-deficient mutants showed normal virulence in different host plants, probably due to functional redundancy (Gueneron *et al.*, 2000; Cunnac *et al.*, 2004) Five candidate effector proteins were shown to be translocated into the host cells by the *R. solanacearum* T3SS and 48 novel *hrpB*-regulated genes have been identified, with half of them encoding novel classes of probable effector proteins with no counterparts in other bacterial species (Cunnac *et al.*, 2004). It is estimated that *R. solanacearum* exports large repertoires of pathogenicity effectors through the T2SS and the T3SS (Poueymiro and Genin, 2009).

4.5 OTHER PATHOGENICITY DETERMINANTS

Lipopolysaccharide (LPS) and lectins are the recognition between *R. solanacearum* and the host has long been thought to implicate an interaction between *R. solanacearum* LPS and plant lectins, so involving LPS in the pathogenicity of the bacterium (Baker *et al.*, 1984; Hendrick and Sequeira, 1984). Bacterial LPS is a component of the outer membrane and has three parts: the lipid A, the oligosaccharide core and the O-specific antigen (Baker *et al.* 1984). The core structure is composed of rhamnose, glucose, heptose, and 2-ketodeoxy-octonate, whereas the O-specific antigen is a chain of repeating rhamnose, N-acetylglucosamine, and xylose in a ratio of 4:1:1 (Baker *et al.*, 1984.) Presence or absence of the O-specific antigen differentiated respectively between smooth and rough LPSs, which were respectively negative and positive HR-inducers (Baker *et al.*, 1984). However, a specific interaction between *R. solanacearum* rough LPS and a plant cell wall receptor was not enough to iniciate the HR, although many of the mutations in the LPS also affected virulence (Hendrick and Sequeira, 1984). In *R. solanacearum*, smooth LPS apparently is required to prevent agglutination by certain plant lectins (Sequeira and Graham, 1977). *R. solanacearum* LPS and EPS are somehow related, since a gene cluster was found to be required for the biosynthesis of both cell surface components (Kao and Sequeira, 1991). Two genes encoding lectins have been characterised in *R. solanacearum* (Sudakevitz *et al.,* 2004; Kostlanova *et al.,* 2005), presumably with a function in adhesion to plant surfaces, which is important for *R. solanacearum* pathogenicity during the early stages of infection. In fact, it was found that these lectins bind L-fucose and interact with the plant xyloglucan polysaccharide belonging to the hemicellulose fraction of plant primary cell walls (Kostlanova *et al.,* 2005). Milling *et al.* (2011) reported that *R. solanacearum* produces copious extracellular polysaccharide (EPS), a major virulence factor.The function of EPS in wilt disease is uncertain. Leading hypotheses are that EPS physically obstructs plant water transport, or that EPS cloaks the bacterium from host plant recognition and subsequent defense. Tomato plants infected with *R. solanacearum* race 3 biovar 2 strain UW551 and tropical strain GMI1000 up regulated genes in both the ethylene (ET) and salicylic acid (SA) defense signal transduction pathways. The horizontally wilt-resistant tomato line Hawaii 7996 activated expression of these defense genes faster and to a greater degree in response to *R. solanacearum* infection than did susceptible cultivar Bonny Best. However, EPS played different roles in resistant and susceptible host responses to *R. solanacearum*. In susceptible plants the wild-type and eps2 mutant strains induced generally similar defense responses. But in resistant Hawaii7996 tomato plants, the wild-type pathogens induced significantly greater defense responses than the eps2 mutants, suggesting that the resistant host recognizes *R. solanacearum* EPS. Consistent with this idea, purified EPS triggered significant SA pathway defense gene expression in resistant, but not in susceptible tomato plants. In addition, the eps2 mutant triggered noticeably less production of defense-associated reactive oxygen species in resistant tomato stems and leaves, despite attaining similar cell densities *in planta*. Collectively, these data suggest that bacterial wilt-resistant plants can specifically recognize EPS from *R. solanacearum*. Nordeen *et al.* (1992) reported that activity of cecropin SB37 against protoplasts from several plant species and their bacterial pathogens. Saile *et al.* (1997) reported that *R. solanacearum* normally invades host through their roots and then systemically colonies aerial tissue. Previous research using stem infection found that the major factor in causing wilt symptom was the high molecular mass acidic (EPS-1) and that B-1, 4-endo glcanase (EG) also contribute to virulence. The importance of EPS-1 and EG for invasion and colonization of tomatoes was investigated by either a wild type derivative or genetically well-defined mutant lacking EPS-1, EG or both. Bacteria of all strains were rendered from surface disinfected roots and hypocotyls as soon as 4h after inoculation. The EPS-minus mutants did not colonize stem as rapidly as the wild type and the EG-mutant. Inoculation of wound petioles showed that the even though the mutant multiplied as well as wild type *in planta*, the EPS-minus strain did not spread as well through the plant stem. It was concluded that poor colonization of stems by EPS-minus strain after petiole inoculation or soil infection was due to reduced bacterial movement within plant stem tissues. Pun and Bhorali (1997) have reported that aqueous solution of (EPS) extracted from the isolates induced with symptom in tomato leaf stem cuttings. Disease development increased as EPS concentration increased. EPS produced typical wilting in 12 species of solanacearum and non-solanacearum plants. Mc Garvey *et al.* (1999) have tested one susceptible and two resistant cultivars of tomato for the differences in infection by *R. solanacearum* and for the subsequent multiplication, colonization and production of the wilt inducing virulence factor, extra polysaccharides 1(EPS1). They reported that bacterial ingress into tap root was faster in the susceptible cv. Marison, followed by resistant cvs L 285 (fivefold slower) and Hawaii 7996 (fifteen-fold slower). Once inside the taproot, *R. solanacearum* colonized to some extent, almost all region of the resistant and susceptible plants. However, colonization occurred sooner in the susceptible than in the resistant cultivars, as measured by viable cell counts of bacterial in the mid-stem. Rates of multiplication and maximum bacterial cell densities were also greater in the susceptible than in the resistant cultivars. Bacterial population in susceptible cultivars produced greater amount of EPS per plant than in the resistant cultivars. Immunofluorescence microscopy using antibodies against either EPS 1 or *R. solanacearum* cells revealed that bacteria and EPS 1 were distributed throughout the vascular bundles and intercellular spaces of the pith in the susceptible cultivars, whereas in the resistant cultivars, bacteria and EPS 1 were restricted to the vascular tissues. Bacterial colonization and multiplication in susceptible and resistant genotype were studied by Ansari *et al.* (1996) and reported that in the susceptible variety, the incubation period for development of symptom did not vary with growth stages of the host plant and the bacterial population grew with increase in the duration of post-inoculation period. However, the tolerant cultivar, although colonization occurred, the bacterial population did not reach the critical level, *i.e.*, 9×10^6 cfu/ml for wilting. Hückelhoven (2007) had reviewed about the cell wall associated mechanism of disease resistance and susceptibility and reported that the plant cuticle and cell wall separate microbial pathogens from the products of plant metabolism. While microbial pathogens try to breach these barriers for colonization, plants respond to attempted penetration by a battery of wall-associated defense reactions. Successful pathogens circumvent or suppress plant non-self recognition and basal defense during penetration and during microbial reproduction. Additionally, accommodation of fungal infection structures within intact cells requires host reprogramming.

5
Losses and Epidemiology

Plant Disease Epidemiology is the study of diseases in plant populations. Plant diseases occur due to pathogens such as bacteria, viruses, fungi, oomycetes, nematodes, phytoplasma, protoza and parasitic plants Agrios (2008). Plant disease epidemiologists strive for an understanding of the cause and effects of disease and develop strategies to intervene in situations where crop losses may occur. Typically, successful intervention will lead to a low enough level of disease to be acceptable, depending upon the value of the crop. Plant disease epidemiology is often looked at from a multidisciplinary approach, requiring biological, statistical, agronomical and ecological perspectives. Biology is necessary for understanding the pathogen and its life cycle. It is also necessary for understanding the physiology of the crop and how the pathogen is adversely affecting it. Agronomic practices often influence disease incidence for better or for worse. Native species of plants may serve as reservoirs for pathogens that cause disease in crops. Statistical models are often applied in order to summarize and describe the complexity of plant disease epidemiology, so that disease processes can be more readily understood (Arneson, 2001; Madden *et al.* 2007). The development depends on the four factors shown as disease pyramid (Fig. 5.1b).

1. Initial amount of pathogen inoculum (infective structures)
2. Rate of disease increase
3. Duration of crop development

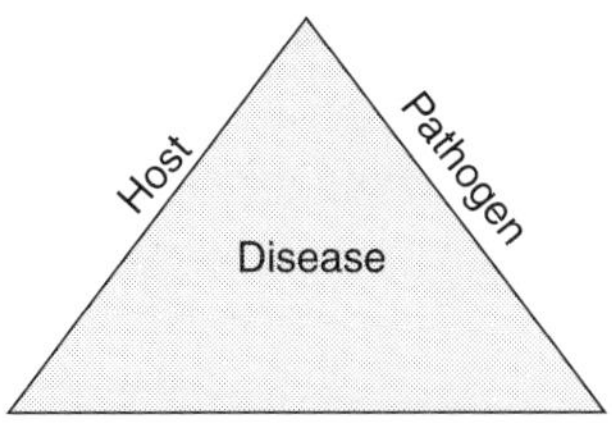

Fig. 5.1a Disease triangle indicating interaction of the pathogen, host, and the environment leading to a plant disease (Stevens,R.B.1960)

These parameters interact to produce a rapid increase in pathogen populations, which manifests as exponential disease development in many production systems. The rate of disease increase over time is dependent upon the interactions of the pathogen, host plant, and the environment. For disease management purposes, the biggest concern for growers is the interaction of the pathogen and host and the ideal environmental conditions, which plays a critical role in determining the nature of plant disease epidemics. This set of interactions is known as the disease triangle (Fig. 5.1a), which determines the fate of a disease on a crop.

Van der Plank (1963) published "Plant Diseases: Epidemics and Control", a seminal work that created a theoretical framework for the study of the epidemiology of plant diseases, this framework is used for modeling and determining thresholds for epidemics that take place in a homogeneous environment such as a mono-cultural crop field (Drenth, 2004).

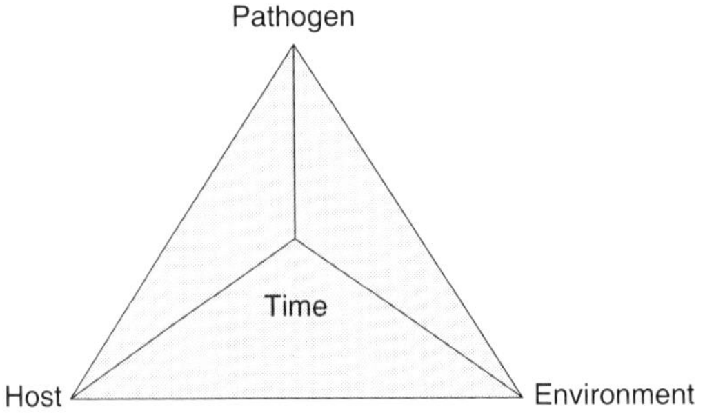

Fig. 5.1b Epidemic pyramid

Disease development is a monocyclic process, while epidemic growth is a polycyclic process. The monocyclic process involves the different stages, *viz*, infection, incubation period and sporulation. Since the bacterial wilt is a polycyclic process involves overlapping and repetive monocyclic stages and covers, in addition, spore liberation, transport and deposition on vulnerable sites. Nagrajan (1983) stated that any biological agent that causes disease through its parasitic habits can be called a pathogen. For a disease to occur coincidence of three parameters is necessary, namely, the susceptible host, virulent pathogen and favourable environment (Fig. 5.1.b) but epidemic develops only when a pathogen completes a few repeated cycles on the same host. As each generation takes a few days, time factor is involved in epidemic buildup. In other words, epidemic growth is a temporal and spatial process. The early growth curve has a lag phase, when the incubation period is longer and infection conditions are less favourable. Subsequently, disease growth is rapid and explodes as time bomb. It is at this stage that the farmers get panicky over this crop's prospects. Later, severity tapers either due to unfavourable weather or crop maturation, or both. A pathogen, or infectious agent, is a biological agent that causes disease or illness to its host.

5.1 SURVIVAL

5.1.1 Soil

Soil is a heterogeneous habitat of beneficial, plant pathogenic and saprophytic microbes, which play an important role in the spread of soil-borne diseases (Samaddar *et al.*, 1998). Infested soil and surface water, irrigation water, are the main sources of spread of inocula. The pathogen can infect undisturbed roots of susceptible hosts through microscopic wounds caused by the emergence of lateral roots. The widespread distribution of this pathogen has been explained by various workers (Titatarn and Toong Jeam, 1985a; Mc Carter, 1968) seed transmission has been reported (Singh *et al.*, 1995). The vertical distribution of *P. solanacearum* was seen at 30-45 cm soil depth (Mc Carter *et al.*, 1969) resulted more wilting with high population in 30 cm top soil. The survival of pathogen depends on soil type and soil origin was found to be stronger determinants. Higher diversity of pathogen was found in sandy than in clayey soils in the Netherlands (Messiha *et al.* 2009; van Diepeningen *et al.*, 2006).

5.1.1.1 *Tomato and Brinjal*

There are various reports on survival of *R. solanacearum* in soil for variable periods (Kelman, 1953; Graham *et al.*, 1979; Graham and Lloyd 1979; Shamsuddin *et al.*, 1979;. Shekhawat *et al.* 1986, 1990; van Elsas *et al.*, 2000, 2005). It survived up to one year in agricultural soil even after treatment with an herbicide (van Elsas *et al.*, 2005a), two years after crop removal (Shamsuddin *et al.* 1979), four years (Graham *et al.*, 1979). Long-term survival in deeper soil layers (Graham and Lloyd, 1979). Survival up to 673 days in naturally infested soils stored in plastic bags at 40°C has been reported (Granada and Sequeira, 1983a). Mc Carter (1968) developed a procedure for infecting field soil and reported that the pathogen persisted thoughout 4 years in soils artificially infected by clipping beds of young tomato plants with a contaminated mower and ploughing in disease plants. Kishun and Sohi (1982) reported that *P. solanacearum* was recovered from soil from cultivated as well as fallow land, but its population decreased with increase in depth.The average number of bacterial cells were 38 $\times$ 10^6, 27.5 $\times$ 10^6, 15 $\times$ 10^4, 1.5 $\times$ 10^4, 5.5 $\times$ 10^4 3.75 $\times$ 10^4, 3 $\times$ 10^2, 2.5 $\times$ 10^2 and 1.5 $\times$ 10^4 cfu/g soil cultivated lands, whereas it was 32.5 $\times$ 10^6, 18.25 $\times$ 10^6, 8.25 $\times$ 10^6, 6.75 $\times$ 10^4, 6.25 $\times$ 10^4, 4.25 $\times$ 10^4, 3.5 $\times$ 10^4, 2.5 $\times$ 10^2 1.75 $\times$ 10^2 and 1.25 $\times$ 10^2cfu/g soil in fallow land at ground level at, 5, 10, 15, 20, 25, 30, 35, 40 and 45 cm below the ground level, respectively. In general, the number of bacterial cells was higher in both the soils up to 30 cm depth, wilting of tomato cv Pusa Ruby observed from 10 to 70% in cultivated land upto 40 cm and 10-30% up to 20 cm depth in fallow land. Granada and Kelman (1983) reported that the *P. solanacearum* survived in soil, rhizosphere and plant roots. The bacteria survive much longer in moist soil than in dry or flooded soil and low pH is less favourable for survival than high pH (Hsu, 1991; Yedesha *et al.*, 2010). The race 1 of pathogen can survive for more than 6 months on seed surface and in the soil (Samaddar *et al.*, 1998) and also reported to spread through seed transmission (Singh *et al.*, 1995; Roopali, 1994).

5.1.1.2 *Potato*

Sunaina *et al.* (1989) reported about the latent survival of *P. solanacearum* in potato tubers and weeds. The infected potatoes (seed tubers, harvest leftover and infected plants) or infested soil, or both of *R. solanacearum* are the source of inoculums. The pathogen can survive in soil (mostly on plant debris) and in the rooting system and rhizosphere of many hosts (weeds, other host crops, potato volunteers). The pathogen can also be introduced into an area by planting infected potato tubers, disseminated, and survive in contaminated surface water (Janse, 1996; Persson, 1998; Schans and Steeghs, 1998; Wenneker *et al.*, 1999) and weed hosts such as *Solanum dulcamara, S. nigrum, Portulaca oleracea* in Europe (Elphinstone *et al.*, 1998) and *Rumex dentatus* and *Solanum nigrum* in Egypt (Farag *et al.*, 2004). Volunteer plants or (in colder climates: perennial) weeds can be a reservoir, and responsible for transmission of the pathogen through successive seasons (Janse, 1996; Lopez and Biosca, 2004). The pathogen can persist for a long time in soil, in infected host plant debris or by colonizing potato volunteer plants, alternative hosts or even non-host plants (Graham *et al.*, 1979; Granada and Sequeira, 1983a). After establishment of the disease in a growing area in temperate zones it can survive for periods between 12 months to 3 years in the absence of a potato crop as reported from New South Wales in Australia (Graham *et al.*, 1979).

5.1.1.3 *Pepper*

Infested soil is the main source of inoculum. The pathogen survived in the rhizosphere of non-host plants, including weeds (Momol *et al.,* 2001). Soil factors influenced the survival of the bacterium. It is not uncommon to find bacterial wilt in the first crop in recently cleared land in tropical and subtropical regions. In disease-free areas soil can be infested through infected planting material (tomato and pepper transplants, potato tubers), contaminated water, or machinery and labourers that are contaminated from infested fields. Transmission by seed is not considered important in pepper, even though *R. solanacearum* has been reported to be seed-transmitted in tomato, eggplant and peanut. Epiphytic colonization of the true leaves of pepper has been shown under high relative humidity. The most significant method of pathogen dissemination is through contaminated irrigation water flowing among wilted plants. *R. solanacearum* can infect undisturbed roots of susceptible hosts through microscopic wounds caused by the emergence of lateral roots. The bacterium is soil and water borne, waterways being major dissemination routes in the environment (Elphinstone, 2005). After destruction of the plant, an inter-host life period has been reported for this pathogen (Denny *et al.*, 1994), where it would have to cope with the stress of oligotrophic habitats and occasional capacity to wilt susceptible hosts (Caruso *et al.*, 2005; van Overbeek *et al.*, 2004). The survival of the pathogen was seen in sterile pure water (Kelman, 1956; Wakimoto *et al.*, 1982; van Overbeek *et al.*, 2004) or sterile natural water only for several months (Kelman, 1956; van Elsas *et al.,* 2001) to a maximum of 1 year (Tanaka and Noda, 1973). *R. solanacearum* filamentous cells were visualized by immunostaining as well as bacillar and coccoid cells reported by Thomas *et al.* (1999) which appears to be the first report of filamentation and budding phenomena occurring in a plant-pathogenic bacterium. Many of the coccoid forms were viable, although non-viable cells were also observed. *R. solanacearum* coccoid forms showed a tendency to aggregate under starvation, which can be considered a survival mechanism, since cell aggregation enables utilization of the content of lysing cells and protects the cells from environmental hazards such as free radicals, toxins and predators (Blat and Eisenbach, 1995).

5.1.2 Survival Forms in Environment

There are several forms of resistance of *R. solanacearum* by which the pathogen can survive in non-favourable environmental conditions, e.g., viable but non-culturable (VBNC) forms, starved cells, PC-type and biofilms (Fig. 5.2).

5.1.2.1 *Viable But Non-culturable Forms*

Bacterial cells in the VBNC state, although metabolically active (viable), are unable to grow on the solid media usually used for their culture (Oliver, 1993), and form dormant cells. Xu *et al.* (1882) have first observed VBNC which is the physiological state has been considered a survival mechanism that some non-sporulating bacteria exhibit under adverse environmental conditions become adverse (Roszak and Colwell, 1987; Grey and Steck, 2001; Oliver, 2005). In soil, *R. solanacearum* became VBNC in less of a month by exposure at 4°C (van Elsas *et al.*, 2000) and within three days without dilucidating the inducing factor (Grey and Steck, 2001). In liquid microcosms containing cupric sulfate, the percentage of VBNC cells depended on copper concentration (Grey and Steck, 2001). In pure water, maintenance at 4°C led to VBNC *R. solanacearum* (van Elsas *et al.*, 2000; van Overbeek *et al.*, 2004) with cold-stressed cells progressively losing wilting capacity (van Overbeek *et al.*, 2004). In environmental water, induction of *R. solanacearum* VBNC state was caused by exposure at 4°C (Caruso *et al.*, 2005; Alvarez *et al.*, 2010b) and by extended oligotrophy (Alvarez *et al.*, 2008). The state has also been reported to occur *in planta* during infection, where the proportion of cells becoming VBNC increased after the plant underwent extensive necrosis (Grey and Steck, 2001).

5.1.2.2 Starved Cells

Several starvation-survival responses have been described, with starved bacterial populations maintaining their numbers over time in a non-growing but culturable state at different levels depending on the species (Morita, 1997). The starvation-survival response constitutes a physiological state different from those of active growth and the VBNC (Heim *et al.*, 2002). It is considered a strategy for survival in energy-deficient systems, in which bacterial populations manage to maintain in a non-growing but culturable state (Morita, 1997). It was found that *R. solanacearum* displayed this strategy in environmental water microcosms at permissive temperatures (Alvarez *et al.*, 2008a; Alvarez *et al.*, 2010a), and the state was not induced by low temperatures (Alvarez *et al.*, 2010b). Starved *R. solanacearum* kept pathogenic in the water microcosms over four years (Alvarez *et al.*, 2008b).

5.1.2.3 PC-type

The physiological characteristics (PC) of this form have the biological significance (Poussier *et al.,* 2003). PC-type variants may have some selective advantage over the wild type, like a higher motility favouring aerotaxis under oxidative stress (Clough *et al.*, 1997). In a reversible switching between wild and PC types when moving from host to non-host environments, PC-type variants specifically adapted to endure harsh conditions (Denny *et al.*, 1994). The reversion from the PC to the pathogenic form observed *in planta* (Poussier *et al.,* 2003) seems to confirm the existence of this phenotypic conversion/reversion cycle (Denny *et al.*, 1994; Poussier *et al.,* 2003).

5.1.2.4 Biofilms

Although little is known on *R. solanacearum* capacity for biofilm formation and the factors involved in the process (Kang *et al.*, 2002), this kind of communities may also have a role in its survival (Yao and Allen, 2007) because *R. solanacearum* forms biofilm like structures in contact with abiotic and biotic surfaces. Different strains formed biofilms on PVC wells at the liquidair interface (Kang *et al.*, 2002; Yao and Allen, 2007), and also biofilm-like aggregations on the surface of tomato seedling roots (Yao and Allen, 2006). Moreover, since aerotaxis-deficient mutants over produced biofilms on abiotic surfaces; aerotaxis might lead cells to avoid toxic oxygen levels at the liquid-air interface by forming protective thicker biofilms, and therefore regulating biofilm formation to facilitate survival (Yao and Allen, 2007).

5.1.2.5 *Soil Type*

Kelman (1953) reviewed *R. solanacearum* and reported that the bacterial wilt is more severe in sandy loam soil than red loam soil. Soil factors prevented persistence of the pathogen from season to season. Soil type, which influenced soil moisture, determined the size of antagonistic populations which, in turn, affected the survival of *R. solanacearum* in soil (Nesmith and Jenkins, 1985). Some soils in wilt endemic areas have never supported the disease where the disease has been introduced there are contrasting reports on persistence in the soil. (Kelman, 1953; McCarter, 1976; Nesmith and Jenkins, 1983, 1985). In Surinam, bacterial wilt of tomato and related hosts was absent on the sea shell ridges of the coastal plain in contrast to sandy ridge and clay soils where disease outbreaks occurred frequently. Nishiyama *et al.* (1999) reported that to identify the sites responsible for the suppressiveness of tomato bacterial wilt in a suppressive soil, population dynamics of *R. solanacearum* in non-rhizosphere soil, roots and stems of tomato plants was compared between a wilt-conducive soil and a suppressive soil both of which were artificially infested with the pathogenic strain SL8. Rhizobacteria were recovered as two fractions; root fraction-l obtained after four washings in water, which was assumed to correspond to the rhizoplane, and root fraction-2 that was macerated after the washings and was assumed to be the fraction corresponding to the part inside of the roots and the firmly attached cells to the rhizoplane. In the conducive soil, all the tomato seedlings wilted while none wilted in the suppressive soil during the 3l-day period of cultivation after transplanting to soils infested with the pathogen at an initial density of 10^3 cfu per g dry soil. The number of pathogens in the conducive soil increased first in root fraction-I, then in root fraction-2 and in stems up to 10R-10^{10} cfu per g wet plant tissue during the l6-day period of cultivation after transplanting. In contrast, in the suppressive soil, the number of pathogens remained below 10^6 cfu per g tissue both in the two root fractions and in stems for 40 days. Survival rate of the pathogen inoculated into the tomato-free soils was higher in the suppressive soil than in the conducive soil. These results suggest that prior multiplication in root fraction-l at a density above 10^6 cfu per g root is necessary for the pathogen to multiply in root fraction-2 and stems at a density above 10^6 cfu per g for the occurrence of wilt, and that the primary site for suppressiveness of the disease in tomato plants cultivated in the suppressive soil in this report was located in root fraction-I.

5.2 SOIL MOISTURE

Soil moisture and antagonistic microorganisms were the most important factors. Some soils are conducive to bacterial wilt and others suppressive. High soil moisture usually favour development of bacterial wilt (Kelman, 1953; Abdullah *et al.,* 1983). Survival of the pathogen is greatest in wet but well drained soils, whereas survival is affected adversely by soil desiccation and flooding (Kelman, 1953; Buddenhagen and Kelman, 1964; Ngoc and Kim, 2010). Soil moisture significantly affected reproduction and survival of *R. solanacearum* in unsterile soils, regardless of soil type, colony counts increased within 7-10 days of introduction into the soil at the highest soil moistures. The survival of *R. solanacearum* was declined in the driest soil. The driest soil contained generally higher number of viable pathogens than wet soil. Some strains of *R. solanacearum* in certain soils are not sensitive to dry soil conditions. Under natural conditions soil organisms are subjected to the stress of cyclic wetting and drying. Dry conditions are a factor contributing to reduction in numbers of *R. solanacearum* in fallow land (Sequeira, 1962). The bacteria can survive for a long time in water (up to 40 years at 20-25°C in pure water) and the bacterial population is reduced in extreme conditions (temperature, pH, salts). Infected land sometimes cannot be used again for susceptible crops for several years. *R. solanacearum* can also survive in cool weather and enter a state of being viable but not culturable because the bacteria usually become avirulent after recovering. *R. solanacearum* has ability for survival in aquatic environments (Kelman, 1956; Kang *et al.*, 2002; van Elsas *et al.*, 2005a,b; Caruso *et al.*, 2005; Hong *et al.*, 2005; Alvarez *et al.*, 2007, 2008a, 2010b), being able to multiply in pure water in the absence of nutrients (Wakimoto *et al.*, 1982). In freshwater it was reported variable periods (Kang *et al.*, 2002; van Elsas *et al.,* 2005; Kelman, 1956; Caruso *et al.*, 2005; Hong, *et al.*, 2005; Alvarez *et al.*, 2007, 2008a, 2010a,b) with pathogenic populations surviving up until four years keeping pathogenic even by watering (Alvarez *et al.*, 2008a). The persistence in water was favoured at permissive temperatures and by the presence of sediment but, it was unfavoured in sediment subjected to drying and by levels of seawater salts similar to those of water in coastal areas, due to osmotic tensions (Kang *et al.*, 2002; van Elsas *et al.*, 2005a,b). The populations depend on a seasonal variation over years, consisting of relatively high levels during spring and summer, and unsuccessful recovery of cells in autumn and winter (Caruso *et al.,* 2005; Hong *et al.*, 2005), with cold stressed cells able to wilt host plants (Caruso *et al.*, 2005). Field irrigation with contaminated water has been associated to most outbreaks; waterways can be considered major dissemination routes of this pathogen (Kang *et al.*, 2002; van Elsas *et al.*, 2005; Caruso *et al.,* 2005, Hong, *et al.*, 2005; Alvarez *et al.*, 2007, 2008a, b). Moderate differences in moisture did not drastically affect populations; contrarily to severe drought (Kelman, 1953; van Elsas *et al.*, 2000, 2005). Alvarez *et al.* (2008a) reported four years survival of *R. solanacearum* phylotype II biovar 2 was seen which caused outbreak. Outbreaks originated from dissemination of the pathogen in water courses, where it has to cope with prolonged nutrient limitation. During the whole period, the bacterium remained pathogenic on host plants and underwent a transition from typical bacilli to small cocci which tended to aggregate. Some starved *R. solanacearum* cells filamented and formed buds, was found viable but non-culturable state, morphological changes and aggregation have not previously been reported for this pathogen as survival mechanisms induced in oligotrophic conditions. The potential existence of long-starved pathogenic cells in environmental waters may raise new concerns about the epidemiology of bacterial wilt disease. Survival of the pathogen was found in sterile pure water (Kelman, 1956; Wakimoto *et al.*, 1982; van Overbeek *et al.*, 2004) or sterile natural water only for several months (Kelman, 1956; van Elsas *et al.*, 2001) to a maximum of 1 year (Tanaka and Noda, 1973). During the starvation period in river water microcosms, proportions of the viable *R. solanacearum* populations progressively lost the ability to form colonies on plates, as already observed in starved bacteria (Besnard *et al.*, 2000; Colwell and Huq, 1994), becoming VBNC as another survival strategy under prolonged oligotrophy. Nutrient deprivation has been recognized as an important stimulus to induce the VBNC state, which represents a common strategy for bacterial survival in energy-deficient environments (Colwell and Huq, 1994; Oliver, 2005). *R. solanacearum* retained virulence during the 4 years of nutrient scarcity in river water microcosms. The virulence of *R. solanacearum* biovar 2 strain 1609, kept in water at 4 and 20°C, were studied by van Overbeek *et al.* (2004) and reported the at 20°C, total cell and plate count (colony forming units; CFU) numbers were similar, between log 5.03 and log 5.55 CFU, and log 5.03 and log 5.51 cells per ml, at days 0 and 132, respectively. However, CFU in the cultures kept at 4°C dropped from log 6.78 CFU/ml at day 0 to below detection after 84 days. The similar disease incidence on plants stem inoculated with *R. solanacearum* either starved (from microcosms) or freshly grown (controls) during the 4 years demonstrated the great capacity of this pathogen to cause disease. Interestingly, bacterial wilt was also observed in host plants after watering with long-starved *R. solanacearum*, which constitutes a more natural way of transmission of this pathogen. The influence of temperature, incident light, presence of sediment, seawater salts and aquatic microbiota on *R. solanacearum* populations has been monitored, also in water microcosms (van Elsas *et al.*, 2001; van Overbeek *et al.*, 2004; Alvarez *et al.*, 2007). Among these factors, only temperature has been assessed in natural water systems, causing a seasonal variation of *R. solanacearum* populations in temperate areas (Caruso *et al.*, 2005; Hong *et al.*, 2005) which ability observed for *R. solanacearum* to persist and cause disease is in agreement with studies in open natural waters, where the bacterium was detected for several consecutive years (Caruso *et al.*, 2005; Hong *et al.*, 2005). Pradhanang (1999) reported that *R. solanacearum* transmission is through drainage water.

5.3 MODE OF INFECTION AND SPREAD

Infection occurs typically through the roots and wounding in root areas, a normal physiological process as rootlets grow. Bacteria are normally concentrated in the lower stem portions of the plant (Buddenhagen and Kellman, 1962). Infection always occurs through wounds caused by transplanting, cultural operations or nematode infection (Nirmala Devi and Tikoo, 1989; Keshwal *et al.*, 2000). Root knot nematodes (*Meloidogyne incognita*) infection predispose the tomato plants to bacterial wilt by interfering with nutrition of plants and root wounding to provide entry to the pathogen. *M. incognita* and *R. solanacearum* complex increased the severity of bacterial wilt in brinjal (Husain and Bora, 2008b). After infection, the bacterium colonizes the cortex and makes its way towards the xylem vessel; from there it rapidly spreads throughout the plant (Momol *et al.*, 2005). The infected roots released vast number of *R. solanacearum* cells into the rhizoshere and invades subsequent crops to cause the spread of the disease (Mc Carter, 1976). Vast number of cells is released into the surrounding soil from in-‸cted roots and from there they invade subsequent crops.

The pathogen spreads through irrigation or rainwater, infected plants, infested seeds of tomato and brinjal and the soil. It has been reported that tomato transplants carrying *R. solanacearum* moved from Georgia, USA to Kent County, South-Western Ontario, Canada and spread the disease (Layne and McKeen, 1967). A new *R. solanacearum* strains 20 exhibiting a new pathogenic potential was found. Since the pathogen is soil borne, it spreads through irrigation water, soil movement, implements and infected plant material (Wicker *et al.*, 2007). *R. solanacearum* is mainly spread by movement of infected seed tubers. Contaminated surface water used for irrigation and infested soil adhered to farmers' shoes and tools also contribute to disease dissemination. Bacterial entrance into potato roots is facilitated by wounds made by tools during post-emergence cultivation, and by nematodes and insects in the soil.

5.4 FACTORS RESPONSIBLE FOR SURVIVAL AND SPREAD OF PATHOGEN

The bacteria of all races of *R. solanacearum* are transmitted through contaminated soil, irrigation water, equipment, or personnel and spread by the following means:

- Shared water irrigation systems
- Transplanting and propagating infected plants
- Taking cuttings without disinfecting implements between plants
- Pinching buds of plants without sanitizing
- Transfer of contaminated soil (on equipment or shoes) to disease-free areas

The pathogen is not readily spread between plants by the following means:

- Splashing of water
- Leaf to leaf contact
- Aerial transfer

5.5 DISEASE CYCLE

The pathogen moves to the host plant, attaches to the plant roots, infects the cortex and colonizes the xylem (Kelman, 1953; Vasse *et al.*, 1995; Yao and Allen, 2006a, b) which requires secretion of cell wall-degrading enzymes and EPS controlled by a regulatory network that uses PhcA (Schell, 2000a,b). After destroying the host, the bacterium returns to the environment and is likely to survive in soil, water or reservoir plants (Denny *et al.*, 1994). Within plant tissues, high densities of the pathogen increase expression of pathogenicity genes, repressed by low bacterial densities in non-host environments (Schell, 2000a, b).

Romantschuk (1992) reported about attachment of plant pathogenic bacteria to plant surfaces. Further, Romantschuk *et al.* (1996) reported that plant aerial surfaces are colonised by a variety of microbes, including bacteria, fungi and yeasts. Some of these microbes are opportunistic pathogens, whereas others are specialised epiphytes. The outcome of the interaction between the microbe and the plant host is dependent on the plant-microbe combination. Often the host range for disease is more narrow than that for epiphytic colonisation.

5.5.1 Directed Motility to Host in Environment

Ralstonia solanacearum senses specific stimuli and moves towards plants by swimming motility (Tans-Kersten *et al.*, 2001) to find more favourable conditions (Yao and Allen, 2006a, b, 2007). The pathogen remains active attracted by chemotaxis to diverse amino acids and organic acids, and specially to host root exudates, whereas those from a non-host were less attractive (Yao and Allen, 2006a). The ability of the pathogen to locate and interact with the host was dependent on aerotaxis or energy taxis (Yao and Allen, 2006b, 2007). Several aerotaxis-deficient mutants have been reported impaired in either localizing on host roots or moving up an oxygen gradient (Yao and Allen, 2006b; Yao and Allen, 2007). Swimming motility, chemotaxis and aerotaxis seem to have a role in the early stages of host invasion (Tans-Kersten *et al.*, 2001; Yao and Allen, 2006a, b, 2007).

5.5.2 Life Within Susceptible Hosts

Three main stages were distinguished in descriptions of the *R. solanacearum*-host pathogenic interaction: root colonization, cortical infection and xylem penetration (Vasse *et al.*, 1995, 2005) (Fig. 5.2).

5.5.3 Root Colonization

After localizing the host roots, *R. solanacearum* can enter through physical wounds and/or natural openings (Kelman, 1953), and attaches at two precise root sites: root elongation zones and axils of emerging or developed lateral roots (Vassee *et al.*, 1995), probably due to the fact that the epidermal barrier is usually weaker in them. Moreover, root elongation zones are major sites of plant root exudation (Forster, 1986). In the attachment to roots, pili and/or LPS seem to have a role (Denny, 1994; Sequeira, 1985), and the implication of flagella has also been demonstrated (Tans-Kersten, 2001). The pathogen moves to the host plant, attaches to the plant roots, infects the cortex and colonizes the xylem (Kelman, 1953; Vasse *et al.*, 1995; Yao and Allen, 2006a), which requires secretion of cell wall-degrading enzymes and EPS controlled by a regulatory network that uses PhcA (Schell, 2000a,b). After destroying the host, the bacterium returns to the environment and is likely to survive in soil, water or reservoir plants (Denny *et al.*, 1994). Within plant tissues, high densities of the pathogen increase expression of pathogenicity genes, repressed by low bacterial densities in non-host environments (Schell, 2000a,b).

5.5.4 Plant Root Cortical Infection

It starts at the sites previously colonized, *i.e.*, root extremities and axils of secondary roots. Due to the infection of pathogen, the root cortex of these zones has the intercellular spaces invaded and filled with bacteria (Vasse *et al.,* 1995). In the intercellular spaces the bacterium is likely to obtain nutrients from pectic polymers ofthe middle lamella by action of pectinolytic enzymes (Schell, 2000a,b) and also folate concentration in the spaces seems to contribute to vigorous proliferation of the pathogen (Hikichi *et al.,* 2007). Infection proceeds to the inner cortex level of primary roots, with bacteria forming large intercellular pockets, and cortical cells next to them displaying features of degeneration (Vasse *et al.*, 1995). As disease progresses, swimming motility may help the invasive cells go through the cortex (Tans-Kersten *et al.*, 2001).

5.5.5 Vascular Cylinder Infection and Xylem Penetration

Bacterial advance from cortex to vascular parenchyma implies crossing the endodermis, a cell layer with suberized walls and phenolic compounds, thought to be a barrier to vascular pathogens (Vasse *et al.*, 1995). The pathogen bypass the endodermis, the bacterium reaches the vascular cylinder at sites where this barrier is compromised. These zones would be the root extremities, where the endodermis is not fully differentiated, and the axils of secondary roots, where it is reoriented by the outgrowth and development of lateral roots, of all them, root exudation sites are preferably colonized by the pathogen (Vasse *et al.*, 1995). Once in the vascular cylinder, *R. solanacearum* infects the intercellular spaces of vascular parenchyma adjacent to xylem vessels, which eventually are also invaded. The pathogen can then be observed breaking into and filling xylem vessels, with the surrounding parenchyma cells being highly degraded (Saile *et al.*, 1997; Vasse *et al.*, 1995). Cell walls are destroyed by the hydrolytic enzymes secreted by *R. solanacearum*. Within the xylem vessels, the pathogen moves throughout the stem to the upper parts of the plant while it is multiplying, being reported to reach even more than 10^{10} cells per cm of stem in tomato plants (Saile *et al.*, 1997). It has been suggested that some *R. solanacearum* cells might form biofilms on host xylem vessel walls, which would protect them from host defenses (Tans-Kersten *et al.*, 2001) and could filter nutrients from the flow of xylem fluid (Yao and Allen, 2007). Although motility could help the pathogen spread out of infected vessels into adjacent uninfected ones, *R. solanacearum* is effectively non-motile in xylem vessels (Tans-Kersten *et al.*, 2001). Extensive multiplication and EPS production taking place in the water-conducting system lead to wilting of the host due to clogging of the vessels. The plant collapses and dies and *R. solanacearum* is released (Kelman and Sequeira, 2008) to a saprophytic life in the soil or other environments where it should survive until contact with a new host. According to these stages, a classification system based on histological localization and isolation described the susceptible hosts for *R. solanacearum* as those in which root and stem xylems were highly invaded (Alvarez *et al.*, 2008, 2009).

5.5.6 Life Outside Susceptible Hosts

In the absence of a host *R. solanacearum* can survive in natural habitats such as reservoir plants, soil and water, where populations can be affected by prevailing abiotic and biotic factors whose combination determines the fate of the pathogen in the environment (Fig. 5.2).

5.5.7 Reservoir Plants

Association of *R. solanacearum* with either reservoir plants or plant debris has been frequently suggested to promote survival of the pathogen in soil and water, and favour overwintering in temperate regions (Hayward, 1991; Graham *et al.*, 1979; Granada and Sequeira; 1983b; Wenneker *et al.*, 1999; van Elsas *et al.*, 2000). Reservoir plants wouid group together resistant/tolerant hosts and some of the non-hosts for the pathogen. Tolerant hosts have been described as those whose xylems can be invaded in the roots and weakly colonized at stem level, whereas non-hosts would not be invaded in plant xylem but there would be in them occasional presence of the pathogen in root cortex or on surface (Alvarez *et al.*, 2008a). From a comparison between the *R. solanacearum* root infection process in susceptible and resistant plants (Vasse *et al.*, 2005) a higher number of potential infection sites was found in the resistant, with two types of lateral roots differently colonized by a virulent strain, and where a brownish coloration in zones proximal to those colonized by *R. solanacearum* was frequently observed, probably due to compounds from secondary metabolism of polyphenols, responsible for the limited root infection (Vasse *et al.*, 2005). In fact, phenols induced by a fungus decreased *R. solanacearum* populations in the rhizosphere, on root surfaces and in the xylem of host roots (Zhu and Yao, 2004b). In less resistant or tolerant plants further movement of *R. solanacearum* onto the stem can be like that in susceptible hosts, with the difference that tolerant plants limit the spread to the upper parts by defence mechanisms such as inhibitor extracts, tyloses and gums that occlude vessels, whose success will depend on the rapidity of the plant response (Prior *et al.*, 1994). Among reservoir plants, a number of weeds were found to harbour *R. solanacearum* cells often as latent infections (Wenneker *et al.*, 1999; Pradhanang *et al.*, 2000), and several of them were considered potential hosts (Pradhanang *et al.*, 2000). In Europe, the most frequently reported carrier is bittersweet nightshade (*Solanum dulcamara*) a common perennial semi-aquatic weed inhabiting river banks, and also black nightshade (*S. nigrum*) and stinging nettle (*Urtica dioica*) (Hayward, 1991; Wenneker *et al.*, 1999). After death of the plant, the bacterium is released into the environment, where it seems to survive in reservoir plants, soil and/or water, through diverse strategies, such as the VBNC state (Fig. 5.2) the starvation survival response, the PC process, or the biofilm formation, until contact with a new host shelter *R. solanacearum* cells and be continuously releasing them into the water (Hayward, 1991; Janse *et al.*, 1998; van Elsas *et al.*, 2001; Elphinstone, 2005). Although these infesting populations are not as high as those from wilted plants (Pradhanang *et al.*, 2000), the continuous flow would contribute to the persistence of the pathogen in the environment.

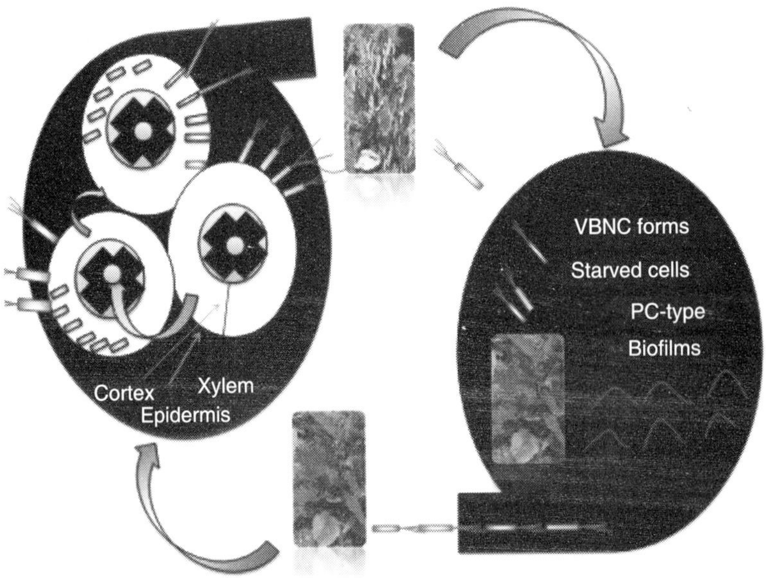

Fig. 5.2 Life outside susceptible hosts: reservoir plants soil and water

5.6 FACTORS RESPONSIBLE FOR DISEASE DEVELOPMENT

5.6.1 Temperature

Temperature, light intensity, soil pH, and moisture play an important role on growth of pathogen and disease devlopment. High soil temperature levels effect disease intensity in four ways: (1) increasing survival of the pathogen in soil; (2) increasing infection; (3) increasing disease development after infection; and (4) increasing the release of bacterial cells from the host and their spread through the soil (Buddenhagen and Kelman, 1964). Major factors favouring *R. solanacearum* survival in the field are permissive soil temperatures and high moisture contents (Kelman, 1953; Hayward, 1991). Bacterial wilt is significantly correlated with temperature, relative humidity and mainly rainfall (Bora *et al.*, 1996). No infection occurs at or below 50 per cent soil moisture. Disease does not envelop at 15°C. The optimum temperature for the growth of the bacterium is 35-37°C and most rapid development at 37°C. Devi and Menon (1980) studied the seasonal incidence of bacterial wilt and reported that high temperature has a significant role in disease development (Singh *et al.*, 2013). The disease mostly developed at mean temperature above 20°C accompanied with the plentiful supply of moisture. Small variation in an environmental parameter caused variation in wilt incidence. The pathogen can grow over a wide range of pH (6.2-7.4) but alkaline soils are generally unfavorable for disease development. There is decrease in infection with the increase in the age of the plants (Singh, 1995). Gallingly and Walker (1949b) reported that host inoculation showed that the disease increased with increase in soil temperature and the increase in air temperature up to 36°C and 28°C, respectively. In pre-inoculation stage, environmental factors were not so effective. Gallegly and Walker (1949b) reported that light period have significant role in development of bacterial wilt the winter type reaction seen when day length was maintained at 12 hrs during summer, and summer type reactions were reproduced at 18 hrs. During the summer disease development in the unbalance solution was increased at the low K level and decreased at the high N Level as compared with that in the basal solution. The inability of resistance genotype to perform consistently in different locations has been attributed to environmental influence such as temperature (Mew and Ho, 1977). They futher reported in 1977 that the resistance to bacterial wilts (*R. solanacearum*) in some tomato cultivars was influenced by change in soil temperature. VC 48 cultivar of tomato was not affected by the different treatment and maintained a moderate resistance at 26, 30 and 32°C. High temperature (30-35°C) and high soil moisture are the main factors associated with high bacterial wilt incidence and severity. Under these conditions, high populations of bacteria are released into the soil from the roots as the plant wilts (Momol *et al.*, 2001). Temperature is the most important factor affecting the host-pathogen interaction as well as survival in soils. In general, increase in ambient temperature to between 30-35°C increases the incidence and rate of onset of bacterial wilt on hosts such as tomato, for many but not all strains of the pathogen. The plants are resistant at moderate temperature may become susceptible at a higher temperature. Resistance is temperature-sensitive and strain-specific (Krausz and Thurston, 1975; Martin and Nydegger, 1982; Mew and Ho, 1977; Tung *et al.*, 1990b). Krausz and Thurston (1975) showed that temperatures of 32°C in controlled environment chambers significantly increased severity of bacterial wilt in two tomato lines resistant to bacterial wilt. The degree of resistance of cultivar Venus to isolate K-60 from the United States was not significantly affected by temperature, but this line expressed no resistance to isolate LB-6 from the Philippines at either 26 or 32°C. Similar effects have also been observed in tomato cultivars grown at different soil temperatures (Mew and Ho, 1977). Soil temperatures at different locations are also a factor in survival. In the cool hill of Shri Lanka, soil temperatures are such that no great stress is imposed and there is prolonged survival of *R. solanacearum*, whereas in the lowland plains near to see level soil temperature rises above 43°C and remains above that temperature for several hours daily. The soils exposed to 43°C continuously for periods of four days and over were free of the pathogen was reported by Seneviratne (1988). The effect of light intensity and photoperiod on host resistance has been reported to bacterial wilt by Kelman (1953). Reduced light intensity did not reduce resistance to isolate LB-6 in tomato line 1169 at 26.6°C but significantly decreased resistance at 29.4°C, and reduced photoperiod significantly decreased resistance of line 1169 to isolate LB-6, independent of temperature (Krausz and Thurston, 1975). Bora *et al.* (1996) reported that bacterial wilt incidence was significantly correlated with soil temperature and total rainfall. Small variations in an individual parameters caused variation in wilt incidence.The soil temperature from 25-30°C accompanied by a maximum air temperature of 26-30°C and monthly rainfall ranges from 200-300 mm favoured bacterial growth and multiplication resulting in severe wilt incidence. Ngoc Hung Tran and Kim Byung So (2010) reported that degree of severity increased with temperature after artificial inoculation from 15°C to 35°C. *R. solanacearum* disease screening, pepper seedling should be inoculated at 30 days old with inoculums density of 10^7-10^8 cfu/ml (OD around 0.14 at 600 nm) cultured on TZC for 48h. Inoculated seedlings are required to remain in a greenhouse at around 30°C and water accordingly. Root wounding methods make the disease symptom developed quickly. Disease evaluation of inoculated seedling is recorded around 20-30 days after inoculation. This disease is severe in summer and rainy season (Kishun, 1987; Hsu, 1991). Several factors are involved in virulence and pathogenecity of *R. solanacearum* (Kang *et al.*, 1994a,b).

5.6.2 Nutrient

Gallegly (1949) reported that the development of wilt (*P. solanacearum*) in sand culture at nutrient concentration 0.1, 0.5, 1, 2, and 3 times during summer month was greatest at 0.1 and decreased with an increase in salt concentration during late autumn and early spring disease development was greater at 0.5 and 1.0. Gallegly and Walker (1949b) further reported that wilt development under drip sand culture, with varying concentration (0.1H-3H) of the basal salt and unbalanced solution, disease development was greatest at 0.1 and decreased with an increase in salt concentration during summer, whereas in early spring and late autumn, disease development increased with increase in salt concentration up to 0.5H and 1H, but decreased sharply with further increase in nutrient concentration. During the early spring disease development was lower at both high N and low N levels, but in the low K solution it was not different from that in the basal. Disease development in the unbalanced solution at 32°C and temperature was decreased at the high N Level and increased at the low N level, while at 24°C and temperature the reverse was true. The winter type reaction in unbalance N solution were reproduced at 24°C, while the summer type results were reproduced at 32°C. The two sand temperatures had no effect on disease development in the low K solution. Disease development was reduced in low K solution at low light intensity. *R. solanacearum* retained virulence during the 4 years of nutrient scarcity in river water microcosms. It had been reported by van Overbeek *et al.* (2004) that the presence of catalase in the agar resulted in higher CFU of *R. solanacearum* biovar 2 strain 1609 and at day 84, log 1.95 CFU/ml still was detectable. No colonies were observed at day 125. The presence of viable-but-nonculturable (VBNC) cells in the 4°C cultures was confirmed using SYTO9 viability staining. Viable cell numbers were log 1.77 higher than CFU on plates with catalase. At day 84 and after 125 days, log 3.70 viable cells per ml still were present. Shifts in subpopulations differing in viability were found by flow cytometric sorting of 4°C treated cells stained with SYTO9 (healthy) and propidium iodide. The SYTO9-stained cell fractions dropped from 99 to 39%, and the PI-stained fractions increased from 0.7 to 33.3% between days 0 and 125. At 20°C, the SYTO9-stained fraction remained stable at 99%

until day 132. SYTO9-stained cells sorted from 4°C cultures at day 100 were injected into tomato plants. Upon incubation for 30 days, these plants did not show wilting. However, more than log 4.19 CFU and log 8.17 cells were recovered from these plants. Cells from colonies isolated from the non-wilted plants did not regain their virulence as demonstrated by subsequent injection into several new sets of tomato plants. Cells from 4°C cultures injected at day 125 were not able to cause wilting of, or proliferate in, tomato plants. The threat posed by VBNC *R. solanacearum* cells upon incubation at 4°C was thus ephemeral because cells lost their capacity to cause disease after 125 days. Interestingly, bacterial wilt was also observed in host plants after watering with long-starved *R. solanacearum*, which constitutes a more natural way of transmission of this pathogen. Bacterial growth, maceration of tissues and symptoms development in the early stages of bacterial wilt of tomato was primarily in plant tops of low calcium content (Power, 1983). Survival of pathogen was 180 days in Egyptian sandy soil (van Elsas *et al.*, 2000). Pathogen survival was positively correlated with dissolved organic carbon (DOC) in soil, negatively with organic matter (OM) content at similar DOC content, and negatively with bacterial diversity. It was found that *R. solanacearum* survived least in loamy soil with relatively high organic matter content (4%), whilst survival was highest in soil with lower organic matter content of 2.0-2.5%. Messiha (2006a) reported that the pathogen survived longer in Dutch than in Egyptian soils and in clay than in sandy soils. Messiha (2006b) reported that survival and suppressive activity of *S. maltophilia* were positively correlated with OM, which was highest in Egyptian organic clay soil. The effect of biological soil disinfestation (BSD) was tested in glass vessels, microplots, and in an accidentally infested commercial field. BSD is based on production of toxic organic acids through anaerobic digestion of fresh organic matter. BSD was accomplished by incorporating grass or potato haulms in soil and covering the soil with airtight plastic. Survival of *R. solanacearum* was found to be affected by soil type (texture and organic matter), temperature and moisture content (van Elsas *et al.*, 2000) in potato.

5.6.3 Nematode Population

The synergistic interaction between root-knot nematode (*Meloidogyne* spp.) and *R. solanacearum* on a variety of hosts is widely recognized. Root infection by nematodes as expressed by root galling index generally correlates with bacterial wilt symptoms as expressed by percent plants wilted The nematode may also act as a modifier of plant tissue in such a way that the tissue becomes more suitable to bacterial colonization. Breakdown of the resistant eggplant culivar Pusa Purple Cluster in India was attributed, at least in part, to such an effect (Reddy *et al.*, 1983).

5.6.4 Hosts Range

There are thousands of species of plant pathogenic microorganisms, but only a small minority of these pathogens has the capacity to infect a broad range of plant species. Most pathogens instead exhibit a high degree of host-specificity. Non-host plant species are often said to express non-host resistance. The term host resistance is used when a pathogen species can be pathogenic on the host species but certain strains of that plant species resist certain strains of the pathogen species. There can be overlap in the causes of host resistance and non-host resistance. Pathogen host range can change quite suddenly if, for example, the capacity to synthesize a host-specific toxin or effector is gained by gene shuffling/mutation, or by horizontal gene transfer from a related or relatively unrelated organism. Earlier over 25 separate botanical families (Sequeira and Kelman, 1976) been reported as host range of *P. solanacearum*. Various workers (Pegg *et al.*, 1974; Titatarn and Tong Jean, 1985b) reported that the pathogen survive on the weed host act as primary source of inoculums for the crops of the following season. A list of host range has been summarized in Table 5.1 Alternative weed host and non-host plant play an important role in survival of *P. solanacearum* strain in the absence of susceptible crops. Granada and Sequeira (1983b) suggested that the ability of strain of *P. solanacearum* to invade the roots of resistant cultivars of host species such as tomato and presumed non-hosts such as bean and corn, may be critical to this survival of the pathogen. The strain of *P. solanacearum* race 1 survived in soil upto 20 weeks while race 2 and 3 survived less than 20 weeks. This strain was able to invade the roots of non-host such as corn and bean, and this extended their ability to survive in absence of a susceptible host for at least 33 weeks. In general, the *P. solanacearum* (race 1) has a wide host range, the *P. solanacearum* (race 2) has host range restricted to *Musa* and a few related hosts, and the potato strain (race3) is restricted to few other alternative host. Race 4 from ginger and race 5 is restricted to mulberry (Persley, 1985). Kishore *et al.* (1994) have isolated *P. solanacearum* from several weed and non-host of rhizosphere, rhizoplane soil sample and they found that low population were in rhizosphere, rhizoplane and root tissue of symptomless host, compared to high population observed in roots of plants showing visible symptom. *Datura stramonium, Solaum xanthocarpum, Antirrhinum* sp. (weed host) and *Capcium baccatum* (non-host) showed visible symptom on inoculation while all the species carried out the pathogen population at 10^2-10^6cfu/g soils in the rhizoplane and 0-10^6cfu/g per root tissue without showing symptoms. Soil-borne pathogens have a broad host range and persist for longer periods in soil by resistant resting structures. The species as a whole has a very broad host range and infects hundreds of species in many plant families. The majority of hosts are dicots with the major exception being bananas and plantains. The association of the pathogen with wide range of hosts has epidemiological significance. They can survive on crop residue, seed, or in soil and water; they may be spread by plant or plant cuttings transfer, mechanical means, insects, and seeds. Now, 250 plant species belonging to 53 different botanical families (Kelman, 1953; Buddenhagen and Kelman, 1964; Hayward, 1994; Algam *et al.,* 2004; Elphinstone, 2005; Aliye *et al*., 2008) have been reported as host range of *R. solanacearum* and this unusually wide host range is continuously expanding. The most important widespread hosts are banana and plantain (*Musa paradisiaca*), eggplant (*Solanum melongena*), groundnut (*Arachis hypogaea*), *Heliconia* spp., potato (*S. tuberosum*), tobacco (*Nicotiana tabacum*), and tomato (*Lycopersicon esculentum*) (EPPO, 2004). The majority of them mostly belong to the *Solanaceae* and *Musaceae* families. According to the host range, *R. solanacearum* strains have been classified into five races (Buddenhagen and Kelman, 1964; Hayward, 1991). The most frequently reported hosts for each of them have been listed (Elphinstone, 2005). For race 1, which has the highest number of host species, solanaceous crops like chilli and sweet pepper, eggplant, potato, tobacco and tomato; non-solanaceous crops like bean, groundnut and sunflower; ornamental plants like *Anthurium* spp., *Dahlia* spp., *Heliconia* spp., *Hibiscus* spp., *Lesianthus* spp., *Lilium* spp., marigold, palms, *Pothos* spp., *Strelitzia* spp., *Verbena* spp. and *Zinnia* spp.; trees like *Eucalyptus* and fruit trees as black sapote, custard apple, and neem. Also, although of minor importance, abaca, cowpea, cucurbits, hyacinth beans, jute, moringa, mulberry, nutmeg, patchouli, *Perilla crispa*, sesame, strawberry, water spinach, wax apple and winged bean.

Table 5.1 Host range of *Ralstonia solanacearum*

Family	*Common Name*	*Scientific Name*	*References*
Annonaceae	Custard Apple	*Anions squamosa*	Hayward, 1985
	Tail flower	*Anthurium* spp.	Kishore *et al.*, 1994
	Weed	*Anthurium audreum*	Velupillai, 1995
Araceae	Pothos	*Epipremnum aureum*	Norman and Yuen, 1998
	Anthurium	*Anthurium* sp.	Mian *et al.*, 2002
Asteraceae	Herb	*Acanthosperma hispidum*	Rao, 1976
	Davanus (New host from India)	*Artemisia pallens*	Chandrsekaran and Prasantkumar, 2010
	Perfoliate blackfoot	*Melampodium perfoliatum*	Champoiseau *et al.*, 2009b
	Zinnia	*Zinnia* spp.	Elphinstone *et al.*, 2005
Araceae	*Pothos spp*	*Epipremnum pinnatum*	
Capparidaceae	Weed	*Cleome monophylla*	Rao, 1976
Cannabinaceae	Hemp	*Cannabis album*	Kishore *et al.*, 1994
Chenopodiaceae	Lamb's quarters	*Chenopodium album*	Kishore *et al.*, 1994; Champoiseau *et al.*, 2009b
Caryophyllaceae	Herb	*Spergula arrensis*	Kishore *et al.*, 1994
	Common chickweed	*Stellaria media*	Champoiseau *et al.*, 2009b, Pradhanang *et al.*, 2000
	Sticky chickweed	*Cerastium glomeratum*	
	Whitesnow or Summer Weed	*Drymaria cordata*	Pradhanang *et al.*, 2000
Caricaeeae	Papaya	*Carica papaya*	Seshadari *et al.*, 1977
Casuarinaceae	Horsetail Beet wood	*Casuarina equisetifolia*	Hayward, 1985
Commelinaceae	Patchouli	*Commelina communis*	Li *et al.*, 2012
Compositae	Goat weed or White weed	*Ageratum conyzoides*	Titaturn and Tong Jeam, 1985b; Kishore *et al.*, 1994
	Davana	*Artemisia pallens*	Chandrsekaran and Prasannakumar, 2010
		Perilla crisspa	Elphinstone *et al.*, 2005
	Ornamental Plant	*Bidens pilosa*	Titaturn and Tong Jeam, 1985b; Kishore *et al.*, 1994
	Pyrethrum	*Chrysanthemum coronarium*	Yik *et al.*, 1994

Contd...

Contd...

	Cosmos	*Cosmos caudatus*	Abdullah, 1983; Ansari, 1990
		Cosmos sulphureus	Gunawardena. *et al.*, 1980; Ansari, 1990
	Mung wart	*Artemisia(Artemisial)* spp.	Kishore *et al.*, 1994
	Dahlia	*Dahlia bariavilum*	Chatterjee *et al.*, 1997
	Bhongra (H)	*Eclipta prostrata*	Titaturn and Tong Jeam, 1985b
	Assam lata (H)	*Eupatorium odoratum*	
	Marigold	*Tagetus erecta*	Navaratnam, 1969
	Zinnia	*Zinnia* sp.	Chatterjee *et al.*, 1997
	A tree in Uganda	*Crassocephalum crepidoides*	Titaturn and Tong Jeam, 1985b
	Purple Fleabane	*Vernonia cineria*	
Convolvuceae	Sweet potato	*Ipomea batata*	Abdullah, 1983; He, 1985
Cruciferae	Chinese cabbage	*Brassica chinensis*	Valdaz, 1985
	Cabbage	*Brassica pekinensis*	
		Brassica oleracea var capitata	Kishore *et al.*, 1994
	Mouseear cress	*Arabidopsis thialiana*	
	Squash	*Cucurbita maxima*	Valdaz, 1985
	Mustards	*Brassica* spp.	Champoiseau *et al.*, 2009c
Euphobiaceae	Croton	*Croton sparsifolium*	Chatterjee *et al.*, 1997
	Weed	*Croton hirtus*	Abdullah, 1980a
	Cassava	*Manihot esculentus*	Nishiyama *et al.*, 1980
	Weed	*Phyllanthus niruri*	Rao, 1976
	Eucalypus young tree	*Eucalyptus urophylla*	Ran *et al.*, 2005
Fabaceae	Groundnut	*Arachis hypogaea*	Hsu, 1991
Gentianaceae	Texas bluebells, Anthurium Flowering Pot Plant	*Lesianthus* spp.	Elphinstone *et al.*, 2005
Geraniaceae	Geranium	*Pelargonium hortorum*	Pradhanang *et al.*, 2002
Graminae	Maize	*Zea mays*	Kishore *et al.*, 1994

Contd...

Contd...

Heliconiaceae	Wild plantains or false bird-of-paradise	*Heliconia* spp.	Elphinstone *et al.*, 2005
Labiatae	Bantulsi	*Perilla crispa*	Hsu, 1991
	Wilayati tulsi (H)	*Hyptis suaveolens*	Titaturn and Tong Jeam, 1985b
	The Scarpet	*Scutellaria scandes*	Kishore *et al.*, 1994
	Sage Nepata (Nepatal) sp.		
	Coleus (New host from India)	*Coleus forskohlii,*	Chandrsekaran and Prasannakumar, 2010
Liliaceae	Flower	*Lilium* spp.	Elphinstone *et al.*, 2005
Lamiaceae	Common Weed in China	*Salvia dugesii*	Li *et al.*, 2012
Leguminoeceae	Cluster bean	*Cyamoposi tetragonoloba*	Uematsu *et al.*, 1983
	Dolichos	*Dolichos lablab*	Keshwal, 1975; Keshwal *et al.*, 1976
	Mung bean	*Phaseolus mungo*	Uematsu *et al.*, 1983
	Garden pea	*Pisum sativum*	
	Bush bean	*Phaseolus vulgaris*	Velupillai, 1985
	Winged bean	*Psophocerpus tetragonoloba*	Abdullah, 1980a,b, 1982
	Cowpea	*Vicia* spp.	Kishore *et al.*, 1994
	Stylosanthes	*Stylosanthes humilis*	Aldrick, 1971
	Chickling Vetch	*Lathyrus*	Kishore *et al.*, 1994
Musaceae	Abaca	*Musa texilis*	Zehr, 1970
	Banana	*Musa paradisica*	Chatterjee *et al.*, 1997
	Bird of paradise	*Strelitzia reginae*	Hsu, 1991
		Strelitzia spp.	Elphinstone *et al.*, 2005
Malvaceae	Flower	*Hibiscus* spp.	Elphinstone *et al.*, 2005
Moraceae	Mulberry	*Morus alba*	Mathew *et al.*, 1993b
Myristicaceae	Nutmeg	*Myristica fragrans*	Mathew *et al.*, 1993a
Oleaceae	Olive	*Olea europeae*	He *et al.*, 1983
Ongraceae	–	*Jassiaea linifolia*	Titaturn and Tong Jeam, 1985b

Contd...

Contd...

Oxalidaceae	Oxalis	*Oxalis* spp.	Kishore *et al.*, 1994
Padaliaceae	Sesame	*Sesamum indicum*	Hsu, 1991
Piperaceae	Piper	*Piper nigram*	Bigat and Escuie, 1967
Polygonaceae	Toothed dock	*Rumex dentatus*	Farag *et al.*, 2004
Polygalaceae	Herb	*Polygala* spp.	Kishore *et al.*, 1994
	Herb	*Polygonum* spp.	
Polygonaceae	Weed	*Polygonum capitata*	Pradhanang *et al.*, 2000
Portulacaceae	Common purslane	*Portulaca oleracea*	Titaturn and Tong Jeam, 1985b
	Purslane	*Portulaca oleracea*	Champoiseau *et al.*, 2009; Elphinstone *et al.*, 2005
Ranunculaceae	Herb	*Ranunculus scleratus*	Kishore *et al.*, 1994
Rosaceae	Strawberry	*Fragaria vesca*	Hsu, 1991
Rubiaceae	Herb	*Galium aparine*	Kishore *et al.*, 1994
Solanaceae	Chilli	*Capsicum annuum*	
	Pepper	*Capsicum annuum*	
	Non host	*Capcium baccatum*	
	Dayking	*Cestrum diurnum*	Chatterjee *et al.*, 1997
	Datura	*Datura metal*	
	Jimson weed	*Datura stramonium*	Kishore *et al.*, 1994
	Jimson weed	*Datura tramonium*	Champoiseau *et al.*, 2009
	Tomato	*Lycopersicon esculentum*	Bigat and Escuie, 1967
	Tobacco	*Nicotiana tabacum*	Chatterjee et *al.*, 1997
	Tultipati(H)	*Physalis minima*	
	Prison berry	*Solanum indicum*	

Contd...

Contd...

	Tomato fruited egg plant	*Solanum integrifolium*	Ozaki and Kimura, 1989
	Brinjal	*Solanum melongena*	Bigat and Escuie, 1967
		Solanum xanthocarpum	Kishore *et al.*, 1994
	Common night shade	*Solanum nigrum*	Chatterjee *et al.*, 1997
	Black nightshade	*Solanum nigrum*	Champoiseau *et al.*, 2009
	Bittersweet or climbing nightshade	*Solanum dulcamara*	
	Weed	*Solanum cinercum*	Graham and Lloyd, 1978
	Potato	*Solanum tuberosum*	Bigat and Escuie, 1967
Tiliaceae	Jute	*Corchorus capsularis*	Sharma and Mukharjee, 1970, Hazarika *et al.*, 2006
		Corchorus olitorium	Chatterjee *et al.*, 1997
Tropaelaceae	Garden nasturtium	*Tropaeolum majus*	Champoiseau *et al.*, 2009
Urticaceae	Stinging nettle	*Urtica dioica*	
Verbenaceae	Teak	*Tactonia grandis*	Mitchell, 1962
	Vervain	*Verbena* spp.	Elphinstone *et al.*, 2005
Zingberaceae	Ginger	*Zingiber officinale*	Chatterjee *et al.*, 1997
	Alien Kahili Ginger	*Hedychium gardnerianum*	Andersona and Gardner, 1999
	Turmeric	*Curcuma domestica*	Gunawardena *et al.*, 1980; Kishore *et al.*, 1994

5.6.5 Symtomless Carrier

In this context, unrecognized symptomless hosts may also be important. Some weed hosts serve as sources of infection as reported from virgin land in the low land tropics is cleared for agricultural use (Buddenhagen and Kelman, 1964). There are some studies observed critically at the rhizosphere of weeds as a protected survival. In the Philippines, the population of *R. solanacearum* gradually declined in both rhizosphere and non-rhizosphere soils of various crop and weed hosts, except in the rhizosphere soil of *Portulaca oleracea*, where it gradually but slowly increased. Suppression of the population was greater with rice and corn. The bacterium may survive between successive plantings of a susceptible crop. These include plant debris, and latently infected plants, the deeper soil layers and rhizosphere of roots of weed hosts (Quimio and Chan, 1979; Jackson and Gonzalez, 1981; Granada and Sequeira, 1983). The range and variety of weed hosts is very extensive (Kelman, 1953). Some symptomsless carriers have been reported (Zehr, 1969; Granada, 1988). In India *Ageratum conyzoides* and *Ranunculus scleratus* showed no symptoms of the disease but the pathogen was readily isolated from surface sterilized roots. In Sweden, *Solanum dulcamarc* serves as a symptomless carrier of *R. solanacearum* biovar 2 (Olsson, 1976a,b). Other weed *Solanum carolinense* has also been reported to act as symptomless host of *R. solanacearum* in USA. In India, *Phyllanthus niruri* (Rao and Sohi, 1976) and *Lagasca mollis* (Kishun *et al.*, 1980) also are hosts for this pathogen. In Philippines, *Portulaca oleracea* was reported to be a symptomless host and in Queensland it is a susceptible host to *R. solanacearum* biovar 3 (Quimio and Chan, 1979; Moffett and Hayward, 1980). In lowland tropics, annual weed hosts are likely to be available year round and thereby increase the population of bacteria in the soil. These weeds may also serve as alternate hosts whenever non-host plants are cultivated, thus making crop rotation ineffective in reducing the bacterial population in the soil (Abdullah, 1982). Pradhanang *et al.* (2003) reported that some thyme oil-producing plants such as thyme (*Thymus vulgaris*) cv. German winter, Creeping thyme (*Thymus serpyllum*), and Greek oregano (*Origanum vulgare* subsp. *hirtum*), while remaining symptomless, became systemically infected by *R. solanacearum* and were therefore identified as hosts of *R. solanacearum.*

Pradhanang *et al.* (2000) reported that common agricultural weeds and crops that grow in the high hills of Nepal were examined after artificial inoculation and under natural conditions in UK and Nepal to determine whether such plant species could act as hosts to biovar 2 of *R. solanacearum*. Bacterial populations in the roots were determined 1 and 2 months after inoculation and at various intervals after harvesting infected potato crops under natural conditions. Inoculated roots of the summer weeds *Drymaria cordata* and *Polygonum capitata* and the winter weeds *Cerastium glomeratum* and *Stellaria media* yielded 10^2-10^7 colony-forming units per g root. High populations of the bacterium were recovered from these plants even after partial surface sterilization, indicating that systemic infection had occurred. *R. solanacearum* populations were recovered from root extracts of 75% of naturally growing *D. cordata* plants when sampled 3 months after harvest of a potato crop with bacterial wilt. Similarly, root extracts of 25% of *P. capitata* plants carried the bacterium. No potential winter weed hosts were infected under natural conditions when sampled 5 and 6 months after harvest of infected potato, indicating that winter conditions in the high hills of Nepal are not conducive to infection. Among crops, mustard (*Brassica juncea* cv. Fine White) developed typical wilt symptoms after artificial inoculation in warm glasshouse conditions (20–28°C). Mustard and barley are winter crops in Nepal. However, neither mustard (*B. juncea* var. Lumle Tori) nor barley (*Hordeum vulgare* cv. Bonus) was infected when planted into heavily infested plots under natural conditions. The results indicated that the role of non-solanaceous summer weeds in the persistence of biovar 2 of *R. solanacearum* in the environment may have been previously underestimated.

5.7 ECONOMIC IMPORTANCE

Bacterial wilt caused an economic importance which infects over 250 plant species in over 53 families as it cause economic damage to diverse agricultural crops. Since the 1950s, scientists have recognized that infections of *R. solanacearum* races 1 and 3 have hindered many solanaceous crops in both tropical and temperate regions. Race 1 has caused the greatest economic damage to potato, tobacco, and tomato crops in the southeastern United States. In Indonesia, Brazil, Colombia, and South Africa, races 1 and 3 caused the greatest economic damage (Hayward, 1991; Janse, 1996. CABI, 2006). It has been reported that in the United States the disease is very difficult to control (Champoiseau *et al*., 2009a,b). Race 3 biovar 2 of disease affect on eggplant, geraniums, potatoes, and tomatoes and race 1 is endemic to the southeastern of United States and currently not present in Canada and United States. Its latent and cold tolerant features have made this pathogen even more difficult to control in areas where the geranium showing early wilting symptoms caused by *R. solanacearum* is known to occur (Gorissen *et al*., 2004) and is considered of quarantine importance (Milling *et al*., 2009). Race 1 causes losses to tomato in Florida (Momol *et al*., 2003); race 3 on tomatoes and other solanaceous crops in the United States is unknown. A total destruction of plant has been reported during rainy season in Asia and Pacific region (Persley, 1985). In India, tomato and brinjal cultivation has been abandoned due to wilt in certain regions of Madhya Pradesh and losses of 25–100% were recorded so the entire communities were affected (Melton and Shew, 2000). Bacterial wilt of potato has been estimated to affect 3.75 million acres in 80 countries. The extent of losses due to bacterial wilt of solanaceous vegetables have been reported from various places are summarized in Table 5.2.

Table 5.2 Extent of loss in solanaceous vegetables

Crop	*Race*	*Biovar*	*Place*	*Loss%/ Amount*	*References*
Solanaceous			Southeast Asia	2%, or 100%	CABI, 2006
Solanaceous	1		North Carolina	25–100%	Melton and Shew, 2000
Tomato	1 & 3	3, 4	Philippines	15%	Zehr, 1969
				75%	Cook and Sequeira, 1994
			Taiwan	$496000/ha	Hartman *et al*., 1991
			Taiwan	15-95%.	AVRDC
			Philippines	80-90%	Agati 1949
			Karnataka	31.47 to 81.7%	Kishun, 1987
			Orissa	10.83 to 92.62%	Mishra *et al*., 1995; Mallikarjun *et al*., 2008,
			Karnataka	90.6 %	Kishun, 1987
			Kerala	15-95%	Anandaraj, 2009

Contd...

Contd...

			Uttarakhand	4-60%	Mishra *et al.*, 1995; Singh *et al.*, 2010
			Madhya Pradesh	100%	–
				Greater	Gunawan, 1988c
Brinjal	1 & 3	3, 4	Uttar Pradesh, Assam	4.24 to 86.14 %	Sabita *et al.*, 2000
			Philippines	10%	Zehr, 1969
			Madhya Pradesh	100%	–
				60 – 80%	Hutchinson, 1913a; Harikishore and Pushkarnath, 1963
			Goa	30 to 100%	Ramesh, 2006; Ramesh, *et al.*, 2009
			West Bengal, Orissa	3 – 40%	Singh *et al.*, 2010; Mishra *et al.*, 1995
Chilli		3, 4	Jharkhand	1 – 3%	Singh *et al.*, 2010; Mishra *et al.*, 1995
			Taiwan	15 to 55%	Wang and Lin, 2005
			Bangladesh	100%	Ali *et al.*, 1994
Potato	3	2	Greece	Extensive	Zachos, 1957
			Israel	Greater	Volcani and Palti, 1960
			–	$950 million	Ephinstone, 2005
			Netherland	U.S. $2.7 million	Meiners, 2000
			United States	$950 million	Milling *et al.*, 2009
			United States	$10 million	Swanson *et al.*, 2007
			–	50%	Hutchinson, 1913a; Harikishore and Pushkarnath, 1963
			Kenya	50-100%	Muthoni *et al.*, 2012

6
Management

The main strategy for controlling bacterial wilt disease consists of phytosanitation, modified cultural practices, crop rotation and intercropping with non-host plants, host resistance, and the use of chemical or biological control methods. These methods should be used in judicious combinations for best management of bacterial wilt. Among them host resistance is the economical stretegy. Shoda (2000) has reviewed control of bacterial plant diseases. Momol *et al.* (2005) have recommended management strategies in tomato. Muthoni *et al.* (2012) have given a general overview of bacterial wilt of potatoes and its management and highlighted the potential of host resistance as an important component of integrated management of bacterial wilt. This disease has limited means of control because crop protection chemicals are less effective and expensive and biological control agents are slow. In addition, phytosanitary methods such as quarantine are either expensive or difficult to apply and cultural methods such as crop rotations are largely impractical because the farms are too small to allow effective rotation, the pathogen has a wide host range, and it persists for long in the soil. Development of resistant cultivars could therefore play an important role in managing the disease. More resistant potato clones have recently been identified by scientists of CIP, and this resistance needs to be incorporated into the popular but susceptible Kenyan potato cultivars so as to increase potato production in Kenya. For better results, use of high resistant varieties may be coupled with use of disease-free tubers and clean fields.

6.1 CULTURAL MANAGEMENT

Farming in some societies is kept on a small scale, tended by peoples whose culture includes farming traditions going back to ancient times. (An example of such traditions would be lifelong training in techniques of plot terracing, weather anticipation and response, fertilization, grafting, seed care, and dedicated gardening.) Plants that are intently monitored often benefit from not only active external protection but also a greater overall vigor. While primitive in the sense of being the most labour-intensive solution by far, where practical or necessary it is more than adequate.

6.1.1 Quarantine

Wherein a diseased patch of vegetation or individual plants are isolated from other, healthy growth. Specimens may be destroyed or relocated into a greenhouse for treatment/study. Another option is to avoid introduction of harmful non-native organisms by controlling all human traffic and activities. The Australian Quarantine and Inspection Service (AQIS) is the Australian government agency responsible for enforcing Australian Quarantine laws although legislation and enforcement are key in order to ensure lasting effectiveness.

McCarter (1991) reported that effective cultural sanitation practices are critical to keep non-infested areas clean which include planting only certified disease-free plantlets, disinfesting all equipment before moving it between fields, controlling flood water flow, and never using surface water for irrigation in the greenhouse, sanitary practices for tomato transplant production may include avoidance of sub-irrigation, wide separation of greenhouses from field production areas, disinfestations of all frames, trays and tools, use of pathogen-free soils or potting mix, control of weeds, and limited handling of plants from infested area. Michel *et al.* (1997) reported that inter-cropping tomato with cowpea (*Vigna unguiculata*) significantly reduced the wilt.

Several workers (Messiha, 2006; Mallikarjun *et al.*, 2008; Champoiseau and Momol, 2009a) reported that destruction of weeds, collateral hosts and other off season hosts will reduce the inoculums potential and instructed the following precautions:

- Avoid contaminated land.
- Eradicate weed hosts.
- Remove wilted plants, root debris and volunteer hosts, and burn them to reduce spread of the disease from plant to plant.
- Disinfest tools when used in an infested field.
- Wash with water or bleach or sterilize by flame.
- Wash the soles of shoes after working in an infested area.
- Work in the infested portion of a field after working in the non-infested areas.
- For transplant production, use disease-free transplants, pasteurized soil medium, or fumigated plant beds.
- Use proper sanitation measures for transplant production, and avoid damage to roots during transplanting.
- Grafting susceptible tomato varieties onto the rootstock of resistant eggplant lines is widely practiced in developed countries of Asia.
- There are tomato varieties with some tolerance to bacterial wilt.
- Variations in race and strain of *R. solanacearum* make it difficult to utilize these varieties in some regions.
- Control root-knot nematodes and root-feeding insects since they may help the disease to establish and spread.

Priou *et al.* (1999) reported the following precautions should be followed under sanitation and cultivation practices for bacterial wilt infection to manage other potato diseases:

1. **Crop sanitation:** Cultivation measures aim to avoid or limit pathogen survival and dissemination.
2. **Removal of potato haulms:** Potato haulms must be removed from the field and buried deep, down-slope and far from irrigation canals after harvest of a bacterial wilt-infested crop, or they can be burned.

3. **Removal of rotten tubers:** Sorting out and leftover diseased tubers must also be removed from the field and destroyed using the same procedure as for haulms after harvest.
4. **Weeding:** *R. solanacearum* survives in weeds, so weeding must be done before planting potato and any other crop used in the rotation.
5. **Roguing volunteer potato plants:** Volunteer potato plants are another means of survival of *R. solanacearum* and must be removed in the subsequent crop (potato or any other crop) soon after their emergence.
6. **Roguing wilted potato plants:** If the incidence of BW is low in a field, wilted potato plants must be removed as soon as they are observed to avoid contamination of healthy neighboring plants. They must be destroyed carefully as previously for potato haulms and sorted-out tubers. Tubers from neighboring diseased plants should not be used for seeds.
7. **Tool decontamination:** To prevent movement of soil from an infested to a disease-free field, all tools must be decontaminated by washing with water and calcium hypochlorite or sterilized by flame.
8. **Tractor decontamination:** Machinery, vehicles, hooves of animals used for traction, and shoes of the personnel coming from an infested field must be washed at least with water before entering another field.
9. **Uncontaminated water:** The flow of water from an infested field to adjacent fields must be avoided. In infested areas, irrigation by well water is preferred over surface water from rivers or irrigation canals.

Other cultivation practices:

(i) **Minimum post-emergence cultivation:** Hilling only at planting time to avoid wounding roots with tools. Subsequent hand weeding is preferable for the same reason.

(ii) **Flooding of paddy rice:** In Asian countries flooding of paddy significantly decreases soil populations of *R. solanacearum*.

(iii) **Exposing soil to heat:** In very warm areas, exposing ploughed soil to summer heat also decreases soil infestation.

Tahat and Sijam (2010) suggested disease management for bacterial wilt on tomato, potato and geranium could be developed by including the following components.

1. Screen additional chemical and biological control products.
2. Exclude the pathogen from tomato transplants, potato seeds and geranium cutting and develop vegetative plant material certification schemes.
3. Study the effects of crop rotation, cover crops and mulches on pathogen dynamics and disease incidence.
4. Develop additional DNA and immunological detection tools that can be used reliably to distinguish subgroups of *R. solanacearum*, especially Race 3 biovar 2, from other endemic strains of *Ralstonia*.
5. Educate country extension, growers and crop advisors in sampling, monitoring and management of related diseases.

6.1.2 Integrated Disease Management

Integrated disease management (IDM), which combines biological, cultural, physical and chemical control strategies in a holistic way rather than using a single component strategy proved to be more effective and sustainable. Plant diseases are considered an important biotic constraint, which leads to signifi cant crop losses worldwide.The adoption of multidisciplinary integrated pest management (IPM) in seven states in the Southeastern United States (Alabama, North Florida, Georgia, Kentucky, North Carolina, South Carolina, and Tennessee) was studied to develop standards for adopting of IPM in fresh-market tomato production and found twenty-six percent of the producers throughout the region felt that the development of insect-and disease-resistant varieties would be most helpful to increase production Bauske *et al.* (1998). Integrated means the use of two or more of these methods in combination offers a higher chance of effectiveness. Plant diseases are considered an important biotic constraint, which leads to significant crop losses worldwide. In practice and in the majority of cropping systems today, emphasis is still being placed on a single technology. Nevertheless, the use of IDM strategy is gaining momentum, but in developing countries it often lacks the enabling environment for its successful implementation. Success requires appropriate policies in place that cover a wide range of themes such as plant protection, private sector investment, trade and export, food safety, land use, education and awareness, and agriculture extension.Wide adoption of IDM practices is a prerequisite for achieving impact at the country level. Experience over the last few decades clearly showed that adoption and support for using participatory approaches help farmers improve their overall field management, including disease management, reducing costs and improving production efficiency (Khoury and Makkouk, 2010). The farmers can control plant insect pests and diseases across the globe by the following strategies "Successful IPM strategies combine use of resistant varieties, cultural practices, monitoring, mechanical and biological controls, and responsible pesticide use for a more ecologically sound approach to agriculture (Chakrabarty *et al.*, 1993a; Wang and Ramasamy, 2011). In South Asia, eggplant is the only vegetable available at an affordable price for the rural and urban poor during the monsoon season. In Bangladesh and India, farmers sprayed insecticides up to 180 times during a 6-7 month growing season to control pest and diseases. An IPM strategy promoted in the two countries involved better field sanitation, prompt removal and destruction of infested plant from insect and diseases. By following the IPM strategy, farmers were able to reduce pesticide use by 65-75%. In Bangladesh, farmers sprayed pesticides only 21-22 times per season compared with 90-110 times in non-IPM fields. The strategy has been adopted by many communities beyond the initial implementation locations, with a significant impact on food security. Several non-chemical options have been reported (Arora *et al.*, 2008) to manage soil- and tuber-borne diseases which can be used in integration to minimize our dependence on pesticides. This can assure the future generations with safe food and a cleaner environment. Arora (2012) reported that environment friendly technologies have potential to reduce our dependence on pesticides and help in conservation of natural balance between microorganisms. An integrated approach involving the use of pathogen free seed, reduction of field inoculums, growing crop under right

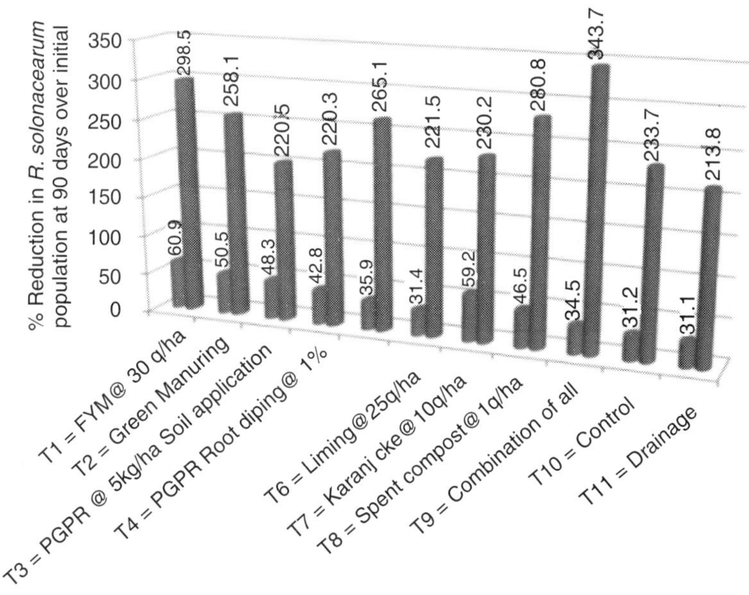

Fig. 6.1 Effect of IDM on *R. solanacearum* population and yield (Swarna Lalima)

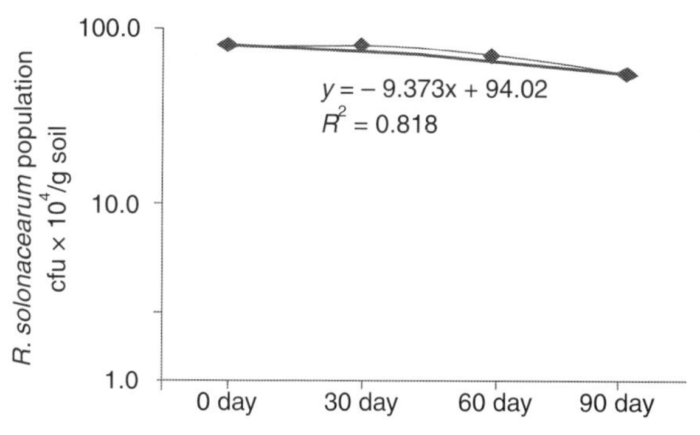

Fig. 6.2 Response curve in combination of IDM

environmental conditions and the biological control can help in the management of the disease in an environment-friendly way (Tahat and Sijam, 2010). This approach is beneficial for a sustainable agriculture and a mean of returning functional evenness to ecosystem (Crowder *et al*., 2010; Nazareno, 2009; Kaushik *et al.,* 2009). These technologies have potential to increase soil microorganisms and earthworms which have potential to increase fertility of the soil. Conventional chemical disease control should be applied only in exceptional cases where mechanical and natural ecological means do not work (Bareisis *et al*., 2002). Integrated pest management can be applied to manage vegetable diseases means using all the tactics available to the grower (cultural, biological, host-plant resistance, field scouting, chemical) that provide acceptable yield and quality at the least cost and are compatible with the tenets of environmental stewardship. Hussain and Bora (2008a) reported that integration of summer ploughing in chilli, half dose of carbofuran 3G, neem cake, streptocycline and full dose of *T. harzianum* were found superior treatments against *M. incognita* and *R. solanacearum* complex in brinjal under field conditions. The treatment recorded effectively improved all the plant growth parameters and yield of crop with corresponding decrease in nematode reproduction rate. The treatment also recorded final bacterial population in the soil along with less percent wilt incidence. Hussain and Bora (2008b) further reported that in the presence of *R. solanacearum*, *M. incognita* activities including population buildup in soil and root in brinjal crop were reduced compared to absence of the *R. solanacearum*. Dutta *et al.* (2009) reported 9.5-fold decrease in *R. solanacearum* population and were observed in the treatment, where seedlings were treated with bioagent community and the soil was amended with lime + bleaching powder + neem cake +vermicompost. Singh *et al.* (2012) reported that an IDM practice consisting of FYM (30 q/ha) + green manuring + soil application of PGPR (5 kg/ha) + PGPR root dipping (1%) + liming (25q/ha) + Karanj cake (10q/ha) + spent compost (10q/ha) found effective against bacterial wilt of tomato under Ranchi conditions. A 60-day glasshouse experiment was conducted to assess the influence of two strains of *P. fluorescens* (GRP3 and PRS9), organic manure, and inorganic fertilizers (urea, diammonium phosphate (DAP), muriate of potash and monocalcium phosphate) alone and in combination on the multiplication of *M. incognita* and growth of tomato. *P. fluorescens* GRP3 was better at improving tomato growth and reducing galling and nematode multiplication than PRS9. Anandaraj (2009) reported that seed treatment and soil application of bioagents *T. viride* and *P. fluorescens,* application of neem cake and vermicompost at the time of planting are essential for management of wilt. Carputo *et al.* (2009) reported that an integrated control strategy can succeed in reducing bacterial wilt incidence or eradicating it, which is a site-specific strategy to control and eradicate bacterial wilt; the main components are the use of healthy seed and planting in clean soils. However, many additional factors influence the incidence of the bacterium, such as environmental conditions (temperature and soil moisture), rotation with non-host plants, the use of less susceptible varieties and cultural practices crop sanitation and nematode control. Sharma and Kumar (2011) studied the integrated disease management to manage bacterial wilt in tomato at Ranchi and found maximum response (Fig. 6.1 and Fig. 6.2) in combination of FYM, Green manure, PGPR soil application along with root dip, PGPR soil application along with drenching, liming, karanj cake and spent mushroom substrate in cultivar Swarna Lalima. It was noted that *R. solanacearm* population was reduced 34.5 % reduction at 90 day over initial and trend was linear (R^2 = 81.8%). It has been reported that increase in total population of bacteria in green manure amended soil could might be the controlling factor in suppressing *R. solanacearum* (Kelaniyangoda, 1992). Further Kelaniyangoda *et al.* (1994 and 1990) reported from SriLanka that environment-friendly integrated management of bacterial in potato with soil amendment with Walsuriyakantha (*Tithonia diversifolia*) (10t/ha + CaO (2t/ha) + urea (200 kg N/ha) or Walsuriyakantha (10t/ha) + cowdung (10t/ha) or walsuriyakantha (10t/ha) + poultry manure (10t/ha) or walsuriyakantha (10t/ha) alone has suppressed the bacterialwilt. The current status of integrated management of bacterial wilt in potato from China was reported by Liyuan (1993), Chakrabarti *et al.* (1993b)Hayward (1987) in tomato. Further, Khoury and Makkouk (2010) also reported the prospects for the integrated control of bacterial wilt (*P. solanacearum*) and Naik (2009) reported in chilli.

6.1.3 Cultural Early Transplant

In USA, the incidence of bacterial wilt was recorded highest in late transplanted (August) crop. The average disease incidence being 3.7, 9.6 and 25.6% for early (May)-mid (June-July) and late season (August) transplanted crop, respectively as reported by Mc Carter *et al.* (1971). A virgin-soil technique for control of bacterial wilt was developed by Kratky and Ko (1974). Tomato plant grown on infected soil bed in greenhouse wilted and died whereas all those grown in soil free from the pathogen in plastic containers placed directly over the infected soil remained healthy and shifting planting dates to cooler periods of the year can help escape the disease. Barksdala *et al.* (1971) reported that early transplant escape the wilt in tomato.

6.1.4 Soil Solarisation

Soil solarization was first used in the fields of Israel (Kartan *et al.*, 1976). It is a method of controlling soil-borne diseases. Waggoner *et al.* (1960) reported that plastic mulches have been used predominantly as a pre-plant treatment to increase soil temperature for early season growth of crops and a post-plant treatment for management of soil moisture, temperature, nutrients and weed control. The benefits of plastic mulches for increased production of horticulture crops have been known for years. Recently, disease control by mulching of soil with transparent plastic during summer has been developed. Uehara (1988) used plastic sheet for soil solarisation for control of bacterial wilt of tomato. He found that without solarisation in the late crop, wilt destroyed the crop in September, whereas no wilt appeared till December in solarised plots. Similarly, in early crop, if not solarised the wilt destroyed the crop in May, whereas wilt did not appear until late June in the event of solarisation. Chellemi *et al.* (1994b) have reported that *R. solanacearum* can be controlled by soil solarisation whereas the effect of fumigation was noted to be highly variable in combined treatment (*i.e.* solarisation and fumigation) in reduction of the population of *R. solanacearum*. Soil solarisation has been included as one of the components in the integrated management in controlling the soil-borne disease (Sood *et al.*, 1997; 1998). Dazomet combined with soil solarisation gave better control of tomato bacterial wilt as compared to methyl bromide. One month after treatment, residue of dazomet in the soil reduced to 0.2% of that of the immediately after treatment (Yamada *et al.,* 1997). Soil solarisation combined with fumigation reduces the pathogen considerably. Kline *et al.* (1999) have used solarisation for control of soil-borne disease. A heated greenhouse, with soil surface covered with 150 µm-infrared PE (Polyethylene) film or a 100 µm clear film was compared with an untreated greenhouse covered with 150 µm infrared film plus a clear 100 µm film on the soil surface or 150 µm black film on the soil surface and a tent made from 100 µm clear film stretched above the black plastic. The highest soil temperatures reached in the heated greenhouse were 63-64.5°C for both system at depth of 2.5 and 10 cm. At a depth of 30 cm the temperature was 57.3-58°C. Soil temperature in the untreated greenhouse varied between 56-57.5°C at 2.5 cm, 50-52.5°C at 10 cm, and 39-41°C at 30 cm. At these depths, the soil borne pathogens were absent as compared to control. Mallikarjun *et al.* (2008) reported that the disease can be kept under soil solarisation combined with fumigation. Kumar and Sood (2001) have studied soil solarization for 8 and 10 weeks during March to May and noted increased soil temperature at 5cm depth in wilt sick soil at Palampur by 5.2°C and 6.2°C during 1998 and 1999, respectively. The increased soil temperature promoted abundant germination and profuse growth in weeds in the solarized plots during early stages of traping followed by their complete death by burning later on in both the 8 and 10 weeks solarised plots. The bacterial wilt incidence in such plots was significantly suppressed by 65.9% and 71.6% during 1998 and 68.9% and 73.4% in 1999 in 8 and 10 weeks of solarization, respectively as compared to 97.7% and 86.9% wilt in non-solarized plots. The significant reduction in wilt incidence is attributed to the incorporation of antagonistic rhizobacteria (*P. fluorescens* and *B. cereus*) in soil prior to solarization as the population of the antagonists was considerably enhanced after solarization. *R. solanacearum* was significantly reduced and resulted in increased vigour of tomato plants in solarized plots as compared to non-solarized plots. Sharma *et al.* (2004) studied effect of soil solarization on *R. solanacearum* population in (Fig. 6.3). Chellemi *et al.* (1997) reported that soil solarization as integrated management tool of soil-borne pests of tomato under humid conditions.

6.2 CROP ROTATION

Crop rotation may be an effective means to prevent a parasitic population from becoming well-established, as an organism affecting leaves would be starved when the leafy crop is replaced by a tuberous type, etc. Other means to undermine parasites without attacking them directly may exist Lemega *et al.*, 2001. Thorold (1949) has recommended from crop rotation studies that neither tomato nor brinjal plant nor any other solanaceous crop should immediately precede tomato, when it is necessary to grow tomatoes on land, which has already carried solanaceous crop, than cauliflower should be intercropped before tomatoes. Keshwal *et al.* (1975) reported that the wilt was controlled by 3-4 cultivation to a depth of 25-30 cm, 20 days apart by rotation involving sunhemp-wheat, guar-garlic, pea-rice or mungbean-cauliflower and by removing weed host of pathogen. The following crop rotations are suggested by Priou *et al.* (1999) and Messiha (2006) in potato cultivation. Kishore *et al.* (1992) reported the relative effectiveness of various crop rotations on the incidence of bacterial wilt in potato.

- **Rotation with non-host plants** is an effective means of decreasing the level of *R. solanacearum* populations in soil provided volunteer potato plants are continually removed by uprooting the plants from the field as they emerge.
- **Rotation with cereal crops and gramineous pastures** rapidly decreases soil inoculums potential. However, the time necessary for eradication has to be determined in each location.
- **Rotate with liliaceous and brassicaceous crops *Allium* spp. (onion, garlic and leak) and brassicaceous** (cabbage and cauliflower) can be planted on infected soil. Furthermore, these crops help the elimination of weeds and volunteers that maintain the pathogen.

(a) Solarization

(b) Raising nusery covered with mosquito net after solarization

Fig. 6.3 Soil Solariztion

- **Rotate with legumes:** Legumes, e.g., fabaceous (pea, bean), can be used as rotation crops for potatoes. Moreover, they have the additional benefit of increasing soil fertility by fixing atmospheric nitrogen. A bean/maize rotation has reportedly reduced soil inoculum potential.
- **Rotate with cucurbitaceous crops:** Cucurbitaceous crops (pumpkin, cucumber, and zucchini) are non-host for *R. solanacearum* and can thus be used in the rotation.
- **Avoid tomato crop:** Tomato crop should be avoided, since it is highly susceptible to bacterial wilt caused by either race 1 or 3 of *R. solanacearum.*
- **Avoid other solanaceous crops:** Eggplant, pepper and tobacco must never be planted after a bacterial wilt infected potato crop.
- **Avoid ginger and groundnut:** Ginger and groundnut must never be planted after a bacterial wilt infected potato crop in infected areas, where race 1 of *R. solanacearum* are prevalent (low elevation tropics).
- **Nematode control:** Nematodes make a path for the bacterium to enter the plant through root injuries. Therefore, they must be controlled to avoid their interaction with *R. solanacearum.* Major nematode control components are soil fumigation, rotation with cereals, application of high quantities of organic amendments (free of *R. solanacearum*), and planting nematode-resistant varieties.

Regular rotation with paddy and other non-host plants (Table 6.1) have been reported to reduce the disease incidence. Rice and corn are reliable rotation crops with susceptible species (Quimio and Chan, 1979; Wang and Lim, 2005). Michel *et al.* (1997) reported that inter-cropping tomato with cowpea (*Vigna unguiculata*) significantly reduced the wilt.

Table 6.1 Non-host crops of *R. solanacearum* suited for rotation with tomato crops

Crop Group	Crop Name
Field Crop	Maize, mung beam, rice, sorghum, vegetable
Root Type	Carrot, yam, yam bean
Bulb Type	Chive, garlic, onion, spring onion
Stem Type	Asparagus, celery, lotus, taro
Leaf Type	Amaranth, cabbage, coriander, leafy brassica, lettuce, mustard, spinach

Contd...

Contd...

Flower Type	Broccoli, cauliflower
Fruit Type	Gourd, pea, pumpkin, yard long bean
Green Manure	Brassica, sesbania, sun hemp

Kloos *et al.* (1991) studied crop rotation for four successive seasons in Mindanao, Philippines against *P. solanacearum*, race 1, in potato with maize, beans, and sweet potato rotated with potato in eight different combinations. They reported that the disease remained at a significant level in all treatments, but all rotations reduced disease symptoms, compared with potato monoculture. A dry period following the third season, significantly reduced pathogen populations in the soil. Short crop rotations and the use of dry season fallow appeared to be useful tools in an integrated approach to control bacterial wilt, race 1, in potatoes.

Bang and Wiles (1995) conducted a trial to compare the effect of different crop rotations on the level of bacterial wilt in succeeding potato crops. Land severely infected with bacterial wilt (*P. solanacearum* biovar 2, race 3) at Tambul, Western Highlands Province (Papua New Guinea), was selected for the trial. Potato crops were followed by maize, sweet potato, bare fallow and weed fallow. Occurrence of wilt in succeeding potato crops was then observed. At the third crop (*i.e.* after one break crop) wilt infection remained high in all treatments. However, infection level appeared lower in potatoes planted after maize or bare fallow than after weed fallow or continuous potato. After the fourth crop, the trend was more pronounced. Percentage of wilting plants was significantly lower ($P_{0.01}$) after two maize crops, or 12 months bare fallow, than after continuous potato or potato-bare fallow-potato. Marketable and total potato yield were significantly higher ($P_{0.01}$) after two maize crops than after continuous potato or potato-weed fallow-potato. These results suggest that weed fallow permits survival of the pathogen. Sweet potato was less effective than maize as break crop to suppress bacterial wilt infection. Sohi *et al.* (1981) reported that two years crop rotation with *Vigna* sp. followed by maize and cabbage or okra followed by *Vigna* sp. and maize gave effective control of *P. solanacearum* whereas cropping sequence of cowpea-maize–cabbage, bhendi-cowpea-maize, maize-bhendi-radish, maize-cowpea-maize, ragi-brinjal (tolerant variety)-bean was very much helpful in reducing the soil-borne inoculums as reported by Rao and Kishun (1986). Effect of previous crop on soil population of *B. solanacearum*. The study revealed that the bacterial population was declined after cowpea and rice but not after brinjal (Michel *et al.*, 1996). The population also declined when soil was left fallow, suggesting that a suitable host plant is required to maintain the bacterial population. Rotation with maize, okra, cowpea and resistant tomato line (CL1131) caused delayed onset of wilt by 1-3 weeks and reduced severity by 20-26% in Nepal (Adhikari and Basnyate, 1998). Lemaga *et al.* (2001) reported from Uganda that one-season rotation in heavily infested field (90% wilt) with wheat and maize did not significantly reduce wilt but increased potato tuber yield, while the effects of *Phaseolus* beans were not significant. On the contrary, a one-season rotation in mildly infested field (about 15-20% infestation) resulted in a significant reduction in wilt incidence as compared to the control. The highest wilt incidence under rotation was below 13% as compared to 62% in the control. Finger millet and sweet potatoes reduced wilt the most, while carrots and onions reduced the least. There were no significant yield differences among the rotation treatments except for the control. A two-season rotation on heavily infested soil reduced wilt to 22% - 49% as compared to 81% in the control.

The highest wilt reduction was obtained with potato-beans-maize-potato treatment, while the lowest reduction was with potato-maize-maize-potato. Planting different crops in two consecutive seasons performed better than planting the same crop. Interestingly, sweet potato, a major staple food, was a good rotation crop for reducing wilt. Beans, the most important food crop and a protein source, and a symptomless carrier of *R. solanacearum* suggested a good rotation crop if used before or after cereals in a potato rotation scheme. Sharma and Kumar (2004) reported from Ranchi (Jharkhand), the best crop rotation was ragi-french bean-okra in which 74.6% reduction in native bacterial population was seen. Rotations with other non-hosts also caused reduction in the population of *R. solanacearum* to varying levels, *viz.*, maize-spinach-watermelon (71%), okra-maize-radish (68.7%), maize-carrot-cucumber (55.9%), maize-ragi-maize (31.6%), ragi-radish-french bean (14.8%). Solo cropping of tomato caused increase in the population of *R. solanacearum* by 162.8% in 3 years.

Mallikarjun *et al.* (2008) reported that the disease can be kept under check if crop rotations are followed with crops such as maize, ragi and okra, since they can reduce the pathogen population significantly. Consider using infested fields (after 3-4 years rotation) during cooler months for tomato production in infested fields to reduce *R. solanacearum*, weeds and nematodes. Suitable rotations can only be determined through local experience because of the diversity of *R. solanacearum* strains and races, and the many agro-climatic zones where reports occur. In spring season of North Florida and similar condition was reported by (Wang and Lin, 2005) from Taiwan and further reported that the plots that have no history of bacterial wilt and have not been planted previously to other susceptible crops like pepper are less likely to have the pathogen.

Champoiseau *et al.* (2009) reported that cover crops, *i.e.*, rye for winter, sudan-sorghum for summer was suitable rotation in North Florida to avoid weeds that support *R. solanacearum* populations where race 3 biovar 2 has not to be established in the United States. Carputo *et al.* (2009) reported that rotate with cereal crops, *viz*; wheat-maize-other cultivation practices. Minimum post-emergence cultivation hilling only at planting time to avoid wounding of roots with tools. Subsequent hand weeding is preferable for the same reason. Hiddink *et al.* (2005) reported the effect of mixed and single crops on disease suppressiveness of soils whereas Aaen (1997) reported the effect of crop rotation on bacterial wilt incidence and potato tuber yield.

6.3 SOIL AMENDMENT

Soil amendments are elements added to the soil, such as compost, peat moss, or fertilizer, to improve its capacity to support plant life. While fertilizer improves soil by adding nutrients only, soil amendments such as peat moss improve soil by making its texture or drainage more conducive to plant health. Peat moss adds no nutrients to soil. Meanwhile, compost enhances soil both through adding nutrients and through improving texture and drainage. Amendments can also change soil pH.

There are two broad categories of soil amendments: organic and inorganic.

6.3.1 Organic Amendment

6.3.1.1 Green Manure

In agriculture, organic matter is considered the backbone of cultivated soils. In the early 1970s the organic matter in the soil was around 0.7 per cent, which has now declined to 0.2 to 0.3 per cent, particularly in the intensive crop cultivation region like northern India. The continuous use of fertilizers and pesticides has altered the biological properties of soil. Microbial activity in soil has decreased. Soil aeration, porosity and water-holding capacity have been affected due to the reduced organic matter. Chemical fertilizers only provide one or two major nutrients, while the soils are becoming deficient in other major, secondary micronutrients. Our efforts should be on gradual reduction of chemical use while sustaining productivity. Use of bulky organic manures, which can improve the organic content of soil and thereby enhancing its physical, chemical and biological properties is important for fertility management. Organic manures improve the nutrient balance. Bulky organic manures can be supplied through farmyard manure, farm compost, town compost, night soil sludge, green manure, etc. Out of these sources only farmyard manure and green manure are widely used. Farmyard manure is prepared from dung, yet about 60 to 70 per cent of dung is used as fuel in rural areas. Also, the dung is not managed properly and the quality of manure is not good. Thus, one practical way left for supplying organic matter is green-manuring. Green-manuring is ploughing into soil live plant material for improving the physical structure as well as fertility of the soil (Fig. 6.4). Green-manuring can be done in two ways depending on the situation (a) green-manuring *in situ* (b) green-leaf manuring *ex situ* (any green materials turned into soil, *viz*, leaves or twigs collected from shrubs and trees grown on bundhs, wastelands or nearby forest). This practice adds organic matter to the soil, improves the physical properties of soil in terms of aeration, porosity, water-holding capacity, and facilitates penetration of rainwater. Green-manure crops hold plant nutrients that would otherwise be lost by leaching and leguminous green-manure crops add nitrogen to the soil. This practice increases the availability of nutrients like phosphorus, calcium, potassium, magnesium and iron. Dhaincha (a leguminous plant) is one the most valuable medicinal herb and ideally suitable as green manure crop. Several species, *viz*, *Sesbania aculeate, S. randiflora, S. sesban* and *S. speciosa* are grown in which nodules born both on roots and stems. The nodules are

Fig. 6.4 Green manuring by use of Dhaicha

homes of nitrogen fixing symbiotic bacteria. The nitrogen fixing capacity of some species are very high and even in case of waterlogging; it performs well also in Chhattisgarh (Oudhia, 2001, 2002 and 2003). Cardoso *et al.* (2006) studied the use of organic matter that improves the physical, chemical and biological soil properties and also as an inducer of soil suppressiveness to plant pathogens. Incorporation of organic compost and freshly cut aerial parts of pigeon pea (*Cajanus cajan*) and crotalaria (*Crotalaria juncea*) in concentrations of 10, 20 and 30 % (v/v), into soil infested with *R. solanacearum* on tomato gave encouraging results on seedlings of cv. Santa Clara transplanted into polyethylene bags with 3 kg of the planting substrate (infested soil + organic matter). All evaluated concentrations with incorporation and incubation for 30 days of aerial parts of pigeon pea and crotalaria controlled 100% tomato bacterial wilt. However, with 60 days of incubation, the 10% incorporation of pigeon pea and crotalaria shoots did not control the disease. Yadessa *et al.* (2010) reported effect of amending top soil with three different levels (1, 5 and 10%) of cocopeat, farmyard manure (FYM) and green compost, and two levels of bacterial inoculations on infection of tomato by *R. solanacearum* compared to non-inoculated treatments. Non-amended topsoil with and without *R. solanacearum* were included as control treatments, which resulted in changes in physico-chemical properties (such as electrical conductivity, organic matter content, total carbon, dissolved organic carbon, NH_4^+, NO_3^-) and microbial activity of the amended soil and the effects were found to be higher at the higher rate of application. Effects on disease suppression and survival of the pathogen in the soil differed depending on amendment type and application rate. Higher disease severity was recorded in soil amended with 10% green compost compared to the control treatment. Complete suppression of *R. solanacearum* was observed in pots amended with 5 and 10% FYM, 1% green compost and 10% cocopeat. Absence of disease at the highest rate of FYM was supported by a lower number of culturable *R. solanacearum* recovered from rhizosphere soil two months post-inoculation in soil amended with 10% FYM. Soil amended with 10% FYM gave higher root and above-ground dry weight. Moreover, FYM added to top soil at 5 and 10% gave significantly higher above ground fresh weight. This study indicated that amending top soil with different types and rates of amendment can suppress bacterial wilt severity and pathogen survival in the soil. Amendments also enhanced tomato yield, the higher rates of amendments being the most effective except in green compost treated with 10% which gave a 27% lower yield compared with the higher rate of FYM. Kelaniyangoda *et al.* (1994 and 1995) also reported on the effect of green manure, organic manure and chemicals on the incidence of bacterial wilt in potato.

Table 6.2 Crop Suitable for Green Manuring (Yawalkar and Agarwal, 1962)

Crop	*Season*	*Green Matter (q/ha)*	*N added (kg/ha)*
Sunhemp	Kharif	152	84
Dhaicha	Kharif	144	77
Pillipesara	Kharif	132	55
Moong	Kharif	57	37
Cowpea	Kharif	108	56
Guar	Kharif	144	62
Senji	Rabi	206	134
Berseem	Rabi	111	60

Incorporation of fresh aerial parts of pigeon pea and crotalaria is an effective method for bacterial wilt control. This is an efficient disease control strategy. Green manure is an inexpensive source of organic matter and it improves the chemical, physical and microbial soil. Uesugi *et al.* (2002) reported that incorporation of 20 to 100% of organic matter is possible for plant growth substrates during the sowing process for seedling production, while the level of organic matter incorporated into the soil should be around 2%, in order to be considered economically viable. Germani and Plenchette (2004) reported that a significant correlation was obtained between their calculated on shoot dry weight mass recorded in each experiment and suggested that *Crotolaria* spp. could be used as pre-crops for providing green manure while at the same time decreasing the level of detrimental nematodes and increasing the level of beneficial mycorrhizal fungi. Sharma and Kumar (2009b) results of interaction of karanj cake and green manure with *R. solanacearum* propagules in soil conducted in a alfisol during 2004-07 (Table 6.3 and Fig. 6.5). Data revealed that different dose of karanj cake and green manure has significant effect on reduction of *Ralstonia* population. Application of karanj cake @ 0.1 kg/m^2 resulted significant reduction (18.75% over initial) in *R. solanacearum* population in soil (Fig. 6.5a). Similarly, incorporation of different doses of green manure of *Sesbania* leaves resulted in reduction (64.04% over initial) in *R. solanacearum* population in higher dose (10 kg/m^2) after 90 days of application (Fig. 6.5b). In case of green manure *Sesbania* incorporation in the plot, higher dose resulted high yield (77.8q/ha) and it was at par with optimum dose of karanj cake (0.1 kg/m^2).

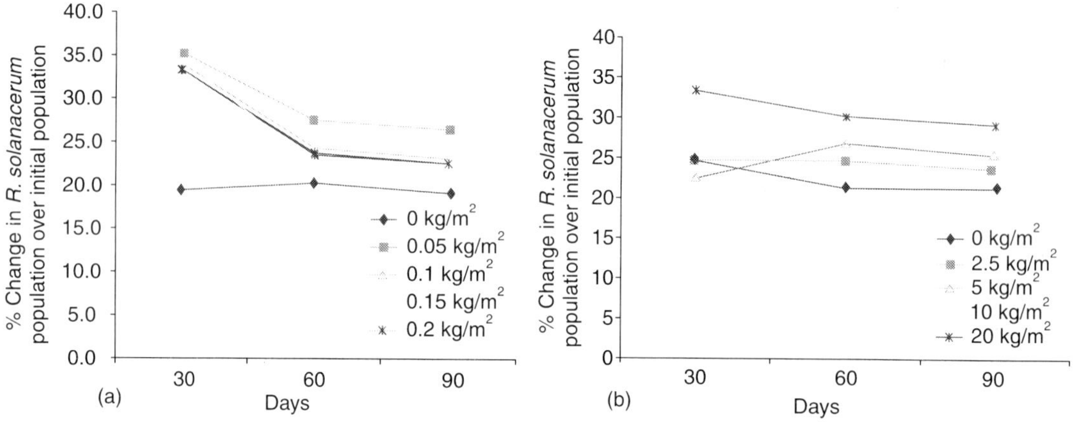

Fig. 6.5 Response of (a) karanj cake and (b) green manure on *R. solanacearum* wilt in tomato (cv. Pusa Ruby)

Table 6.3 Effect of different doses of karanj cake and green manure on *R. solanacearum* population in tomato (cv. Pusa Ruby)

Treatments	*Dose (kg/m^2)*	*Plant Survival (%)*	*Yield (q/ha)*
Karanj cake	0	65.0(56.99)	55.4
	0.05	70.0(53.76)	69.3
	0.1	70.0(57.72)	74.5
	0.15	60.0(55.01)	82.0
	0.2	50.0(45.0)	102.5

Contd...

Contd...

Green manure	0	58.3(61.76)	44.6
	2.5	60.0(51.14)	66.0
	5.0	53.3(55.81)	68.5
	10	48.3(54.14)	41.0
	20	56.6(47.96)	77.6
CD(P0.05)		7.17	23.1
CV%		7.77	19.8

In another experiment they reported in 2009 that use of green manure and FYM reduced the apparent infection rate of bacterial wilt in Swarna Lalima.

Lemaga *et al.* (2001) reported that potato bacterial wilt can be managed by soil amendment. They conducted experiment for 3 consecutive seasons with organic materials: *Sesbania sesban* (S) and *Leucaena diversifolia* (L.) applied in sufficient amounts to supply 100 kg N ha^{-1} either singly or combined with P and PK. Also added were NP and NPK from inorganic sources. The organic materials were incorporated into soil one week before planting, while the inorganic fertilisers were side-dressed at planting all at rates that would supply 100 kg ha^{-1} of N, P and K. Nitrogen in the form of urea was split-applied at planting and one month after. Disease incidence was lowest with treatments NP and S + PK and highest with the control. Application of organic manures alone did not necessarily result in reduced wilt incidence except in a few cases. Both marketable and total tuber yields were consistently highest with S + PK and differed significantly from the control in all seasons. A combined analysis over the three seasons showed that the treatment S + PK gave a significantly higher yield (20.8 ha^{-1}) than all other treatments, while the control yielded significantly lower (9.7 ha^{-1}) than the other treatments. *Sesbania* as an organic manure performed better than *Leucaena* and potassium was found to be a useful nutrient for crop performance. When K was applied with NP, LP and SP, it brought about marketable yield increases of 11, 23 and 37%, respectively. Generally, the rate of wilt development, expressed in wilt incidence per unit time, was highest at early stage of growth, thereafter, it declined and stabilised during much of the tuber bulking stage. The interaction between soil fertility and bacterial wilt incidence merits further studies in different environments.

6.3.2 Manures

The use of soil amendments (Sas) is a widespread means to control diseases caused by soil-borne plant pathogens (Huang and Huang, 1993). Various Sas are reported to suppress bacterial wilt of tomato (Power, 1983; Sun and Huang, 1985; Prior and Béramis, 1990) and potato (Elphinstone and Aley, 1993; Nielsen and Todd, 1945). Organic amendment with compost proved beneficial (Chelleni *et al.*, 1992) and that of made 15 days earlier with soybean flour, rice husk + urea and recommended dose of fertilizer in soil effectively prevented bacterial wilt in susceptible cultivars Roma VF (Miah *et al.*, 1995). The bacterial wilt of tomato was suppressed in the poultry and FYM added soils, and higher microbial activity was likely responsible. Mallikarjun *et al.* (2008) reported that application of sawdust, peat moss is efficient in reducing the incidence. Islam and Toyota (2004) reported that different composts were added to a conducive soil to study their effects on the bacterial wilt of tomato caused by *R. solanacearum*. The soils to which bark

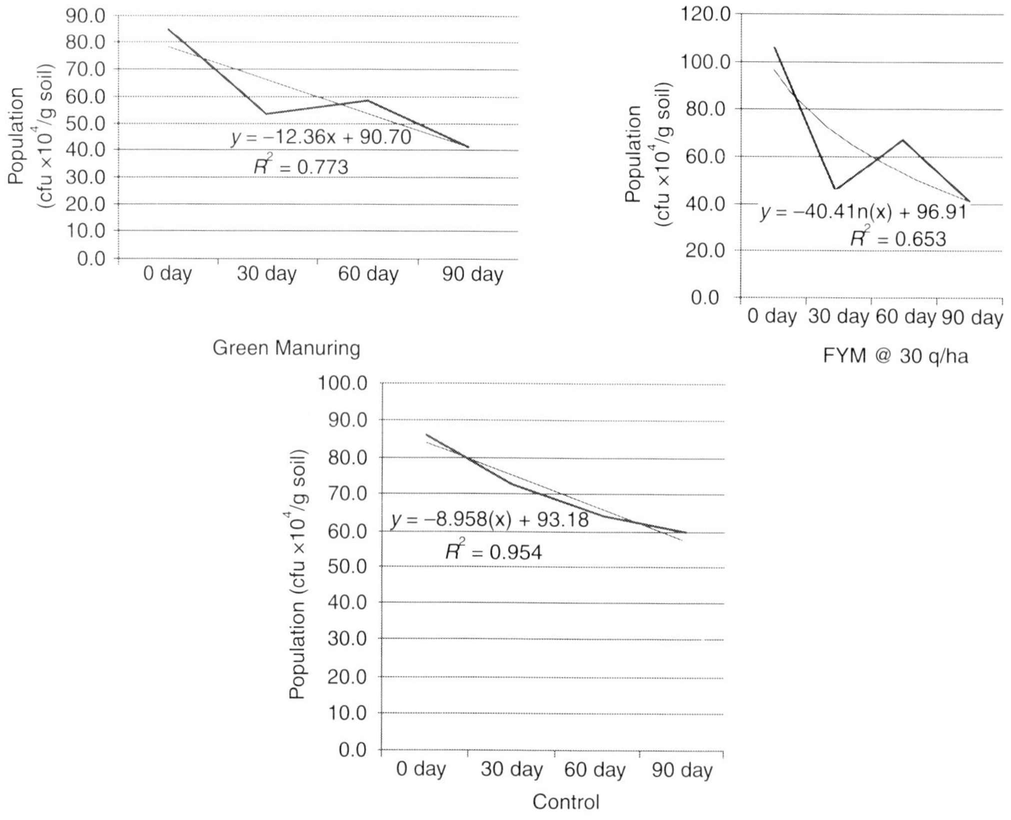

Fig. 6.6 Response of organic manure on apparent infection rate of *R. solanacearum* in tomato (cv. Swarna Lalima)

compost, coffee manure and pork manure were added showed higher wilt incidence (average more than 5 plants out of 6 wilted) after 30 days of cultivation compared to those in the non-treated soil, while the soils with poultry and FYM showed consistently lower wilt incidence (less than one plant out of 6 wilted). Survival of the pathogen was better in the control and in the bark, coffee and pork manure added soils than in the poultry and FYM added soils. A lower C/N ratio and higher amounts of water-soluble organic carbon and nitrogen were found in the poultry and FYM than the other composts, suggesting that the contents of easily decomposable organic matters might be higher in the added soils. Higher numbers of culturable bacteria and fungi and higher microbial activity (measured by CO_2 emission and dehydrogenase activity) in the poultry and farmyard manure added soils than in the other compost added soils have been related to suppress the bacterial wilt of tomato. Islam and Toyota (2004) also reported that bacterial wilt incidence on tomatoes was suppressed in farm yard manure amended soil when the application rate was higher than 4% with 8% application rate being the most effective in terms of disease suppression and growth of tomato plants. The rapid decline in *R. solanacearum* populations observed in amended and unamended soil due to high organic content which was 13.6% in unamended soil and even in amended soils. Several reports are available on amendment of the organic soils with cow manure had an accelerating effect on the decline of *R. solanacearum* in Egyptian sandy soil and Dutch clay soil. The suppressive effect of manure was probably due to the immediate release of ammonia as a result of microbial decomposition (Lazarovits *et al.*, 2001). Michel and Mew (1998) also showed that the suppressive effect of soil amendment on the survival of *R. solanacearum* depends on a particular soil type. It is has been reported that in some cases the disease suppression effect of the compost could be seen only after several years of addition (van Bruggen and Termorshuizen, 2003). Composts may enhance the plant defense mechanism by improving soil structure and waterholding capacity and hence protecting the plant from being susceptible to pathogens when there would be drought stress in non-amended soil (Termorshuizen *et al.,* 2004). Messiha (2006) reported that the amendment with compost or manure generally increased the dissolved organic carbon (DOC) concentrations compared to the respective non-amended organic soils; this did not necessarily leads to increased survival of *R. solanacearum.* Amendment of organic soils with 1.7% compost did increase survival of the pathogen in Egyptian sandy and clay soils, but not in Dutch soils, which were already relatively high in soil organic matter compared to Egyptian sandy soil. Amendments also enhanced tomato yield, the higher rates of amendments being most effective except green manure compost at 10% which gave a 27% lower yield compared with the higher rate of FYM. Among the amendments tested, FYM at 5 or 10% would be an interesting option to manage *R. solanacearum* in tomato growing region of Ethiopia. Further Yadessa *et al.* (2010) reported that soil amended with 5 and 10% FYM, 10% coco peat and 1% green manure compost completely inhibited infection by *R. solanacearum* green compost, while disease severity reached 43% in unamended control soil. Absence of disease, among other factors, could be attributed to improvement in physiochemical and biological properties of amended soils. The higher microbial activity found in soil amended with 10% FYM has contributed to suppressive effect of FYM at that rate.

6.3.3 Cakes

The benefit of neem for crop production has been realized by scientists in agriculture. Neem oil, neem cake, neem leaves and other parts of neem trees are being used quite extensively in the agriculture. It acts as an insecticide in grain storage, as soil amendment, fertilizer efficiency enhancer, and a very effective foliar pesticide. Soil amendments with inorganic and organic mixtures reduce wilt incidence in some locations, but more research is required to provide explicit and economically feasible types of material and doses for reliable disease management. The utilization of several organic and inorganic sources has been compiled for the management of *R. solanacearum* in India and abroad (Table 6.3). The application of sawdust; peat moss, oil cake from rubber or neem, cashew nut shell powder, coconut pith and saw dust is efficient in reducing the incidence when combined with N sources. All reduced population of the bacterial pathogen and increased population of the antagonistic saprophytes (Jay Prakash and Rajan, 1975). Amendment by use of chicken manure resulted in an increase of the bacterial population during the first two days followed by a decline of the pathogen population due to the accumulation of ammonia or the increase in pH (Himathongkham and Riemann, 1999). Ammonium is known to reduce the growth of *R. solanacearum*, but was not found to be suppressive and its effect was found to be pH dependent (Michel and Mew, 1998).The mechanism of suppression of *R. solanacearum* is unknown but it was hypothesized that the effect is related to a shift in the soil microbial community towards a community with enhanced antagonism against *R. solanacearum*. The suppressive effect of soil amendment on the survival of *R. solanacearum* depend on soil type (Michel and Mew, 1998). Sharma and Kumar (2000) reported that cake of Indian beech (karanj) or mustard also reduced the population of pathogenic bacteria in soil and increased the survival of tomato plants. Application of nitrite form of fertilizers is also capable of reducing the bacterial population. Further they reported in 2009 from four years of study that soil application of karanj cake resulted in reduction of initial population of *R. solanacearum* in soil in post-monsoon seasons. The percent reductions in initial population of *R. solanacearum* noted at 90 d after transplanting of tomato were 30.5 and 33.2 at 10 and 20 q karanj cakes/ha; against 14.6% in the natural control process. Based on the linear equation reduction of initial population, it was suggested that karanj cake was a stronger bacteriostatic substance than other chemicals (Fig. 6.7a, b). Further, Sharma and Kumar (2009b) reported that its combination with bleaching powder and lime enhanced the efficacy of control in tomato cv. Pusa Ruby (Fig. 6.7). *In vitro* studies further confirmed that extracts of cakes of karanj and neem and that of green manure posses ability to inhibit the growth of *R. solanacearum* in TCC medium (Fig. 6.7d).

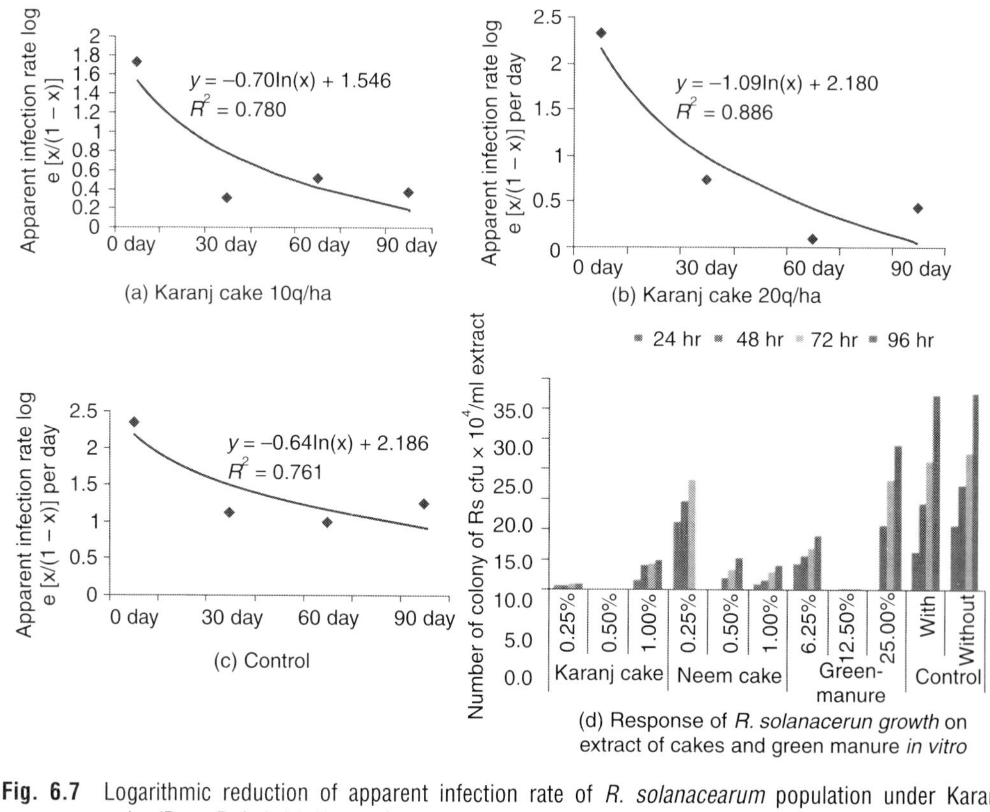

Fig. 6.7 Logarithmic reduction of apparent infection rate of *R. solanacearum* population under Karanj cake (Pusa Ruby) (a) Karanj Cake 10 q/ha (b) Karanj Cake 20q/ha (c) Control (d) Response of *R. solanacearun growth* on extract of cakes and green manure *in vitro*

6.3.4 Inorganic Amendment

6.3.4.1 Lime

Soil treatment with sulphur to produce pH of 4, followed by an application of lime to restore the pH value of 5 failed to give efficient control of bacterial wilt of tomato (Aberdeen, 1946). The disease was reported least likely to occur at high altitude and in calcareous soil (Kaan *et al.*, 1971). Pre-plant application of lime 200 lb/acre in form of $CaCO_3$ augmented Ca accumulation in leaf tissue and soil and reduced the rate of bacterial development on susceptible cultivars soil but had little effect on fruit yield (Lacascio *et al.*, 1988). Use of lime at about 0.5 ton/acre resulted in at least 67% reduction in infection apparently by making soil condition less favorable for the pathogen (Anonymous, 1976). Power (1983) reported that the best control of *R. solanacearum* and optimum plant vigour was obtained with the pre-plant incorporation of sea-shell grit (42% CaO) in a depth of 0.2 m. Spray of Ca_2 (NO_3) at rate of 0.3% at fortnightly interval stimulated plant growth, and increased the incidence of wilting of even on sea shell amended soil, indicating N had a marked antagonistic effect on Ca absorption and translocation. In a manurial experiment with 4 tomato varieties done by Jaworski and Morton (1964) did not found any difference in plant survival in different levels of Ca and Mg. Plant survival was fewer at the medium level of K (50 lbs K per acre) than at the high level of K (199 lbs per acre) or with no K. Yamazaki *et al.* (1996) reported that highly resistant cultivars were noted to be dependent on types of soils. Liming increased the available calcium and raised the soil pH to 7.5. Yamazaki *et al.* (1998) have reported that the resistance was negated at low concentration in Hawaii 7998 (highly resistant) and pathogen population in stem decreased with increasing Ca concentration and affect resistance level in the plants (Yamazakia and Hoshina, 1995). On the contrary to above Jaworski and Morton (1964) did not found any difference in plant survival in a manorial experiment with 4 tomato varieties, and the plants receiving different levels of Ca and Mg. Yuan *et al.* (1997) have shown that the white mud soil had 2-7 times greater calcium concentration than that of in susceptible soil and had pH 7.2 while susceptible soil had pH < 7.0. Application of CaO alone significantly reduced the population of *R. solanacearum* (Michel *et al.*, 1997). Pre-plant application of lime 200 lb/acre in form of $CaCO_3$ augmented Ca accumulation in leaf tissue and soil and reduced the rate of bacterial development on susceptible cultivars and soil but had little effect on fruit yield (Locscio *et al.*,

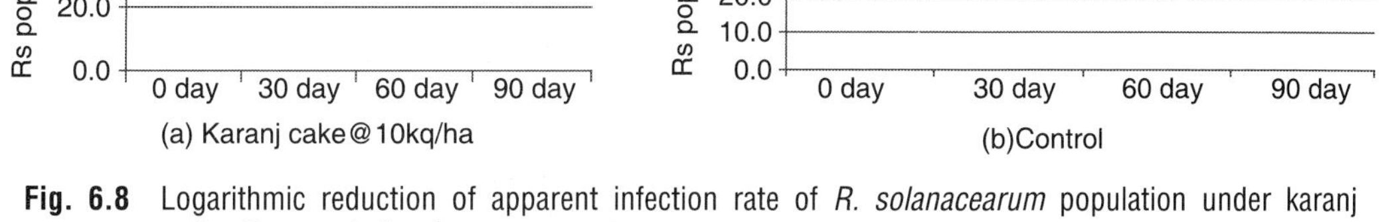

Fig. 6.8 Logarithmic reduction of apparent infection rate of *R. solanacearum* population under karanj cake (Swarna Lalima)

1998). Champoiseau *et al.*, (2009) reported that calcium concentration in soil should be increased to have better control of wilt disease. Sharma and Kumar (2009a) revealed that application of lime ($CaCO_3$) @ 25 q/ha before 15 before transplanting resulted in 18.7% reduction of initial population of *R. solanacearum* in soil at 90 d after transplanting of tomato reported a mild reduction of 18.7%.against the natural reduction of 14.6% in untreated soil. The response of lime has been shown in Fig. 6.9 in susceptible cultivars.

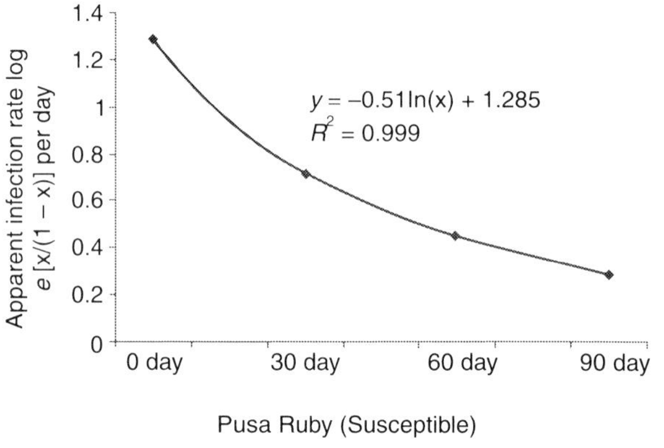

Fig. 6.9 Response of Lime on *R. solanacearum* in Tomato in Pusa ruby

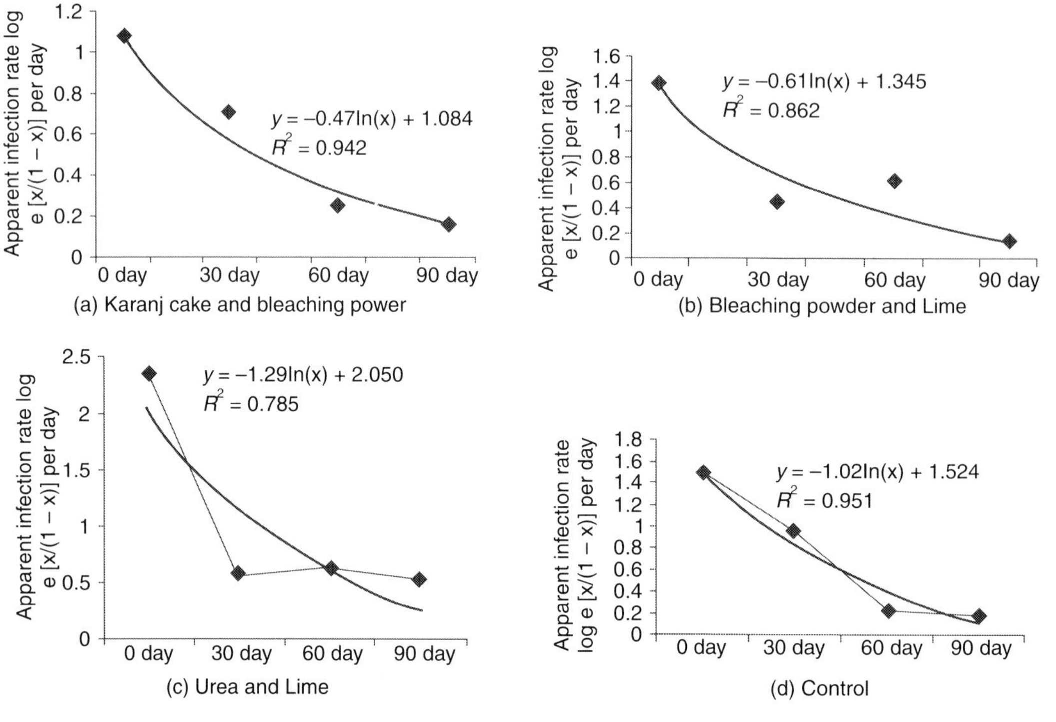

Fig. 6.10 Reduction in apparent infection rate of *R. solanacearum* population in combination treatments in Pusa Ruby (a) Karanj cake and bleaching powder (b) Bleaching powder and lime (c) Urea and lime d) Control

6.3.4.2 *Mineral Nutrition*

Cavalcantle *et al.* (1995) have studied the influence of mineral nutrition on reaction of tomato cv. Yoshmatsu and Santacruz and reported that Yoshmatsu maintained its resistance under different nutritional regimes. In nutrient solution with 1/10 Mg 1/10 Ca and 1/10 P, the level of resistance was altered under the condition of Mg and Ca deficiency, but still remained within the category of resistant disease rating. Height and weight of dry matter of the inoculated plants were maximum in 1M (Hoagland) treatment. Resistance of Santa Cruz was altered at all nutrition levels, particularly 2 H, 1/10 P and 1/10 Mg, and this cultivar was rated susceptible in Ca deficiency regime *i.e.* 1/10 Ca. Disease susceptibility reduced when the nutrient concentration was increased from 1M to 2M. Maximum height and dry weight of inoculated Santa Cruz plant were achieved, respectively in the 1/5N and 1/H treatments. Sharma and Kumar (2000, 2009a) reported that lime is effective in reducing the *R. solanacearum* population. It has been reported that lime ($CaCO_3$) increase the pH towards neutral and increase the availability of nutrients in soil like calcium (Yamazaki *et al.,* 1996) magnesium, phosphorus, boron and molybdenum in soil, which improve the plant health (Singh, 1991; Sharma and Kumar 2009a). Hacisalihoglu *et al.* (2010) worked on physiological response of calcium-efficient and calcium-inefficient in two genotypes, *viz,* PI 117566 [calcium-efficient] and PI 109315 (calcium-inefficient) of tomato and reported that the Ca-efficient tomato genotypes may not play a key role in bacterial wilt resistance under sufficient Ca conditions because Ca efficiency showed no effect on the suppression of bacterial wilt.

6.3.4.3 *Bleaching Powder and Lime*

Sharma and Kumar (2000) reported that bleaching powder @3g /m^2 is effective in reducing the *Ralstonia* population. Bleaching powder ($CaOCl_2$) at the rate of 50 g/m^2 is reported to reduce bacterial wilt by 68.4% in potato (Verma and Shekhawat, 1991), 10 g and 5 g/litre in tomato (Mazumder, 1998). Rao and Kishun (1986) reported that pre-planting soil treatment with bleaching powder 62% reduction in wilt. The apparent infection rate (*r*) reduction in *R. solanacearum* population is shown in Fig. 6.11.

Mallikarjun *et al.* (2008) reported that application of bleaching powder (15 kg/ha) has also been found effective against this disease. Further, Sharma and Kumar (2009b) reported that soil application bleaching powder resulted in reduction of initial population of *R. solanacearum* in soil in post-monsoon seasons over four years mean. The percent reductions in initial population of noted at *R. solanacearum* 90 days after transplanting of tomato were 31.9 and 35.0 at 30 and 45 kg bleaching powder ha^{-1}. The natural control process resulted into 14.6% reduction as evident from untreated control. Incidence of bacterial wilt declined by application of bleaching powder @ 12 kg/h mixed with fertilizer at planting or its soil application at first earthing up (Arora, 2012). Sharma and Kumar (2009a) reported that bleaching powder application the *R. solanacearum* population was (76.7 × 104 cfu/g soil) at initial stage but the population decreased later on to 52.16 × 10^4 cfu/g soil at 90 days. The apparent infection rate – 0.0021 and – 0.0293 unit per day was recorded at 30 day, whereas it reduced to – 0.0039 and – 0.0104 unit/day at 90 day, respectively. The reduction trend was found to be linear nature in both the doses (Fig. 6.11).

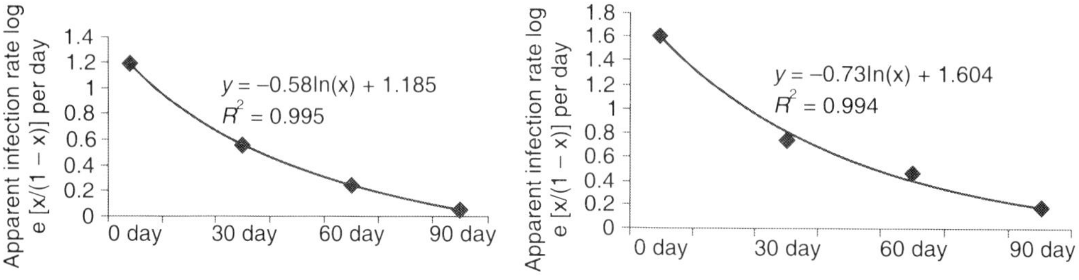

Fig. 6.11 Logarithmic reduction of *R. solanacearum* population under various doses of bleaching powder (1998-99 to 2001-02)

The effect of bleaching powder and lime on the survival of *R. solanacearum* has been studied both separately and in combination (Sharma and Kumar, 2000, 2009b). Bleaching powder caused 53.7% reduction over initial population after three times application in field in three cropping years. In every year bleaching powder was applied once,15 days prior to transplant. In similar set of conditions lime caused reduction of 43.8%. The authors noted that the bacterial population in soil was significantly lowered which lead to higher tomato plant survival (56.68%) than lime (50.79%) however the growth and yield attributes of plants were not satisfactory. The plants raised on bleaching powder treated soil yielded for less than those raised on lime treated soil (35.9 and 128.7% over control) + combined application of bleaching powder and lime resulted in 37.37% reduction in initial population and gave 107.5% higher yield over untreated control. Dutta *et al.* (2009) reported that soil amendments with bleaching powder and lime were found to be highly effective in reduction of bacterial wilt incidence.

6.3.4.4 Urea and Lime

Application of urea and lime resulted peak of nitrite accumulation after soil amendment application which depend on soil conditions (Hauck and Stephenson, 1965). Adding oxide calcium (CaO) oxide or magnesium oxide (MgO) at a rate of 5t/ha affected the survival of *R. solanacearum* to but the pathogen population decreased to undetectablle level only when combined with urea (Michel *et al.*, 1997). Urea alone did not reduce *R. solanacearum* in soil. A similar synergestic effect of CaO and urea was found by Elphistone and Alley, 1993). Percentage of wilt of a susceptible potato line was lower when soil was ammended in combination rather than with CaO and urea separately. Both components alone also significantly reduced wilt compared with non-ammended control. Most effective control of bacterial wilt was achieved when urea and lime is applied together. However, Michel *et al.* (1997) have noted that this effect was not universal and seemed to be linked with the resultant pH of soil after the ammendment. When resultant pH was more than 7.0 then the disease was very low whereas at pH close to 7.0 or low the effect on disease severity was not so marked. The generation of toxic substances after adding two relatively harmless compounds such as urea and CaO was assigned the cause of effectiveness of the soil ammendments. During urea hydrolysis, a soil solution with pH > 9 is formed at the reaction site (Hauck, 1984). This may have led to an important decrease in

R. solanacearum population at the reaction sites, resulting in an average decrease in pathogen population in the whole soil volume during the first week. Depending on the soil properties, the application of urea alone can generate high ammonium and pH conditions, which then can lead to nitrite accumulation (Chapman and Liebig, 1952). Different nitrogenous amendments (including urea) have been reported to increase populations of *Nitrosomonas* spp. But not of *Nitrobacter* spp. (Prior and Béramis, 1990), which could lead to nitrite accumulation. High soil temperature, an important environmental factor in the tropics, was reported to increase nitrite accumulation after urea application and to shorten the half-life of nitrite in the soil (Huang and Juang, 1982).The duration of nitrite accumulation is of great importance because nitrite has been reported to be phytotoxic (Bingham *et al.*, 1954; Hamilton and Lowe, 1981). Sharma and Kumar (2009b) reported that urea (200 Kg/ha N) and $CaCO_3$ (5000 Kg/ha) treated plot resulted reduction in primary inoculums 30.8% at 90 day which was significant reduction in apparent infection rate. The apparent infection rate = –0.059 unit/day at 30 days and r = –0.004 unit/day at 90 days. The apparent infection rate was different from other two combined above treatments (Fig. 6.3.8). The yield was increased to 52% over control. Islam and Toyota (2004) reported that different composts were added to a conducive soil and amendment consisting of urea (200 Kg/ha N) and CaO (5000 Kg/ha) significantly reduced the pathogen population and the tomato wilt. Application of CaO alone significantly reduced the population (Michel *et al.* 1997). Michel *et al.* (1998) reported that the effect of a soil amendment composed of urea (200 kg of N per ha) and CaO (5,000 kg/ha) on the survival of *R. solanacearum* in four Philippines soils was investigated. Within 3 weeks after application, the soil amendment either caused an initial decrease, a final decline, or no change in the pathogen population, depending on the particular soil type. An initial decrease occurred in a soil with a basic pH and resulted in a significantly ($P < 0.001$) lower pathogen population immediately and at 1 week after amending the soil. This decrease was probably due to the high pH in the soil during urea hydrolysis. A final decline in the *R. solanacearum* population after 3 weeks occurred in two soils in which nitrite accumulated after 1 week. In these soils, no decline in acterial levels occurred when nitrite formation was inhibited by 2-chloro-6-trichloro methyl pyridine. In the soil with low pH, no nitrite accumulated and the *R. solanacearum* population did not decline.

6.4 ORGANIC + INORGANIC

Huang (1991) reported that S-H mixture consisting of 4.5% bagasse, 8.4% rice husk, 4.25% oyster shell power, 8.25% urea, 1.04% potassium nitrate, 13.16% calcium super phosphate and 60.5% mineral ash (Slag) developed for the control of soil-borne disease also controlled *P. solanacearum* in Taiwan. Similarly, Yao *et al.* (1994) reported modified S-H mixture consisting of 4.4% bagasse, 8.4% rice husk powder, 4.25% oyster shell powder, 8.25% urea, 2% potassium nitrate, 12% calcium superphosphate, 5% calcium oxide, 0.3% sodium borate, 0.5% zinc sulphate and 54.4% mineral salt ash, with the addition of 20% chicken manure have stimulating growth of saprophytic microorganisms, especially the actinomycetes and *Bacillus* sp. Bacterial wilt can also be controlled by the application of fertilizers and cost-effective soil amendments, such as addition of compost, manure, urea, calcium oxide or solarization, to change soil pH and reduce survival and activity of plant pathogens (Gorissen *et al.*, 2004). Messiha (2006) reported that use of NPK and organic soils with compost or cow manure in conventional soils were compared with non-amended controls. NPK fertilization and cow manure amendment reduced pathogen survival in most soils while compost was not effective. Disease development was highest in sandy soils and similar in all other soil types. It was positively correlated with DOC (substrate availability), and negatively with K and Ca contents in soil, which may increase plant resistance. Further Messiha (2006) reported that amendment of the conventional soils with NPK gave a remarkable decrease in the survival of *R. solanacearum* in all soils tested except for the clay soil, which was already extremely high in nitrate and ammonium without NPK amendment. The effect of NPK was clearest in sandy soil that had the highest pH (pH 7.9). The application of fertilizer increased the pH and concentration of ammonia increased which reduced the pathogen. Ammonium is known to reduce the growth of *R. solanacearum*, but was not found to be suppressive and its effect was found to be pH dependent (Michel and Mew, 1998) and showed that a high pH (similar to the pH in the Egyptian soils) may have a deleterious effect on *R. solanacearum*. In addition, soil pH affects the availability of soil ions a higher pH restricts the availability of many soil nutrients. Positive correlations were found between the availability of potassium, sodium, phosphorus, and nitrogen respectively and the 50% reduction time and negative rate of decline of the pathogen, implying a slower decline at higher concentrations of these nutrients. Indeed, the soil with the highest (most negative) decline rate had the lowest available nutrient contents and the highest pH. However, these correlations may simply be coincidental, and are difficult to interpret. Suppression of pathogens in organically managed soils were reported by van Bruggen and Termorshuizen, 2003. Several reports are available on soil amendment with organic materials or NPK fertilizers or with different combination of these organic and inorganic amendments significantly affected bacterial wilt in potato incidence and increased potato yields (Lemaga, 1999-2000; Lemaga *et al.*, 2005). It has also been reported that soil amendment with compost or manure (Schonfeld *et al.*, 2003) or pig slurry (Gorissen *et al.*, 2004) reduced population of soil-borne pathogen including *R. solanacearum*. Messiha *et al.* (2007) reported that a significant reduction in potato bacterial wilt incidence in cow manure-amended soil. Angonistic microbial population (Schonfeld *et al.*, 2003) and addition of household compost was found to increase disease suppression (Schonfeld *et al.*, 2003; van Elsas *et al.*, 2005b). Pig slurry had a similar suppressive effect (Gorissen *et al.*, 2004). It was found that amendment with chicken manure resulted in an increase of the bacterial population during the first two days followed by a decline of the pathogen population due to the accumulation of ammonia or the increase in pH (Himathongkham and Riemann, 1999). Ayana *et al.* (2011) recommended that use of silicon fertilizer as a soil amendment under field conditions to augment resistance in moderately resistant cultivars where bacterial wilt disease problems prevail. However, a silicon fertilizer or silicon source was not found to substantiate or improve a susceptible cultivar. Sugarcane bagasse was demonstrated to possess a potential as an alternative soil amendment material and as an alternative silicon source. Ayana *et al.* (2011) have recommended that the use of silicon fertilizer as a soil amendment under field conditions to augment resistance in moderately resistant cultivars where bacterial wilt disease problems prevail. However, a silicon fertilizer or silicon source was not found to substantiate or improve a susceptible cultivar. Sugarcane bagasse was demonstrated to possess a potential as an alternative soil amendment material and as an alternative silicon source.

6.4.1 Cake and Chemical

Sharma and Kumar (2000) reported that karanj cake act as organic amendment to reduce the *R. solanacearum* population in soil and bleaching powder act as bactericide which might have reduced the *R. solanacearum* population in soil resulting in higher survival of tomato plants and increased yield. Further Sharma and Kumar (2009b) reported that the soil treated with cake along with bleaching powder reduced the primary inoculum by 32.5% at 90 days whereas the apparent infection rate was recorded r = –0.021 unit/day at initial stage (30 days). The maximum reduction in *r* was recorded –0.152 unit/day at flowering (60 days) and remain effective was recorded (–0.004 unit/day) at 90 days. The trend resulted in logarithmic reduction in the apparent infection rate.

Table 6.4 Organic sources used for management of *Ralstonia solanacearum*

Source	*Name of plant*	*Scientific name*	*Dose*	*References*
Cake	Neem	*Azadirachta indica*	–	Jaya Prakash and Rajan, 1975; Leksomboon, 2009
	Rubber	*Hevea brasiliensis*	–	Jayaprakash and Rajan, 1975
	Indian beech	*Pongamia glabra*	10 q/ha	Sharma and Kumar, 2000
	Mustard	*Brassica compestris*	10 q/ha	Sharma and Kumar, 2000
Straw / pith / saw dust/ bagasse/ leachates	Rice	*Oryza sativa*	–	Miah *et al.*, 1995
	Saw dust	–	–	Jaya Prakash and Rajan, 1975; Mallikarjun *et al.*, 2008
	Coconut pith	*Cocos nucifera*	–	Jaya Prakash and Rajan, 1975
	Sugarcane bagasse	*Saccharum officinarum*	–	Ayana *et al.*, 2011
	Oyster mushroom	*Pleurotus spp*	–	Jaya Prakash and Rajan, 1975
	Shiitake mushroom	*Lentinula edodes*	–	Pacumbaba *et al.*, 1999
	Pig slurry	–	–	Gorissen *et al.*, 2004
S-H mixture	Bagasse, rice husk, oyster shell powder, urea, potassium nitrate, calcium superphosphate and mineral ash	–	–	Anith *et al.*, 2004 Huang,1991

6.5 BIOLOGICAL CONTROL

Biopesticides are certain types of pesticides derived from such natural materials as animals, plants, bacteria, and certain minerals. For example, canola oil and baking soda have pesticidal applications and are considered biopesticides. At the end of 2001, there were approximately 195 registered biopesticide-active ingredients and 780 products. Biopesticides fall into three major classes:

1. Microbial pesticides consist of a microorganism (e.g., a bacterium, fungus, virus or protozoan) as the active ingredient. Microbial pesticides can control many different kinds of pests, although each active ingredient is relatively specific for its target pests.

2. The most widely used microbial pesticides are subspecies and strains of *B. thuringiensis*, or Bt. Each strain of this bacterium produces a different mix of proteins, and specifically kills one or a few related species of insect larvae. While some Bt's control moth larvae found on plants, other Bt's are specific for larvae of flies and mosquitoes. The target insect species are determined by whether the particular Bt produces a protein that can bind to a larval gut receptor, thereby causing the insect larvae to starve.
3. Plant-Incorporated-Protectants (PIPs) are pesticidal substances that plants produce from genetic material that has been added to the plant. For example, scientists can take the gene for the Bt pesticidal protein, and introduce the gene into the plant's own genetic material. Then the plant, instead of the Bt bacterium, manufactures the substance that destroys the pest. The protein and its genetic material, but not the plant itself, are regulated by EPA.

6.5.1 Advantages of Using Biopesticides

- Biopesticides are usually inherently less toxic than conventional pesticides.
- Biopesticides generally affect only the target pest and closely related organisms, in contrast to broad spectrum, conventional pesticides that may affect organisms as different as birds, insects, and mammals.
- Biopesticides often are effective in very small quantities and often decompose quickly, thereby resulting in lower exposures and largely avoiding the pollution problems caused by conventional pesticides.
- When used as a component of integrated pest management (IPM) programs, biopesticides can greatly decrease the use of conventional pesticides, while crop yields remain high.

Biological control is a method of controlling pests (including insects, mites, weeds and plant diseases) that relies on predation, parasitism, herbivory, or other natural mechanisms. It can be an important component of IPM programs. Biological control is defined as the reduction of pest populations by natural enemies and typically involves an active human role. Biological control agents of plant diseases are most often referred to as antagonists. Champoiseau *et al.* (2010) reported that a number of soil bacteria and plant growth promoting rhizobacteria (PGPR) are currently being investigated for their role in the control of *R. solanacearum* now commercially available.

6.5.2 Antagonism

Spadaro and Gullino (2005) reported that technical, economical and environmental factors are forcing to adopt new sustainable methods, such as the use of microbial antagonists, for the control of soil-borne pathogens. Research has mainly focused on antagonistic fungi and bacteria, often not providing consistent or satisfying results. Biocontrol agents can be combined with other chemical products, with physical methods (solarization or steam sterilization) or agronomical practices, such as enhancement of suppressive soils, use of amendments or microbial optimization in the case of soilless systems. Different biocontrol strategies should be developed for different pathogens. The use of microorganisms can play an important role in a more complex vision of crop protection, as a key element of IPM programs. Mixtures of antagonists with complementary activities could be developed, but the need for multiple registration is critical. Genetic manipulation could result in new biocontrol strains with increased production of toxic compounds or lytic enzymes, improved space or nutrient competence, wider host range or enhanced tolerance to abiotic stresses. The potential risks related to the environmental release of genetically modified microorganisms must be carefully assessed. Genes and enzymes involved in the biocontrol mechanism could be applied directly or transferred to crops. Finally, the production system, formulation and methods of application (soil and seed treatment or microbial colonization of the hydroponic nutrient solution) are crucial to maintain and improve the efficacy of microbial antagonists.

Antagonism refers to a type of negative association where one species adversely affects the environment for another species. Such organism may be of great practical importance, since they often produce antibiotics or other inhibitory substances which affect the normal growth processes or survival of other organisms. Antagonistic relationships are quite common in nature. The mechanisms by which antagonistic microorganisms affect pathogen populations are not always clear, but they are generally attributed to one of the four effects:

- Direct parasitism or lyses and death of pathogen.
- Competition with the pathogen for food.
- Direct toxic effects on the pathogen by antibiotic substances released by the antagonist.
- Indirect toxic effects on the pathogen by volatile substances, such as ethylene, released by the metabolic activities of the antagonist.

Although microorganisms from a variety of natural habitats produce antibiotic, soil microorganisms are the most common producers. It is not unusual for one organism to produce five or six different antimicrobial agents. There is some question about the role of antibiotics in nature; production in the laboratory is under conditions quite different from those in nature. Production of antibiotic in soil may enable the antibiotic producing organism to thrive successfully in a competitive environment. Organisms that elaborate antibiotics represent the classic example of this phenomenon; however, antibiotic may result from a variety of other conditions operative in mixed populations. Cyanide is produced by certain fungi in concentration toxic to other microorganisms, and algae fatty acids, which exhibit a marked antibacterial activity. Other metabolic products that may result from microbial activity in the soil which are likely to be inhibitory to other species are methane, sulfide, and other volatile sulphur compounds.The following are the processes through which biological agents are involved in disease management.

6.5.3 Effect of Microbes

Hoitink and Boehm (1999) postulated that manipulation of natural communities of antagonistic microorganisms in soil through organic amendments provides a potentially effective form of biological control of soil-borne plant pathogens. Bio-control agents such as *Pseudomonas fluorescens, P. glumae, Burkholderia cepacia, Bacillus* sp., *Erwinia* sp. and a Hrp-mutant of *R. solanacearum* were found to be relatively ineffective in the control of *R. solanacearum* populations under natural conditions (Shiomi *et al.* 1999; Lopez and Biosca, 2004; Ran *et al.* 2005). Use of fluorescent pseudomonads as biocontrol agents, for example, sometimes succeeded *in vitro* or under controlled conditions but failed under field conditions or even in pots (Ran *et al.*, 2005). Several biological control agents such as *P. fluorescens, B. licheniformis, B. cereus, B. subtilis* and mycorrihiza have been reported by Mallikarjun *et al.* (2008) as very effective in delaying and reducing the wilt development. Several effective biocontrol agents have been listed in Table 6.5. Among them several effective bacteria have been isolated from the rhizosphere of healthy plants grown in infected areas in Egypt which are *P. fluorescens, Streptomyces griseus*, *B. subtilis* and *Trichoderma harzianum* (Tolba, 1998) and more potential antagonists against *R. solanacearum* have been identified and characterized by Messiha (2001). The antagonistic potential of this organism was found not to be affected by incorporation of iron to the medium used for testing the antagonistic properties. Ferric chloride caused a drastic deterioration in the antagonistic activity of *P. fluorescens* that may be attributed to a possible effect of iron on siderophore pigments (Messiha, 2001). The potentially most active antagonist (up to 75% disease control) was recovered from brinjal rhizosphere and was identified as *Stenotrophomonas maltophilia*. It was found that *S. maltophilia* was able to colonize and persist in plant tissue from soil or dipped roots (Garbeva *et al.*, 2001). Three antifungal compounds, designated anthobaccines A, B and C, were isolated from the culture fluid of *Stenotrophomonas* sp. (Nakayama *et al.*, 1999). *S. maltophilia* produces other antibiotics, for example, maltophilin, but this compound was inactive against Gram-positive and Gram-negative bacteria (Berg *et al.*, 1996; Jakobi *et al.*, 1996). Lytic enzymes were also found to be excreted by *S. maltophilia* (Berg *et al.*, 1996; Zhang and Yuen, 1999; Dunne *et al.*, 2000). Biological control is reduction of inoculums' density of disease producing activities of a pathogen in its active or dormant stage, by one or more organisms (Baker and Cook, 1994). This can be accomplished naturally by manipulating the environment, host or antagonists or by mass introduction of one or more antagonists. In case of *R. solanacearum,* pathogenic and antibiotic producing *Pseudomonas* and Bacilli, actinomycetes, extract of mycorrhiza-colonized tomato and the leachates of Shiitake mushroom has been tried (Table 6.5).

Sharma *et al.* (2012) reported that biochemical basis of plant defense responses against *R. solanacearum* bacterial wilt of tomato due to two *Trichoderma* species in resistant (Swarna Lalima) and susceptible (Pusa Ruby) varieties of tomato grown under pot conditions were inoculated with pathogen *R. solanacerum* (OD 600 nm = 0.3 abs unit) simultaneously with the suspension cultures of *T. harzianum* and *T. viride*. Higher content of total carbohydrate, phenols, soluble proteins and free amino acids in the bacterial wilt resistant plants were recorded which could be used as marker for resistance against bacterial wilt. The marked increase in the content of phenol in the stem of *R. solanacerum* infected plants in response to *T. viride* consolidated the hypothesis of antagonist stimulated disease resistance mechanism in tomato against the incidence of bacterial wilt.

6.5.3.1 *Pseudomonas* spp.

Suresh and Rai (1991) reported that *P. fluorescens* was not effective as antagonist. Malya *et al.* (1996) however, studied the mechanism of suppression of bacterial wilt disease by root dipping with *P. fluorescens* pfg-32, and the role of antibiotic substances vis-a-vis siderophore production. Antagonistic substances (Cartin and Wang, 1997). Furuya *et al.* (1991) reported that pre-treatment of the roots of tomato seedling dipped in suspension of antibiotic producing *P. glumae* have reduced the severity of wilt *R. solanacearum* caused by challenge inoculation. Further Furuya and Matsuyama (1991) have also used pre-transplantation dip of seedlings in suspension of antibiotic producing strain of wilt disease. Disease suppression was reported to

Table 6.5 Bioagents used for management of bacterial wilt caused by *R. solancearum*

Name of Bioagents	*Strains*	*Methods of application*	*Varieties*	*Incidence/ Inoculum (%)*	*References*
Tomato					
Actinomycetes	18 isolates	Dipping seeds		100.0	Moura et al., 1998
Bacillus sp.	113-3	Dipped in suspension for 30 min	Koko seeds	65	Kyungseok *et al.*, 2007
Bacillus spp.	BA-46			21.13	Silveire *et al.*, 1995
	BA24			11.29	
Bacillus amyloliquefaciens		Root drenching			Tan *et al.*, 2013
B. cereus	BA –3				Mallikarjun *et al.*, 2008
B. coagulans					Mallikarjun *et al.*, 2008
B. subtillis	BA-2A, NB_2	Poured into roots zone			Phae *et al.*, 1992
	816-6	Dipped in suspension for 30 min	Koko seeds	65	Kyungseok *et al.*, 2007
	BS-5	Soil application	Pusa Ruby	9.27	Singh *et al.*, 2012, 2013 Mallikarjun *et al.*, 2008
	EXTN-1	Root drench (30m l susp. of 10^6 cfu/ ml) + Foliar spray	Poland and PT18	80-90	Thanh *et al.*, 2009
B. pumilus	228-7	Dipped in suspension for 30 min	Koko seeds	65	Kyungseok *et al.*, 2007
	SE 34	BioYield (a formulated PGPR contained two *Bacillus* strains)			
B. licheniformis					Mallikarjun *et al.*, 2008
B. vallismortis	EXTN-1	Dipped in suspension for 30 min	Koko seeds	65	Kyungseok *et al.*, 2007
Bacterio-cinogeeic Strelitzia	Str.10				Arwinyanto *et al.*, 1994

Contd...

Contd...

Pseudomonas aeruginosa	PfD-1				Cartin and Wang, 1997
P. flurescens	Str, 90B4-2-2 Pfg-32	Root dipping-			Wakimoto, 1989; Malya *et al.*, 1996; Mallikarjun *et al.*, 2008; Jaxon, 2007.
		Seed treatments			Vanitha *et al.*, 2009
	PfD-1	Vermi compost + FYM (Substrate as carrier)			Bora and Deka, 2007
	PfD-1	Vermi compost (Substrate as carrier)			Bora and Deka, 2007
	PfD-1	Seed + root + soil method (Carbooxy methye cellulose)		8.83	Bora and Deka, 2007
					Biswas and Singh, 2008
P. glumae					Furuya *et al.*, 1991
P. putida	Race 1, Biovar 1 89B61	At seeding stage and 1 week prior to inoculation	Solar Set		Anith *et al.*, 2004
Paenibacillus polymixa	H32-5.	Dipped in the rhizobacterial suspension for 30 min	Koko seeds	65	Kyungseok *et al.*, 2007
Streptomyces corchorusii					EL-Raheem *et al.*, 1995, 1996
Glomus mosseae	Race 1, Biovar 3	Soil application	Pusa Ruby	41.4	Mallikarjun *et al.*, 2008
		Soil application	Pusa Ruby		Sharma and Kumar, 2009b
G. fasciculatum					Singh, 1994-95
G. fasciculatum		Extract of tomato roots			Suresh and Rai, 1991
Trichoderma viride		Soil application	Pusa Ruby	29.0	Sharma and Kumar, 2009b, Laxmi Kumari *et al.*, 1972
T. harzianum		Soil application	Pusa Ruby		
Azotobacter + *Phosphobactrin*		Soil application	Pusa Ruby	37.3	

Contd...

Contd...

Brinjal					
Bacillus sp	EC4, EC13			70.0	Messiha, 2001
	EC4, EC13			70.0	Ramesh *et al.*, 2011
Bacillus megaterium					Zinniel *et al.*, 2002
B. pumilus					Zinniel *et al.*, 2002
B. subtilis					Zinniel *et al.*, 2002
Pseudomonas florescencs	CVPf	Seed + root + soil (Carbooxy methyl cellulose)		83.33	Chakravarty and Kalita, 2011a
	Hv37				Kloepper, 1993
	CVPf	Vermi Compost + FYM (as carrier)			Kloepper, 1993
	GR3ARS3				Sriram and Rangeshwaran, 2010-11
	CVPf	Seed + root + soil method of application		83.33	Chakravarty and Kalita, 2011a,b
	EB9, EB67				Ramesh *et al.*, 2011
P. mosselii	CK8c,CK24C				Sriram and Rangeshwaran, 2010-11
P. aeruginosa	CK19E, AFP3, AFP4, REP7, OTN8				Sriram and Rangeshwaran, 2010-11
P. plecoglossivcida	BA11D1				Sriram and Rangeshwaran, 2010-11
P. putida.					Assis *et al.*, 1998
P. alcaligenes					Reiter *et al.*, 1998
					Benhamou *et al.*, 1998
Stenotrophomonas maltophilia				75.0	Sriram and Rangeshwaran, 2010-11
Alcaligenes spp.					Zinniel *et al.*, 2002
Kluyvera spp.					Zinniel *et al.*, 2002
Flavobacterium spp.					Zinniel *et al.*, 2002

Contd...

Microbacterium spp.					Zinniel *et al.*, 2002
Clavibacter michiganensis					Zinniel *et al.*, 2002
Curtobacterium spp.					Zinniel *et al.*, 2002
P. fluorescencs	IHRPf-24	Soil drenching @1% solution at 15 days at 20 day of transplanting		15.0	Gopal Krishnan and Krishna, 2010-11
		Seedling root dip		20	
		Soil application @5kg/ha		18.8	
	CVPf	CVPf and seed + root + soil application,Vermi compost and FYM		83.33	Gargi and Kalita, 2011
Chillies					
Trichderma harzianum + *Pseudomonas fluorescens*	Pant bioagent-3	Seedling dip (1.0%) for 30 minutes	Not mentioned	9.94	Singh *et al.*, 2008
Potato					
Bacillus spp.	strain B5	Seed Treatment			Shekhawat *et al.*, 1992a,b Aliye *et al.*, 2008 Sunaina *et al.*, 2006
B. subtilis	B2G			56%	Lemessa and Zeller, 2007
	PFMR1			82.7	
	EXTN-1	Root drench (30 ml suspension of 10^6cfu/ml) + foliar spray	Ackersegan and Diamond	80-90	Thanh *et al.*, 2009
P. Aeruginosa		Soil			Bora and Dika, 2008
P. flourescens	str. BC8	Soil inoculation to seedlings			Ciampi *et al.*, 1989; Gallardo and Panno, 1989
	APF1			60%	Lemessa and Zeller, 2007
P. macerans	BSDFS and PF9			66.2 65.7	
Stenotrophomonas maltophilia	PD3531, PD3532, PD3533, PD3534/ race 3 isolates of eggplant			75.0	Messiha *et al.*, 2006

Table 6.6 List of botanicals used for control of bacterial wilt caused by *R. solanacearum*

	Name of Botanicals	*Botanical Name*	*Methods of application*	*Incidence (%)*	*References*
Extract	*Onion*	Allium cepa	Seedling dip (30 minutes in 1% extract + spray		Pun and Das 1997; Sharma and Kumar, 2009b
		Allium fistulosum	Soil treatment with 100% and 50% extract	6 and 14	Deberdt *et al.*, 2012
	Shallot	*Allium cepa var ascalonicum*	1% extract		Hanudin, 1987; Hutaglung, 1988; Komarova and Korunets, 1997
	Garlic	*Allium sativum*	Seedling dip (1% extract for 30 minutes) + spray		Hanudin, 1987; Hutaglung, 1988; Komarova and Korunets, 1997; Sharma and Kumar, 2009b
			Seed + seedling root dip (10% extract)		Singh, 1995
			Seed + seedling root dip (4% extract)	Excellent control	Singh, 1994
	Brahma manduki	*Centella asiatica*	Seedling dip + soil drenching @20% extracts	69.44	Bora *et al.*, 1997
	Neem	*Azadirachta indica*	Soil + Seed extract 4% + mycorrhiza (*Glomus faxiculatum*)	52-54	Singh, 1995-96
	Dhaincha	*Sesbania cannabina*	Seedling dip treatment + soil drenching @20% aqueous extracts	69.44	Bora *et al.*, 1997
	Marigold	*Tagetus erecta*			Hanudin, 1987
	Harrey	*Terminalia chebula*	Seedling dip treatment + soil drenching @20% aqueous extracts	69.44	Bora *et al.*, 1997
	Arjuna	*Terminalia arjuna*	Seedling dip treatment + soil drenching @20% aqueous extracts	69.44	
	Indian sorrel	*Oxalis corniculata*	Seedling dip treatment + soil drenching @20% Aqueous Extracts	69.44	Bora *et al.*, 1997
		Ipomaae reptans	Seedling dip treatment + soil drenching @20% Aqueous Extracts	69.44	
	Water peper	*Polygonum hydropiper*	Seedling dip + soil drenching @20% aqueous extracts	69.44	

Contd...

Contd...

	Clove oil		5ml /kg soil		Huang and Lakshaman, 2010
Powder	Asafoetida + Turmeric	Ferula foetida + Curcuma domestica Syn. C.longa	1 g + 5 g Seedling dip for 30 minutes	70.3 69.0	Bora, 1995; Mazumder, 1998
		Ferula foetida + Curcuma domestica Syn. C.longa	1 g + 5g Seedling dip for 30 minutes + spray		Sharma and Kumar, 2009b
	Cashew nut	*Anacardium occidentale*			Jaya Prakash and Rajan, 1975
	Chestnut	*Castanea sativum*			Jaya Prakash and Rajan, 1975
Leaf	Arhar	*Cajanus cajan*	Incorporation of 20 to 30%	100.0	Cardoso *et al.*, 2006
	Dhaincha	Crotalaria juncea	Whole Incorporation		Cardoso *et al.*, 2006; Sharma and Kumar, 2011
Plant Essential Oils	Thymol oil	*Thymus vulgaris*	Spray 0.7%	33.1	Ji *et al.*, 2005; Mamol *et al.*, 2005; Ji *et al.*, 2007b
	Palmarosa oil	*Cymbopogon martinii*	Soil treatments 0.04%	48.1	Ji *et al.*, 2005; Mamol *et al.*, 2005; Paret *et al.*, 2010
	Lemongrass	*C. citratus*	0.04%	95 to 100	Paret *et al.*, 2010
	Eucalyptus (*s*)	*Eucalyptus globulu*	0.04 and 0.07%		Paret *et al.*, 2010
	Ginger	*Zingiber officinale*			
	Chinese chive	*Allium tuberosum*	Use of root exudates collected by trapping		Yu, 1999a 1995a,b
Bark	Suji	*Cryptomeria japonica*	Ethanol Extract injecting volatile oil from leaves into root wool		Yu *et al.*, 1999a,b
	Hinoki	*Chamaecyparis obtuse*			
Flour	Soya bean	*Glycine max*			Miah *et al.*, 1995

be positively correlated with length of dipping time and concentration of bacterial suspension. Dipping of roots of tomato seedling into suspension of C_{10}^{10} cfu/ml for 24h was recorded as the most effective treatment. Pretreatment of roots with non-antibiotic producing *P. glumae* strain or heat killed *P. glumae* cell also reduced disease severity suggesting that mechanism involved was other than the antibiotics medicated process. Antimicrobial substances (Furuya *et al.*, 1997) may be involved in antagonism of *P. solanacearum* in tomato. Strain ATCC 7700 of *P. aeruginosa* was highly antagonistic *in vitro* to various plant pathogenic bacteria and fungi, especially infected with the wilt pathogen. Plants grown in non-sterilized soil accumulated more tomatine, than those in sterilized soil. In all treatments more tomatine accumulated in stress than the roots. Vinitha *et al.* (2009) reported that the effects seed treatments with *Pseudomonas fluorescens* revealed that significantly improved the quality of seed germination and seedling vigour, reduced the disease incidence plants raised from *P. fluorescens* treated seeds followed by challenge inoculation with *R. solanacearum.* Use of antagonists such as *Pseudomonas flourescens,* avirulent *P. solanacearum* and actinomycetes has been found effective biocontrol agents in bacterial wilt in potato (Arora, 2012). Burkhead *et al.* (1994) reported that Pyrrolnitrin is producted by biological control agent *P. cepacia* B37w in culture and in colonized wounds of potato. Rashid *et al.* (2004) reported about organic acids production and phosphate solubilization by phosphate solubilizing microorganism under and related it with bio-control *in vitro* condition.

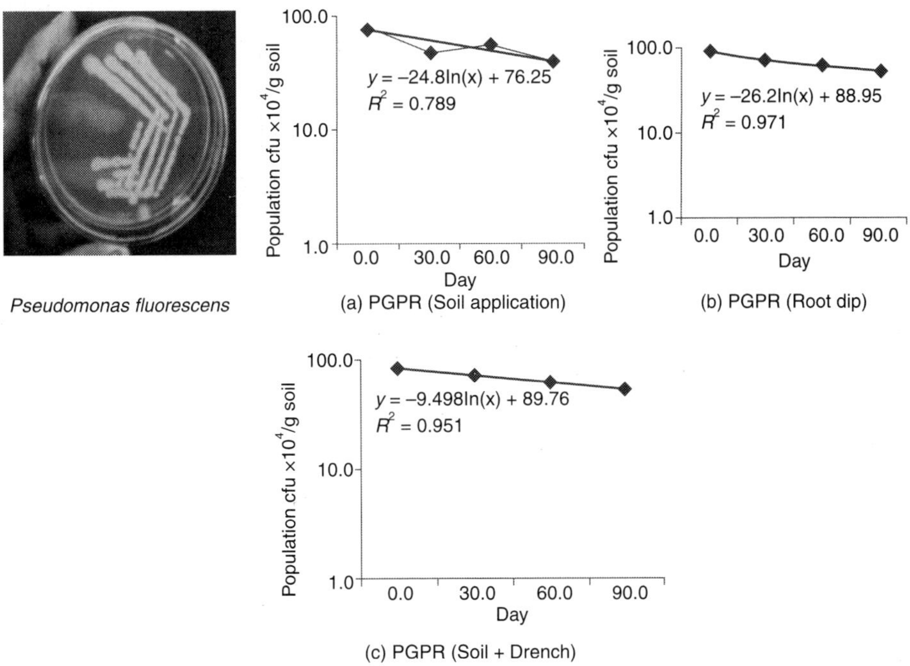

Fig. 6.12 Effect of various treatments of PGPR on *R. solanacearum* population in tomato

The antagonistic ability of the endophytic bacteria towards *R. solanacearum* was assayed *in vitro* indicated that 16 fluorescent and 18 non-fluorescent isolates from brinjal, inhibited the growth of the pathogen. In the streak method, inhibition of *R. solanacearum* at the meeting point of the antagonist and the pathogen was noted. Further, the inhibition zone caused by culture filtrates of endophytic bacteria ranged between 1.5 and 10.0 mm in radius. Although, there is general agreement that suppression *in vitro* does not necessarily relate to the same in field conditions (Kloepper, 1993). It is a commonly followed method in the initial screening against bacterial pathogen (Meenakumari *et al.*, 2003). According to Cook and Sequeira (1991) *R. solanacearum* is a poor competitor outside the plant system, giving opportunity for biological control.When cell-free culture filtrates were used, diffusible metabolites of the isolates inhibited *R. solanacearum* growth indicating the role of secondary of in the pathogen suppression *in vitro*. Xiao and Kisaalita (1998) have reported that *Pseudomonas* exert their beneficial effects in part by producing yellow-green, fluorescent siderophores called pyoverdins, under iron-deficient conditions.

Ramesh *et al.* (2009) have reported for 28 bacterial isolates which effectively inhibited *R. solanacearum* of the brinjal. In greenhouse, the plants treated with *Pseudomonas* isolates (EB9, EB67), *Enterobacter* isolates (EB44, EB89) and *Bacillus* isolates (EC4, EC13) reduced the wilt incidence by more than 70%. Most of the selected antagonists produced antibiotics DAPG (2,4 diacetylphloro glucinol) which inhibited *R. solanacearum* under *in vitro* conditions and might have been responsible for reducing wilt incidence under *in vivo* conditions. Also production of siderophores and IAA in the culture medium by the antagonists was recorded, which could be involved in biocontrol and growth promotion in crop plants. They reported that *Pseudomonas* is the major antagonistic endophytic bacteria from brinjal, which have the potential to be used a bioagent as well as plant growth-promoting rhizobacteria. Large-scale field screening and detailed knowledge on antagonistic mechanism could provide an effective biocontrol solution for bacterial wilt of solanaceous crops and concluded that *Pseudomonas* is the major antagonistic endophytic bacteria from brinjal which have the potential to be used as a biocontrol agent as well as plant growth-promoting rhizobacteria. Duijff *et al.* (1997) reported about involvement of the outer membrane lipopolysaccharides in the endophytic colonization of tomato roots by biocontrol *P. fluorescens* strain WCS417r. Gargi and Kalita (2011) tested the potentiality of an indigenous strain of *P. fluorescens* as a biocontrol agent in the management of bacterial wilt disease of brinjal. Different organic materials were evaluated for their potential as substrate carriers in the formulation of *P. fluorescens* based biopesticide. Vermicompost and farmyard manure were found to support significantly longer shelf life of the biocontrol bacteria. Carboxy methyl cellulose and substrate carrier based formulations of *P. fluorescens* were applied by different methods in pot and field experiments to study their effectiveness in the control of bacterial wilt of brinjal. Seleim *et al.* (2011) have used seven isolates of *R. solanacearum* isolated from naturally wilted roots of tomato. All isolates were pathogenic to tomato plants and produced typical symptoms of wilt. Isolate No. 5 exhibited the highest virulence followed by isolate No. 4. *P. fluorescens*, *P. putida*, *B. subtilis* and *E. aerogenes* were isolated from tomato rhizosphere and tested against *R. solanacearum in vitro* and *in vivo*. All PGPR strains except *E. aerogenes*, increased seed germination up to 15% over untreated control. Under greenhouse conditions, *P. fluorescens* exhibited the highest disease reduction of tomato bacterial wilt disease followed by *P. putida* and then *B. subtilis* while *E. aerogenes* showed the lowest disease reduction percentage. The tomato plants treated with all PGPR isolates significantly stimulated plant growth promotion under greenhouse conditions. In field trials, addition of *P. fluorescens*, *P. putida* and *B. subtilis* at concentration of 10^8 cfu mL^{-1} showed that *P. fluorescens* caused the highest disease reduction percentage of tomato bacterial wilt disease while *P. putida* exhibited the lowest disease reduction percentage. Chakravarty and Kalita (2011) have conducted an experiment in the laboratory and farm to explore the potentiality of various organic formulations of *P. fluorescens* and to manage bacterial wilt disease of brinjal under local conditions. Different organic materials and adhesives were used in the formulation of Pf based biopesticides and stored at room temperature (30°C) and 4°C to evolve a suitable substrate carrier-adhesive based bioformulation on the basis of comparative longer shelf life of the biocontrol bacteria. Storage condition of 4°C was found to be more suitable as it yielded higher viable count of the biocontrol bacteria in all the organic formulations studied. The formulation CVPf was found to give highest viable count of 890.33 × 10^8 cfu/g at 4°C and 45 × 10^6 cfu/g at room temperature, at 120 days after storage. Five best organic formulations of Pf were selected for application by different methods in pot and field experiments to study their effectiveness in the control of bacterial wilt of brinjal. CVPf formulation and seed + root + soil method of application performed significantly better than others with 83.33% control of bacterial wilt of brinjal in field experiment. Effective management of bacterial wilt of brinjal by *P. fluorescens* under local conditions signifies its potentiality and scope as a plant growth promoting rhizobacteria (PGPR) when formulated using effective substrate carrier and adhesive. Vermicompost and farmyard manure used as substrate carrier in conjunction with carboxymethyl cellulose as adhesive in the formulations CVPf and CFPf, respectively might have provided better nutrient sources and congenial microenvironment required for proper growth and subsequent longer shelf life of *P. fluorescens* in the formulated product. Bora and Deka (2007) also found that the biopesticide combination of vermicompost, *P. fluorescens,* carboxymethyl cellulose and mannitol showed best shelf life as it maintained highest population recovery at different days after storage. Islam and Toyota (2004) reported higher microbial activity in farmyard manure due to increased rates of CO_2 evolution and high dehydrogenase activity which was the key factor in the suppression of bacterial wilt of tomato.

6.5.3.2 *Bacillus* spp.

Biological control of bacterial diseases of solanaceous crops caused by *R. solanacearum* through the antagonistic bacteria has been reported earlier (Shekhawat *et al.*, 1992a,b,c). Phae *et al.* (1992) have used culture suspension of strain NB_{22} of *B. subtilis* poured into the roots zone of tomato in heavily infected plants they suggested that suspension of NB_{22} has a wide range of anti-fungal and antibacterial activities and can be used for biological control of soil-borne diseases of tomato. Wei Chun *et al.* (1994) reported that suspension of strain of *Bacillus* sp. Str, 90B4-2-2 accelerated the germination of tomato seed and enhanced the seedling vigor when seeds were soaked in a suspension of 2 × 10^9 cfu/ml for 30 minutes and soil of seed bed was drenched with a 125 ml suspension contains 5 × 10^{10} cfu/ml or 2.5 × 10^{10} cfu/ml as the cells of the antagonist could multiply and accumulate in rhizoshere soil as well as penetrate roots and move up into stems and leaves. Silveria *et al.* (1995) isolated 30 isolates of *Bacillus* spp. and tested for their antagonistic activity against *R. solanacearum* under greenhouse condition. They compared two methods of application, *viz.*, irrigation of seedbed substrate with bacterial suspension 5 days before seedling and seed bacterization with 120 ml suspension of *R. solanacearum* containing 8 × 10^8 cfu/ml at 16h before seedling. Twenty days old tomato seedlings were transplanted into plastic pots containing naturally infested soil 5d before with a 120 ml suspension of *R. solanacearum* containing 8 × 10^8 cfu/ml. Evaluation was performed and bacterial wilt index was noted, lowest in the treatment of BA24 at 1 × 10^7 cfu/ml which corresponded to 11.29% control after seventeen days of transplanting. Among the conductive soil, treatment BA-46 gave the best level of disease control (21.13%). The ability of growth promotion in seedlings was compared among the 3 applied isolates of *Bacillus* spp. and it was recorded that seedling emergence had significantly improved. Among various bacterial antagonists, the endospore forming Gram-positive *B. subtilis* is reported to be a good biocontrol agent and plant growth promoting bacterium (Soohee Chung *et al.,* 2008; Myint and Ranamukhaarachchi, 2006), which rapidly colonizes the rhizosphere and suppresses deleterious microorganisms as well as soil-borne pathogens (Myint and Ranamukhaarachchi, 2006). Besides antagonistic ability, *Bacillus* spp. can either directly or indirectly facilitate rooting (Mayak *et al.*, 1999) and growth of plants (Glick, 1995) possibly due to production of plant growth promoting substances like indole acetic acid, phosphate solubilization, ammonia, and siderophore (Nautiyal, 1999; Mathurot *et al.*, 2009). Lwin and Ranamukhaarachchid (2006) have screenened three selected potential antagonists (*viz.,* LR 3, LR 6 & LR 10) and four commercial antagonistic microorganisms [*viz.*, EM concentrate, EM instant, EM-fermented plant extract (EMFPE) and *B. subtilis*] *in vitro* and *in vivo* as biocontrol agents against the bacterial wilt pathogen. All the biocontrol agents tested reduced the bacterial wilt disease to various degrees. Among the biocontrol agents tested, EM instant was the most effective, while the degree of disease suppression by other microbes varied with the time of application. The effects of EM-FPE and Bokashi were not different from EM instant. Among the antagonists isolated from infested soils LR 10 showed a higher potential of disease suppression. The antagonists isolated from EM sources were effective to suppress against bacterial wilt pathogen. It means that EM contains some microbes, which can suppress the disease. Application method of biocontrol agent should be suitable for antagonist's action and prior to pathogen attack.

Phosphorus solubilizing bacteria has also been isolated from soil by Mehta and Nautiyal (2001). They reported that establishment and performance of phosphorus solubilizing bacteria are severely affected by environmental factors. A significant variation in production of siderophore was observed within the isolates of *B. subtilis*. Role of siderophores in induced systemic resistance (ISR) in plants is also reported (De Meyer *et al.*, 1999) and reported that one of the possible mechanisms of the observed antagonism of *R. solanacearum* could be siderophore production by the isolates. Indole acetic acid (IAA) production by rhizospheric bacteria help in promoting plant growth has also been studied by Loper and Schroth (1986). It has been reported that IAA production by PGPR can vary among different species and strains. It is also influenced by culture condition, growth stage and substrate availability. Many micro-organisms that interact with plants can synthesize hormones similar to those produced by the plant as growth regulator, such as auxins, gibberellins and cytokines. Among them, auxins are one of the most well-known phytohormones because of its important role in the initial processes of lateral and adventitious root formation and root elongation (Bent *et al.*, 2001). Variability in production of PGP attribute was found in *B. subtilis* isolates, which is influenced by various factors including soil conditions (Principe *et al.*, 2007). Beneduzi *et al.* (2008) reported that siderophore production is another plant growth promoting attribute by bacteria. In India, a formulation developed from *B. subtilis* (strain B5) and evaluated as seed treatment was found effective against *R. solanacearum* in potato under different argoclimate conditions. The treatment controlled bacterial wilt and also enhanced tuber yield (Sunaina *et al.*, 2006). *Bacillus* spp. antagonizes *R. solanacearum in vitro* through siderophore mediated competition for iron and through antibiosis which, in turn, suppresses disease in soil infested with *R. solanacearum* (Tomer and Sunaina, 2010). Singh *et al.* (2012) reported that *B. subtilis* isolates from rhizospheric soil of wilt infected tomato plants from different agroclimatic regions of India, which has significantly performed better than untreated control after 35 days of inoculation with *R. solanacearum* and *B. subtilis*. Among the nine isolates of *B. subtilis* tested for plant growth promoting attributes, BS-9 exhibited maximum phosphorus solubilization (64 μg/ml), siderophore production (1.20 cm diameter), indole acetic acid production (0.139 μg/ml) and ammonia production *in vitro,* while under glasshouse condition, it increased height of tomato plants and biomass production. The isolates, those having both antagonistic and plant growth promoting ability were confirmed to be *B. subtilis* by using specific 16S rRNA based marker. The isolates of *B. subtilis* isolated from rhizospheric soil of tomato possessed antagonistic and PGP attributes, preferred to grow under acidic conditions. Among various bacterial antagonists, the endospore forming Gram-positive *B. subtilis* is reported to be a good biocontrol agent and plant growth promoting bacterium (Soohee Chung *et al.*, 2008; Myint and Ranamukhaarachchi, 2006), which rapidly colonizes the rhizosphere and suppresses deleterious microorganisms as well as soil-borne pathogens (Myint and Ranamukhaarachchi, 2006). Besides antagonistic ability, *Bacillus* spp. can either directly or indirectly facilitate rooting (Mayak *et al.*, 1999) and growth of plants (Glick, 1995) possibly due to production of plant growth promoting substances like indole acetic acid, phosphate solubilization, ammonia, and siderophore (Nautiyal, 1999; Mathurot *et al.*, 2009). *B. subtilis* was observed by estimating plant biomass (dry weight) and height showed variation between treated with antagonistic bacterial isolates and control. Isolate BS-9 had better ability to promote growth of the tomato plants, which was significantly higher in height, and dry weight of shoot and root as compared to control and resulted in 25.30% and 82.71% increase in dry weight of shoot and root of tomato plants respectively compared to untreated control. Lemessa and Zeller (2007) reported that dry weight of tomato plant increased when inoculated with strains APF1 and B2G was 96% and 75%, respectively. Moreover, biomass and height of potato plant were also increased by inoculating the antagonistic bacteria in the soil (Aliye *et al.*, 2008). Priou *et al.* (1999) reported that initial symptoms occurred in susceptible plants 3 days after challenge, appearing as one-sided wilting during the day time and recovering at night. Afterwards, the wilt gradually developed throughout the whole plant, until permanent wilting occurred. Sixteen mixtures of PGPR and six individual PGPR strains provided significant disease suppression against *R. solanacearum*, when compared to the SPB control treatment. The plants treated with PGPR were taller and more vigorous than plants not treated with PGPR. Furthermore, the number of diseased plants did not increase over time after the initial disease rating. Dutta *et al.* (2009) reported that integration of biocontrol agents and soil amendments for effective and economic management of bacterial wilt of tomato in a long term sustainable basis. The bio-agents consortium treated tomato seedling when grown in soil amended with vermin compost, neem cake, lime and bleaching powder significantly reduced the bacterial wilt incidence and *R. solanacearum* population as compared to other treatments. Park *et al.* (2007) have studied the biocontrol activities of five strains of selected rhizobacteria in tomato against bacterial wilt caused by *R. solanacearum*. After root bacterization the plants were grown in a perlite-hydroponic system.Upon challenge, inoculation with the pathogen, all of the rhizobacterial strains efficiently suppressed the bacterial wilt in tomato in various rates, at maximum by the strain; *Bacillus vallismortis* strain EXTN-1. While the percentage of infected plants in then on-bacterized control plants was 95%, it was only 65% in plants pre-treated with EXTN-1. It was also demonstrated that the movement of *R. solanacearum* within the stem was significantly hampered when the plants were root bacterized. As EXTN-1 has no antagonistic properties against *R. solanacearum*, the bacterial wilt was probably suppressed by a mechanism other than antibiosis. The srain EXTN-1 has the ability for reducing the pathogen spread in tomato, the strain could be effectively used as a potential biocontrol agent against bacterial wilt. Sági (2003) reported that the development of crops resistant to bacterial pathogens through genetic engineering has not been progressing as rapidly as other areas, which is surprising information that is available on interactions between plants and bacteria. The first avirulence genes were cloned from bacteria (Staskawicz *et al.,* 1984) and the very first plant resistance genes defined and cloned in tomato (Martin *et al.,* 1993; Salmeron *et al.,* 1996) and *Arabidopsis* (Bent *et al.,* 1994; Mindrinos *et al.,* 1994; Grant *et al.,* 1995) conferred resistance specifically to strains of *P. syringae*. Plant-bacterial interactions have thus provided one of the best systems for studying plant-pathogen relationships and the molecular basis of disease resistance in plants isolated from tomato rhizosphere and tested against *R. solanacearum in vitro* and *in vivo*.

6.5.3.3 *Trichoderma* spp.

The best examples of biological agents are *Trichoderma* spp. which act in different ways as:

1. **Plant Growth Promoter:** *Trichoderma* strains solubilize phosphates and micronutrients. The application of *Trichoderma* strains with plants increases the number of deep roots, thereby increasing the plant's ability to resist drought.
2. **Biochemical Elicitors of Disease:** *Trichoderma* strains are known to induce resistance in plants. Three classes of compounds that are produced by *Trichoderma* and induce resistance in plants are now known. These compounds induce ethylene production, hypersensitive responses and other defense related reactions in plant cultivars.
3. **Transgenic Plants:** Introduction of endochitinase gene from *Trichoderma* into tobacco and potato plants has increased their resistance to fungal growth. Selected transgenic lines are highly tolerant to foliar pathogens such as *Alternaria alternata, A. solani,* and *Botrytis cirerea* as well as to the soil-borne pathogen, *Rhizoctonia* spp.
4. **Bioremediation:** *Trichoderma* strains play an important role in the bioremediation of soil that are contaminated with pesticides and herbicides. They have the ability to degrade a wide range of insecticides: organochlorines, organophosphates and carbonates.

5. **Industrial application:** Various strains of *A. niger* are used in the industrial preparation of citric acid. It accounts for over 99% of global citric acid production, or more than 1.4 million tonnes per annum.
6. **Enzyme production:** Many useful enzymes are produced using industrial fermentation of *A. niger*. *A. niger* is also commonly used for the production of native and foreign enzymes, including glucose oxidase and hen egg white lysozyme. In these instances, the culture is rarely grown on a solid substrate, although this is still a common practice in Japan, but is more often grown as a submerged culture in a bioreactor.
7. **Use in biotechnological field:** Another use for *A. niger* within the biotechnology industry is in the production of magnetic isotope containing variants of biological macromolecules for NMR analysis. In 2006, it was reported that a secreted RNase produced by *A. niger* called *actibind* has antiangiogenic and anticarcinogenic characteristics.
8. **Medicines:** Members of the genus are also sources of natural products that can be used in the development of medications to treat human disease.
9. **Biotransformation:** *A. niger* has also been used for waste treatment and chemical modifications known as biotransformation. Since past 20 years, *A. niger* has been used as a host for transformation to food enzymes.

In brinjal, different species of *Trichoderma* are well known to antagonize other pathogenic fungi and success has been achieved in a number of crop diseases. *T. harzianum* has increased activity of polyphenol oxidase and peroxidase in response to infection by the pathogen has been reported by many workers (Jennings *et al.,* 1969; Vidhyasekaran, 1988) and is considered to play an active role in contributing to disease resistance in certain plant host-pathogen interaction following infections. It has been reported that *Trichoderma* spp. have ability to adopt extreme soil conditions, *viz*, soil temperature, moisture and fungicidal load added in soil and that is why, it is importance in the field of biological control of plant diseases. The *T. viride* produced antibiotics, *viz.*, viridin, dermadine and gliotoxis (Sawant, 2002). A vast number of plant pathogens, particularly soil-borne fungi controlled by *Trichoderma* spp. (Anahosur, 1999). Singh *et al.* (2004) has reviewed and reported that *Trichoderma,* a microbe, has multifaceted In brinjal, different species of *Trichoderma* are well known to antagonize other pathogenic fungi and success has been achieved in a number of crop diseases. *T. harzianum* has increased activity of polyphenol oxidase and peroxidase in response to infection by the pathogen has been reported by many workers (Jennings *et al.,* 1969; Vidhyasekaran, 1988) and is considered to play an active role in contributing to disease resistance in certain plant host-pathogen interaction following infections. It has been reported that *Trichoderma* spp. have ability to adopt extreme soil conditions, *viz.*, soil temperature, moisture and fungicidal load added in soil and that is why, it is important in the field of biological control of plant diseases. Bora and Bora (2008) reported that the efficacy of microbial antagonists, *P. fluorescens* and *T. viride* was evaluated as seed, root and soil treatments in lime amended and un-amended soils for management of bacterial wilt (*R. solanacearum*) of brinjal. The antagonists significantly reduced the disease incidence in both unamended and lime amended soils. In addition to a reduction of disease incidence, there was an increase in the recovery of antagonists from rhizosphere and yield in lime amended soil. Seed, root and soil treatments with *P. fluorescens* when applied in a combination effectively suppressed wilt incidence (0.25%) and produced high yield (87.2 q/ha). The antagonists, *P. fluorescens* (178 × 10^6 cfu/g) and *T. viride* (190 × 10^6 cfu/g) survived at maximum levels in this combined application. Sharma and Kumar (2009b) reported that *T. viride* supplemented soil resulted in 29.0% reduction at 90 days the in *R. solanacearum* population. Sharma *et al.* (2009) studied that biochemical basis of plant defense responses against infection of bacterial wilt *R. solanacerum* through interaction with antagonists like *Trichoderma* spp, *in vitro* with seedlings of resistant (Swarna Lalima) and susceptible (Pusa Ruby) varieties of tomato artificially inoculated with pathogen *R. solanacerum*. Simultaneously suspension cultures of antagonists like *T. harzianum* and *T. viride* were inoculated in the same plants. The higher content of total phenol, total soluble protein and total free amino acids in the bacterial wilt resistant plants indicating their efficacy to be used as marker for resistance against bacterial wilt.The marked increase in the content of plant phenolics in the stem of *R. solanacerum* infected plants in response to treatment with *T. viride* consolidated the hypothesis of antagonist-stimulated disease resistance mechanism in tomato against the incidence of bacterial wilt. The study clearly opened up the field for further investigations into molecular strategies for imparting resistance against bacterial wilt in tomato.

6.5.3.4 *Paenibacillus polymyxa*

It is well known that bacteria of the genus *Paenibacillus* produce endospores that can stay dormant for extended periods. Several potential biocontrol agents against bacterial wilt of tomato have been reported (Guo *et al.*, 2004; Lemessa and Zeller, 2007; Xue *et al.*, 2009). However, one of the most important problems in biocontrol using microbial products is the storage time of living microbes. Algam *et al.* (2010) reported that the effects of chitosan and sixteen *Paenibacillus* strains were tested against the wilt pathogen *R. solanacearum* evaluated *in vitro* and under greenhouse conditions. Chitosan and two *Paenibacillus* strains, in particular *P. polymyxa* MB02-1007, were found to have strong antibacterial activities *in vitro* against *R. solanacearum.* In addition, chitosan applied as soil drench or seed treatment significantly reduced wilt incidence by 72% and 48%, respectively while *P. polymyxa* MB02-1007 as a soil drench or a seed treatment significantly reduced wilt incidence by 82% and 88%. In general, regardless of the application method, plant growth parameters as well as the activities of chitinase and b-1, 3-glucanase in tomato plants were significantly increased by chitosan and *P. polymyxa* MB02-1007 as compared to the corresponding control, both in the absence and presence of *R. solanacearum*. The growth of tomatoes, however, was promoted by chitosan more as a soil drench than as a seed treatment, while *P. polymyxa* MB02-1007 as a seed treatment was more effective than as a soil drench. Use of chitosan is the first report for the control of tomato wilt and promotion of tomato plants. They concluded that both chitosan and *P. polymyxa* MB02-1007 show promise for plant growth promotion and control of *R. solanacearum* in tomato. Combining chitosan and *P. polymyxa* may improve disease control.

6.5.3.5 *Mycorrhyza*

Mosse (1981) has established vesicular-arbascular mycorrhizae under aseptic conditions. Several reports on VAM fungi are available that these fungi decrease the severity of many soil-borne fungal diseases as well as that caused by *P. solanacearum,* however; further studies are required

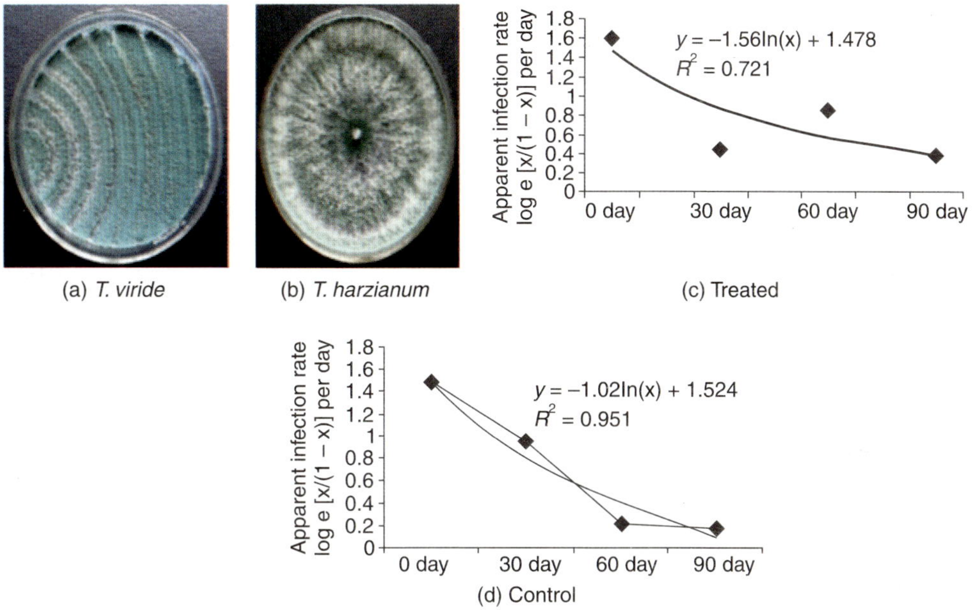

Fig. 6.13 Effect of *Trichoderma* spp. (a) *T. viride* (b) *T. harzianum* (c) Treated (d) Control in Pusa ruby tomato

to establish the mechanism of active principle of VAM fungi inhibiting *P. solanacearum*. (Bagyaraj, 2002). It has been reported that mycorrhizal fungus colonises the cortical cell of the roots, thereby form a symbiotic relationship with the host (Read *et al.*, 1992). Singh (1994) reported that good control was also obtained with mycorrhiza (*Glomus faxiculatum*) on bacterial wilt incidence caused by *R. solanacearum* (race 1, biovar 1) in susceptible tomato cv. Solar Set. Vascular Arbuscular Mycorrhiza (VAM) and charcoal compost caused reduction in the level of bacterial wilt in tomato (Kobayahi, 1999). Extract of tomato roots colonized with *G. fasciculatum* reduced population of *P. solanacearum* in nutrient broth while N-fixing bacteria (*Azotobacter)* was ineffective (Suresh and Rai, 1991). However, further studies are required to establish the mechanism of active principle of VAM fungi inhibiting *P. solanacearum*. However, in the open field where there was a high density of pathogen propagules biological control alone had only a week effect and integrated control was essential to suppress the soil-borne disease. Sharma and Kumar (2009b) reported that there was a significant reduction of *R. solanacearum* population in soil over four years of study. Addition of VAM mycorrhiza (*G. mosseae*) resulted in maximum reduction (41.4%) of *R. solanacearum* population. The VAM was effective at 30 days in which the apparent bacterial growth rate (BGR) was –0.032 unit/day, at flowering stage –0. 011 unit/ day reduced at 90 days –0.009 unit/day. This effect of reduction may be due to colonization by the *G. mosseae* in the root. *Azotobacter chroococcum* and *Phosphobactrin* resulted in reduction (37.3%) at 90 days in the primary inoculum. The apparent bacterial growth rate was r = –0.018 unit/ day at 30 days and at flowering stage (60 days) was r = –0.011 unit/day (Fig. 6.14b) and a trend showed a logarithmic reduction in apparent infection rate (Fig. 6.14b).

Anith *et al.* (2004) reported that application of PGPR, *P. putida* 89B61 significantly reduced bacterial wilt incidence when applied to the transplants at the time of seeding and 1 week prior to inoculation with *R. solanacearum*. BioYield, a formulated PGPR that contained two *Bacillus* strains, decreased disease significantly. Equity, a formulation containing more than 40 different microbial strains, did not reduce wilt incidence compared with the untreated control. With inoculum at low pathogen densities of 1×10^5 and 1×10^6 cfu/ml, disease incidence of Actigard-treated plants was significantly less than with nontreated plants. This is the first report of Actigard-mediated reduction of bacterial wilt incidence in a susceptible tomato cultivar. When PGPR and Actigard applications were combined, Actigard plus *P. putida* 89B61 or BioYield reduced bacterial wilt incidence compared with the untreated control. The *Azotobacter* has ability to produce antifungal antibiotics and also to fix nitrogen, which improves the fertility of soil (Laxmi Kumari *et al.*, 1975) and also acts as bioagent (Wani, 2002). *A. chroococcum* treated plot resulted 89.9% increase in yield increase over control.The trend showed a logarithmic reduction in apparent infection rate (Fig. 6.14c). The use of microbes will be effective for longer period in managing the *R. solanacearum* population in soil. Ambardar (2011) has reported that the application of *G. mosseae* (Gm1, Gm2 and Gm3), *P. fluorescens* and *B. cereus* individually and in combination reduced the wilt disease significantly as compared to control in cv. Solan gola and BWR-5.Interestingly, the antagonist Pseudomanas fluorescens with isolates of vesicular arbuscular mycorrhizae showed synergistic effect against wilt incidence, while no such effect was found with *Bacillus cereus*. The concomitant application of *P. fluorescens* with all the three isolates of *G. mosseae* and that of *B. cereus* with VAM isolates Gm1 and Gm2 synergistically

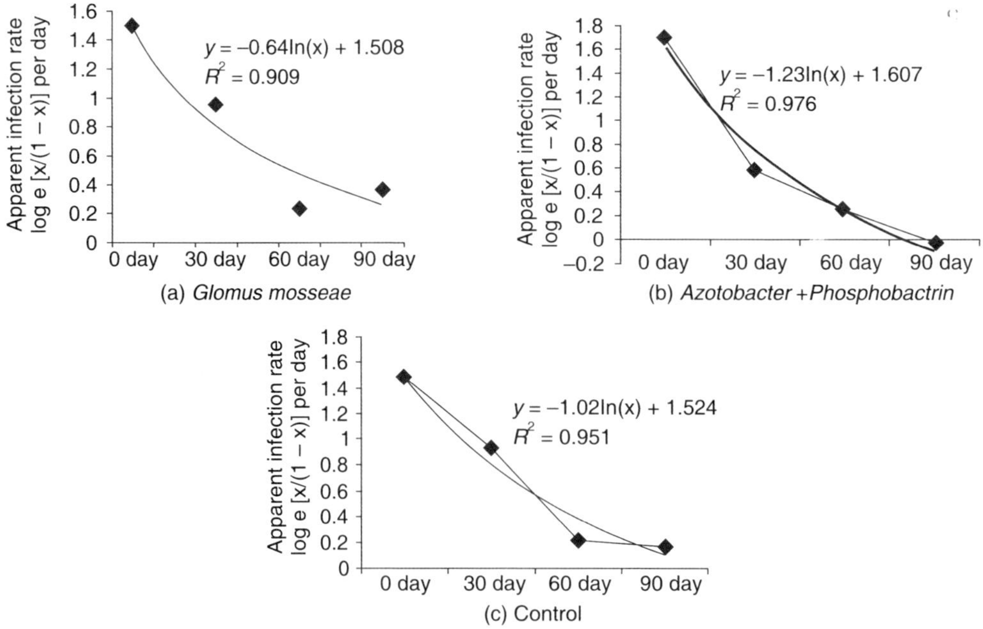

Fig. 6.14 Response of *Glomus mosseae* and *Azotobacter* on *R. solanacearum by (a) Glomus mosseae (b) Azotobacter +Phosphobactrin* (c) Control

reduced wilt, whereas *B. cereus* in combination with Gm3 did not influence wilt incidence. The combined application of vesicular arbuscular mycorrhizae and the antagonist had much more pronounced effect. The plant height, root length and number of fruit per plant were also significantly increased.

6.5.3.6 *Streptomyces*

Eighteen actinomycetes isolated from rhizosphere, rhizoplane and root tissue of tomato showed 100% control of *R. solanacearum*. The most efficient application was dipping the seeds in the propagules suspension of the actinomycetes (Moura *et al.*, 1998). EL-Shanashoury *et al.* (1996) also used the above two species of *Streptomyces* to control tomato wilt pathogen *in vitro* and *in vivo* and reported that *S. mutabilis* inhibited *P. solanacearum in vitro*. Tomato seeds were treated with herbicides and were shown in rows or sterilized soil containing *S. corchorusii* and *S. mutabilis* and were artificially infected with *P. solanacearum*.The combined application of *S. corchorusii* and *S. mutabilis* resulted in better control of wilt than that of the lone application.

6.5.3.7 *Stenotrophomonas maltophilia*

Stenotrophomonas maltophilia was isolated from the rhizosphere of brinjal in the Nile Delta of Egypt, and its antagonistic potential against *R. solanacearum* race 3 biovar 2, the causal agent of potato brown rot, was *in vitro* evaluated on KB agar medium and *in vivo* on potato plants. *In vitro*, four isolates of *S. maltophilia* (PD3531, PD3532, PD3533, and PD3534) appeared antagonistic. The isolate (PD3533) was screened as the most promising antagonist for the *in vivo* tests (Messhha *et al.*, 2006). In the greenhouse, the antagonist was applied directly to soil or by bacterization of potato eyepieces. *S. maltophilia* significantly suppressed potato brown rot in Egyptian clay soil but not in Dutch clay soil. Survival of a rifampicin and chloramphenicol resistant *S. maltophilia* strain (PD4560) was investigated in two pairs of clay soils, conventionally and organically managed, from Egypt and the Netherlands. The survival of *S. maltophilia* was significantly less in Dutch than in Egyptian soils, while this was the reverse for *R. solanacearum*. These results are in agreement with those obtained in the *in vivo* biocontrol tests. In conclusion, *S. maltophilia* may be useful for control of brown rot in the area where it was originally isolated, the Nile Delta in Egypt.

6.5.3.8 *Azotobacter chroococcum*

Azotobacter chroococcum and *Phosphobactrin* resulted in reduction (37.3%) at 90 days in the primary inoculum. The apparent bacterial growth rate was –0.018 unit/day at 30 days and at flowering stage (60 days) was –0.011 unit/day (Fig. 6.14) and remained ineffective in reducing = 0.004 unit/day) at 90 days. The trend showed a logarithmic reduction in apparent infection rate. It has been reported that extract of tomato roots colonized with N-fixing bacteria *Azotobacter* was ineffective (Suresh and Rai, 1991). The *Azotobacter* has ability to produce antifungal antibiotics and also to fix nitrogen, which improves the fertility of soil (Laxmi Kumari *et al.* 1975) and also acts as bioagents (Wani, 2002). The use of microbes would be effective for longer period in managing the *R. solanacearum* population in soil. *A. chroococcum* treated plot resulted in yield of 600.0 kg/ha which was 89.9% increase over control.

6.5.4 Mushroom Leachate

Pacumbaba *et al.* (1999) reported that the disease could be managed by soil drenching with mycelial leachate of shiitake mushroom (*Lentinula edodes*) which prevented wilt in cv Big Boy of tomato.

6.5.5 Mutant of *P. solanacearum*

King *et al.* (1995) and Kang *et al.* (1995) have successfully used extracellular protein-detective mutants of *P. solanacearum* for the biological control of bacterial wilt. An avirulent and bacteriogenic strain Tm3 of *R. solanacearum* gave biological control of tomato bacterial wilt (Dogo *et al.*, 1999) suggested that the antibiotics, bacterocin and induced resistance of hosts are possible mechanisms of this protection.

6.5.6 Spent Mushroom Compost

Water extract filtrate of raw wheat straw and spent mushroom straw of three *Pleurotus* sp. was found to inhibit *R. solanacearum*. The use of spent straw and raw straw powders and their water extracts have a good potential to control the disease and increase plant productivity (El- Fallal and Moussa, 2008). Sharma and Kumar (2011) reported that use of spent mushroom compost (Fig. 6.15) have reduced the apparent infection rate of bacteria in tomato.

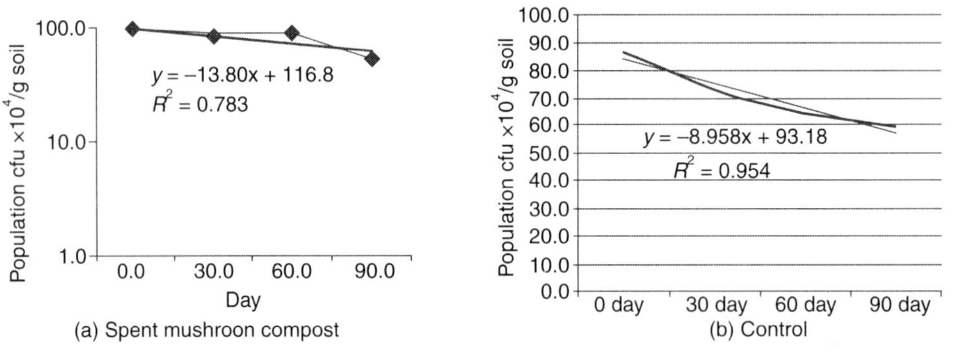

Fig. 6.15 Response of (a) *Spent mushroom compost* and (b) Control on *R. solanacearum* population

6.6 USE OF BOTANICALS

Several botanicals were used for control of bacterial wilt (Bora *et al.* 1997) caused by *Pseudomonas solanacearum.* by *Lantana camara* (Banerjee and Chatterjee) of tomato.

6.6.1 Asafoetida and Turmeric

Sharma and Kumar (2009b) reported that the asafoetida + turmeric powder treated plot also reduced the primary inocula to 32.7% at 90 days in tomato. In the early stage (30 days) the effectiveness was slow. The apparent infection rate was −0.038 unit/day at flowering stage (60 days) and it remain effective in reducing at 90 days – 0.002 unit/day. The trend indicated moving trend which was different from microbes (Fig. 6.16a). Asafetida + turmeric powder treated plot in resulted yield of 540.1 kg/ha which was 71.2% increase over control. Similar result was obtained by Mazumder (1998) who reported from Assam. Seedling treatment with mixture of asafoetida (*Ferula foetida* Ritz.), turmeric powder (*Curcuma longa* Linn.) and water (ATW) in ratio of 1g: 5g: 10 liters controlled bacterial wilt or by spraying 5% ATW solution or soil drenching at 15, 30 and 45 days after transplanting (Pun and Das, 1997) and soil drenching at 20, 50 and 80 days after transplanting with an organic formulation of asafoetida and turmeric powder in water caused a substantial reduction in bacterial wilt (Bora, 1995).

6.6.2 Onion Extract

Sharma and Kumar (2009) reported that the seedling treated with onion extracts (5%) caused a significant reduction in primary inoculum was also found to be 39.3% at 90 days. The apparent infection rate was −0.0106 unit/day at 30 days −0.023 unit/day at flowering stage (60 days) and −0.01 unit/day at 90 days (Fig. 6.16b). The plant extracts can also be used for minimizing the *R. solanacearum* population in soil, which will be cheaper than other treatments. Onion extract was better than garlic extract. Several workers have reported that the extract of onion (*Allium cepa*) reduced the disease (Hutaglung, 1988; Hanudin, 1987; Komarova and Korunets, 1997) *in vitro* and *in vivo.* Deberdt *et al.* (2012) reported that bacterial wilt (*R. solanacearum*, phylotype IIB/4NPB) was controlled by the antimicrobial effect of *Allium fistulosum* aqueous extract applied in a pre-plant soil treatment. Three concentrations of extract (100, 50, and 25%, 1:1 wt/vol) were evaluated by *in vitro* inhibition assay and *in vivo* and in a growth chamber. *In vitro*, *A. fistulosum* (100 and 50%) suppressed growth of *R. solanacearum*. Pre-plant treatment of the soil with *A. fistulosum* extract significantly reduced the *R. solanacearum* populations. No pathogen was detected in the soil after treatment with 100% concentrated extract from the third day after application until the end of the experiment. *A. fistulosum* also significantly reduced the incidence of tomato bacterial wilt. In the untreated control, the disease affected 61% of the plants whereas, with 100 and 50% extracts, only 6 and 14% of the plants, respectively, were affected which suggests that *A. fistulosum* extracts could be used in biocontrol-based management strategies for bacterial wilt of tomato.

6.6.3 Garlic Extract

Singh (1994) reported that seed and seedling root dip of tomato with 4% garlic extract gave excellent control of bacterial wilt and good control was also obtained with heat killed cells, whereas during 1995-96 she reported that seed and seedling root dip with 10% garlic extract or 10^{12} cfu/ml suspension of heat killed cells of *P. solanacearum* gave 64% control in tomato. Hutaglung (1988) reported that the disease decreased by adding a 10 ml suspension of 35g garlic bulb/77 ml sterile water or 6 g of ground garlic bulb to the rhizosphere for normal plant growth and increased the fruit weight. Kamrova and Koruneta (1997) reported that garlic suspension and fungicidal mixture completely inhibited bacterial growth *in vitro.* In the field, the fungicidal mixture, the garlic suspension and peat oxyhumantes gave the best control. Sharma and Kumar (2009b) reported that the seedlings treated with garlic extracts (5%) were less effective in reducing primary inoculum which was found to be 9.97% at 90 days (Fig. 6.16c). The seedling treated with garlic extract (5%) proved more effective at the initial stage (30 days) in which the apparent infection rate −0.082 unit/day at 30 days and −0.01 unit/day

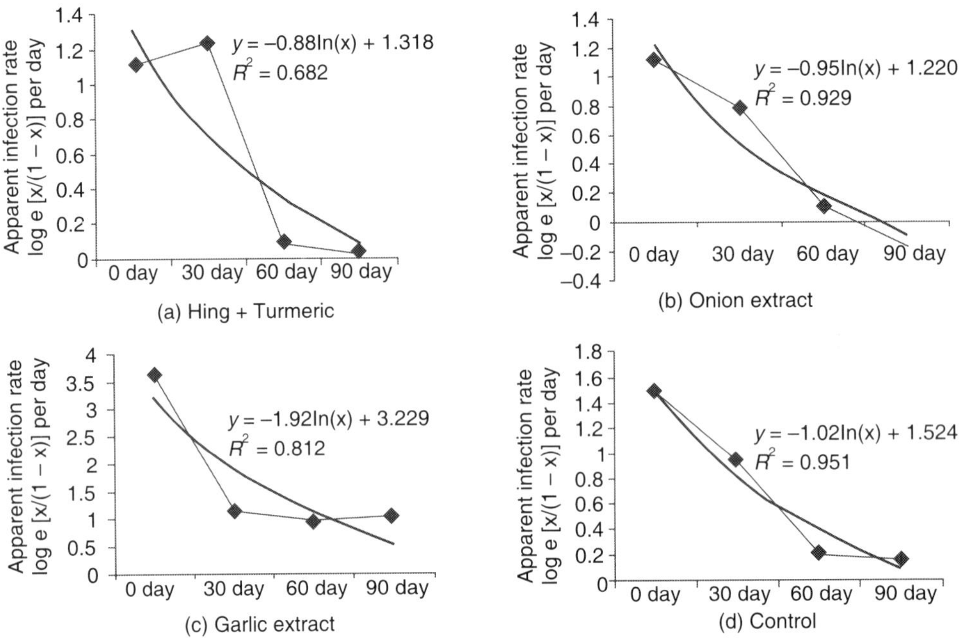

Fig. 6.16 Response of (a) Asafoetida + Turmeric (b) Onion Extract (c) Garlic Extract (d) Control on apparent rate of infection on *R. solanacearum* population

at 60 days and later on its effectiveness was reduced. Several workers have reported that the extract of garlic (*A. sativum*) reduced the disease (Hutaglung, 1988; Hanudin, 1987; Komarova and Korunets, 1997) *in vitro* and *in vivo.* Application of plant extracts is used for minimizing the *R. solanacearum* population in soil, which is cheaper than other treatments.

6.6.4 Fruit Exacts

The fruit exacts of *Terminalia chebula* Ritz showed inhibitory effects to several bacterial pathogens including *P. solanacearum* (Gowda *et al.*, 1975).

6.6.5 Plant Essential Oils

Thymol is a known antibacterial agent produced by thyme (*Thymus* spp.) or other medicinal plants and has been widely used in medical and food sciences (Aeschbach *et al.,* 1994; Shapiro and Guggenheim, 1995; Manou *et al.*, 1998; Momol *et al.,* 1999). Mamol *et al.* (2005) and Champoiseau and Momol (2009) have reported that volatile plant essential oils thymol and palmarosa oil at concentration of 0.7% in tomato were sealed with plastic mulch for 3 or 6 days. Both thymol and palmarosa oil soil treatments reduced bacterial wilt incidence significantly. Thymol was more effective than palmarosa oil based on the final assessment, when 33.1 and 48.1% of the plants had wilted in plots treated with thymol and palmarosa oil, respectively. Soil treatment with either thymol or palmarosa oil produced significantly higher yield of tomato than the untreated control. Earlier they also reported (in 2003) that thymol application significantly reduced bacterial wilt incidence on the susceptible cultivar Solar Set. Disease incidence in untreated plots reached 65.5%, while in thymol treated plots only 12% of plants wilted. Thymol application also increased yield of Solar Set significantly compared with the untreated control. This was the first report on the use of thymol for controlling a plant disease under field conditions, which indicated that this compound provided effective control of bacterial wilt on susceptible tomato cultivars when used as preplant treatment of soils. Ji *et al.* (2007) reported that the combined use of thymol and ASM provided the greatest reduction of root galling in tomato yield (cv. FL47). Thymol-treated plots produced significantly higher marketable yield than untreated plots and the thymol treatment in combination with ASM significantly increased tomato yield compared with thymol or ASM alone. These results indicate that use of thymol and ASM was beneficial in controlling bacterial wilt and root-knot. They developed an effective method for applying thymol through drip irrigation lines for managing these diseases in tomato production. Hacisalihoglu *et al.* (2007) reported that tomato plants treated with ASM by foliar spray (25 $\mu g\ ml^{-1}$) and soil drench (12.5 $\mu g\ ml^{-1}$) and inoculated with different concentrations of pathogen resulted significantly reduced bacterial wilt incidence when plants were inoculated with lower concentrations of *R. solanacearum*. The results suggest that BW could be significantly reduced by ASM under low bacterial wilt inoculum conditions. Pradhanang *et al.* (2003) reported that the potting mixture (soil) infested with *R. solanacearum* was treated with the essential oils at 400 mg or µl and 700 mg or µl per liter of soil in greenhouse experiments and population densities were determined just before and 7 days after treatment. The population was declined in thymol, palmarosa oil, and lemongrass oil treatments at both concentrations, whereas tea tree oil had no effect. Tomato seedlings transplanted in soil treated with 700 mg/liter of thymol, 700 ml/liter of palmarosa oil, and 700 ml/liter of lemongrass oil were free from bacterial wilt and 100% of plants in thymol treatments were free of *R. solanacearum*. Soil amendment with fresh leaves of essential oil-producing plants did not reduce bacterial wilt incidence compared to untreated inoculated control. Paret *et al.* (2010) reported that Palmarosa (*Cymbopogon martini*), lemongrass (*C. citratus*), and eucalyptus (*Eucalyptus globulus*) oils were investigated for their effects on *R. solanacearum* race 4 and their potential use as biofumigants for reducing bacterial wilt disease of edible ginger (*Zingiber officinale*). Three concentrations of the oils (0.04, 0.07, and 0.14% vol/vol) were evaluated by culture amendment assays, epifluorescence microscopy, and studies in potting medium. In culture amendment assays with palmarosa and lemongrass oils at 0.04%, both oils significantly reduced the growth of the bacterium compared with the control, and at 0.07 and 0.14% they showed complete inhibition of bacterial growth. Epifluorescence microscopic observations showed cell deterioration in 95 to 100% of the cells at all concentrations of palmarosa and lemongrass oils, indicating its bactericidal properties. Eucalyptus oil treatments at 0.04 and 0.07% had bacteriostatic effects on the cells. The pathogen was not detected in *R. solanacearum*-infested potting medium after treatment with palmarosa and lemongrass oils at 0.07 and 0.14% in any of the experiments. Bacterial wilt incidence on edible ginger was significantly reduced when planted in essential oil-treated potting medium. None of the essential oil treatments reduced the growth or yield of edible ginger grown for 180 days in 5-liter pots. Isman (2000) also used plant essential oils for pest and disease management. Koul *et al.* (2008) reported that potential and constraint of essential oils are used as green pesticides.

6.7 HOST RESISTANCE

Plant disease resistance is crucial to the reliable production of food which provides significant reductions in agricultural use of fuel, land, water and other inputs. Various reports on devastating plant disease impacts (Irish Potato Famine, Chestnut blight), as well as recurrent severe plant disease issues (rice blast, soybean cyst nematode, citrus canker). However, disease control measures are reasonably successful for most crops. It is estimated that diseases in many crops typically reduce plant yields by 10% every year in more developed counties, but yield loss to diseases often exceeds 20% in less developed countries. Plant disease resistance derives both from pre-formed defenses and from infection-induced responses mediated by the plant immune system. Relative to a disease-susceptible plant, disease resistance is often defined as reduction of pathogen growth on or in the plant, while the term disease tolerance describes plants that exhibit less disease damage despite similar levels of pathogen growth. Although obvious qualitative differences in disease resistance can be observed when some plants are compared (allowing classification as "resistant" or "susceptible" after infection by the same pathogen strain at similar pathogen inoculum levels in similar environments), a gradation of quantitative differences in disease resistance is more typically observed between plant lines or genotypes. Plants are almost always resistant to certain pathogens but susceptible to other pathogens; resistance is usually pathogen species-specific or pathogen strain-specific. Resistance to bacterial wilt in tomato has been studied by Li *et al.* (2001). Rommens and Kishore (2000) have reported that potential disease resistance genes can be exploited for the agriculture use. Muthoni *et al.* (2012) reported that plant resistance is an important component of the integrated disease management.

Plant Resistance: Sophisticated agricultural developments now allow growers to choose from among systematically cross-bred species to ensure the greatest hardiness in their crops, as suited for a particular region's pathological profile. Breeding practices have been perfected over centuries, but with the advent of genetic manipulation even finer control of a crop's immunity traits is possible.

Sources of Resistance: Several sources of resistance of solanaceous vegetables have been reported and summarized in Table 6.7 and they were utilized in breeding programme for development of resistant varieties by plant breeders.

Table 6.7 Wild species of solanaceous vegetables used in breeding bacterial wilt resistance

Name of crops	*Wild Species*	*References*
Tomato	*Lycopersicum pimpinellifolium*	Simpson and Ogorzaly, 1986
	Solanum lycopersicoides	Rick *et al.*, 1990
	Solanum pennellii	
	Physalis ixocarpa	Chellemiet *et al.*, 1994a
	Lycopersicum hirsutum	Zhang *et al.*, 2002
	L. esculentum pimpinellifolium	Asati and Yadav, 2004
	L. peruvianum	Carmelle *et al.*, 2006b
	L. esculentum cerasiforme	
	Physalis ixocerpa	

Contd...

Contd...

	S. lycopersicum var. *cerasiforme* (PI 129080)	
	S. pimpinellifolium (PI 127805A)	
	S. lycopersicum var. *pyriforme* (Beltsville 3814)	
	S. pimpinellifolium X *S. lycopersicum* (199 UPR)	
	S.lycopersicum (Mulua)	
Brinjal	*Solanum torvum*	Gousset *et al.,* 2005
	Solanum anthiocum	
	Solanum integrafolium	Ali *et al.*, 2003
	Solanum macrocarponin	Hermsen, 1987; Madalageriand Gowda, 1988; Yadav *et al.*, 2009
	Solanum khasianum	Yadav *et al.*, 2009
	Solanum mimosum	
	Solanum inasanum	
	Solanum kurzii	
	Solanum diversiforum	
	Solanum melongena	
	Solanum melongena var insanum	
	Solanum lycopersicum	
	S.wrightii	
	Solanum sissymbrifolium (PI337597)	Srinivassam and Gopimony, 1969
Potato	*Solanum commersonii*	Hermsen, 1987; Laferriere *et al.*, 1999
	Solanum sparsipilum	Rowe *et al.* 1972, Sequeira, 1979; Hermsen, 1987
	Solanum multidissectum	
	Solanum stenotomum	
	Solanum chacoense	
	Solanum raphanifolium	
	Solanum pinnatisectunh	
	Solanum jarnestii	
	Solanum phureja	
	Solanum cidiatum (PI196300)	
	Solanum stenotomum	Fock *et al.*, 2001
	S. goniocalyx	Hermsen, 1987
Chilli	*Capsicum annuum*	
	C. baccatum	
	C. chinens	
	C. frutescens	
	C. pubescense	
	C. praetermissum	

6.7.1 Sources of Resistance and Inheritance Patterns

The global survey of the major breeding research for tomato bacterial wilt resistance points out that the main sources of resistance used worldwide are perhaps only half a dozen accessions of *S. pimpinellifolium* (PI 127805A), *S. lycopersicum* var. *cerasiforme* (PI 129080), *S. lycopersicum* var. *pyriforme* (Beltsville 3814), a progeny from a cross between a *S. pimpinellifolium* and *S. lycopersicum* (199 UPR), *S. lycopersicum* (Mulua) the Philippine accession used in Hawaii in the 1970s. This number could be extended to seven accessions if one adds CRA66 by assuming it is a Guadeloupe tomadose and not a progeny from OTB2 Whether there are 6 or 7 main sources of resistance, the genetic basis of tomato resistance mechanisms used worldwide for breeding is quite narrow. North Carolina (NC) material has been integrated in most of the other breeding programs, in particular those of Hawaii, Japan, Philippines and Taiwan. By combining NC material, or not, with other sources of resistance, and breeding in geographical areas where different strains of bacterial wilt are prevalent, the breeders exploited the genetic potentialities at their disposal, and created material resisting a wide range of bacterial wilt strains as exemplified by the results obtained in the worldwide trial carried out by Wang *et al.* (1998). The top nine resistant accessions which had high levels of resistance in almost all 12 locations tested (>90% survival on average) were developed in Hawaii (H7996, H7997 S and L, H7998 S and M), Philippines (TML46 and TML114, R3034), and Japan (BFOkitsu). Other sources of resistance in wild tomatoes have been described sporadically in the literature in accessions of the same species (*S. pimpinellifolium*, cherry and pear *S. lycopersicum*) as well as in other wild relatives of tomato (Jaworski *et al.*, 1987; Mohamed *et al.*, 1997; Carmeille *et al.*, 2006b; Hai *et al.*, 2008). The high genetic diversity has been displayed by *R. solanacearum* complex (Fegan and Prior, 2005) and the strong interactions between strains and resistant materials (Lebeau *et al.*, 2011) suggested that various resistance mechanisms, including strain specific ones, exist in tomato resistant germplasm. Therefore, breeders have some opportunities at their disposal to enlarge the relatively narrow range of resistance sources primarily used so far, and to continue accumulating different mechanisms of resistance in tomato genotypes to obtain better stability of resistance in different environments. However, they might be limited by the fact that some bacterial strains are not controlled by any resistant accession. Inheritance studies have focused mostly on F_2, F_3 and RILs progenies of Hawaii 7996 crossed with the susceptible WVa700 (*S. pimpinellifolium*).

Huckelhoven (2007) reported that the plant cuticle and cell wall separate microbial pathogens from the products of plant metabolism. While microbial pathogens try to breach these barriers for colonization, plants respond to attempted penetration by a battery of wall-associated defense reactions. Successful pathogens circumvent or suppress plant's is nonself recognition and basal defense during penetration and during microbial reproduction.

6.7.2 Improvements Through Breeding for Host Resistance

Breeding for disease resistant vegetables becomes an integral part of IPM strategies. Opena and Tschanz (1987) reported that World Vegetable Centre (WVC), Taiwan (Earlier AVRDC) initiated research on bacterial wilt of tomato in 1973 to understand the population dynamics of the pathogen; develop resistant screening protocols; identify resistant stocks; determine the genetic basis of resistance; and to develop tropically adapted resistant cultivars. More than 100 strains tested all belonged to biotype III or Race 1. A highly reliable field screening method was devised and adopted as the standard protocol since 1986. Disease control through the development of resistance gene is the only method that offers practical solution to most of diseases caused by *P. solanacearum*. Breeding bacterial wilt-resistant varieties has been difficult in tomato, and attempt to develop cultivars with acceptible resistance have been unsuccessful (Ascota *et al.*, 1964; Mew and Ho, 1976; Opena *et al.,* 1987, 1988). One type of bacterial wilt resistance in North Carolina germplasm was found to be olegogenic and recessive (Opena *et al.,* 1987). Several resistants to tolerant lines of solanaceous vegetables, reported from India and various countries, are summarized in Tables 6.7 & 6.11. (Goth *et al.*, 1991) reported that the improvement of levels of bacterial wilt resistance in brinjal through breeding. Düring (1996) reported that resistance in transgenic plants can be introduced through foreign genes in bacteria by genetic engineering. Kumar and Singh (2006) reported about the genetic resources, chromosome engineering, and crop improvement.

6.7.2.1 *Tomato*

The effective means to control bacterial wilt is the development of resistant varieties/hybrids with acceptable quality. The knowledge on genetics of resistance helps in the formulation of the breeding method. Resistant varieties often becomes susceptible when grown in different localities or become susceptible after a few years depending on the ability of the pathogen to overcome resistance. Thus, the life of a resistant variety may be only 3-7 years. (Jay Shree *et al.,* 2002). Most of the popular local varieties are susceptible to this disease and chemical control through treatment of soil is both cumbersome and uneconomical (Madalgeri *et al.,* 1983). Inheritance of resistance to bacterial wilt in tomato has been discussed by Okiveria *et al.* (1999). Neto *et al.* (2002) studied inheritance of resistance in tomato plants using the P_1, P_2, F_1, F_2 and the two backcross generations obtained from the cross between the Drica and Santa Clara cultivars, which are considered standards for resistance and susceptibility, respectively. Inheritance of resistance to bacterial wilt in tomato plants is quantitative with partial dominance of the alleles that express greater disease progress curve (AACPD) and bacterial wilt index which are oligogenic or polygenic traits. *L. pimpinellifolium* a wild tomato (IMB) (Ascota *et al.*, 1964) resistance to bacterial wilt has been controlled by a small gene and associated with small fruit size and intermediate growth habit. Several works have been done to improve wilt resistance in tomato at North Carolina, Florida, Taiwan and Hawaii (Kelman, 1976). Boshou (2005) reported that unfortunately, for most crops susceptible to bacterial wilt, there are almost no source of high level, gene-for-gene type resistance encoded by single dominant genes. Instead, available sources of resistance are usually polygenic and it has been difficult or impossible to transfer all the identified quantitative trait loci (QTL) into desirable cultivars due to their number or linkage to undesirable traits. However, some level of bacterial wilt control is possible using resistant or moderately resistant tomato cultivars. A cultivar is a cultivated plant that has been selected and given a unique name because it has desirable characteristics (decorative or useful) that distinguish it from otherwise similar plants of the same species cultivars, such as FL7514 and BHN 466. However, resistant cultivars may produce fruits that are reduced in size and thus not acceptable to the commercial tomato industry. Additionally, resistance in these cultivars may vary with location and temperature, because of strain differences. The major QTL for resistance to bacterial wilt on chromosome 6 is linked (in repulsion phase) with the Mi 1 gene for resistance to root-knot nematodes (Thoquet *et al.,* 1996a). Opena and Tschanz (1987) reported that the resistant germplasm L 285 (*L. esculentum* var. *cerasiforme*) is a primitive type from WVC collection is more advanced germplasm for Hawaii and North Carolina which showed acceptable levels of resistance have been used in the breeding program. A seven-parent diallel experiment indicated that both additive and non-additive gene actions regulate the genetic system for wilt resistance. In a test of fresh market F_1 hybrids, the proportion of wilt-resistant F_1's was generally higher when•both parents were resistant compared to resistant x susceptible crosses. However, in some processing hybrids with one susceptible parent, the resistance was comparable to the highly resistant stocks, L 285 and CLN 65-349, indicating non-additive action. More than 164 tomato varieties based on the WVC improved breeding lines have been released in 39 countries to help farmers deal with the three most challenging problems of tomato production in the tropics, *viz*, bacterial wilt. In East Africa, WVC improved varieties increased tomato production by almost 40%, reduced production costs by 17%, and increased farm income by 21%. In India, tomato varieties resistant to bacterial wilt produce 87% higher yields and receive 90% better market price than non-resistant varieties. Chellemiet *et al.* (1994a) used several procedures for evaluating the resistance of tomato to bacterial wilt were used to account for diversity in strains of *P. solanacearum* Smith and to approximate resistance under field conditions. Five strains of *P. solanacearum* from Florida and one from North Carolina were inoculated onto 19 tomato genotypes and one tomatillo (*Physalis ixocarpa* Brot.) genotype using a stem-puncture technique. Genotypes were also transplanted as seedlings into naturally infested soil. Resistance was evaluated by comparing the response of each genotype to the susceptible cultivars Bonny Best and Sunny. With the stem-puncture technique, the mean incidence of disease ranged from 30% with the strain from North Carolina to 94% with a strain from Northern Florida. Significant differences in the resistance of genotypes and pathogenicity of strains were observed. However, no interaction between strain and genotype was observed. Using naturally infested soil, the mean incidence of disease was 51% and significant differences in the resistance of genotypes were also observed. Hawaii 7997, Hawaii 7998, and CRA 66 had the lowest incidence of disease, regardless of inoculation method. The results indicate that assessing pathogen diversity and using a combination of resistance screening techniques can facilitate the evaluation of many genotypes, account for potential regional variability in the pathogen, and differentiate levels of field resistance to tomato bacterial wilt. Aarons *et al.* (1993) reported that to locate genes associated with resistance to bacterial wilt (*P. solanacearum)* in tomato, an F_2 population segregating for resistance was analysed using DNA genetic markers. The population was derived from a cross between a resistant tomato genotype, L285, and a susceptible cultivar, CLN286. This cross is intraspecific in nature and therefore low in DNA sequence polymorphism, 100 clones from the established tomato molecular map were surveyed with 30 restriction enzymes to find a total of 46 restriction fragment length polymorphisms (RFLPs). Additionally, seven random amplified polymorphic DNA markers were added to the map. Ten days after inoculation, two genomic regions were found to be associated with bacterial wilt resistance. One of the putative resistance loci, located on a linkage group consisting of RFLPs known to reside on chromosome 6 of tomato, was significantly ($p = 0.004$) correlated with bacterial wilt resistance. Fourteen days after inoculation, three genomic regions were correlated with bacterial wilt resistance. These included the region on chromosome 6, plus two others on linkage groups

7 and 10. The region on linkage group 7 was unusual in that the allele from the susceptible parent (CLN286) associated with higher levels of resistance in the F_2. Together, the three putative resistance loci explained 49% of the total variation in bacterial wilt resistance in the F_2. Five cases of significant gene × gene interactions were also observed. Danesh *et al.* (1994) have studied genetic resistance of olegoresistance to bacterial wilt on tomato and confirmed that the existence of partial resistance gene on chromosome 6, an F_2 individual homozygous for the resistant parents alleles on chromosome 7 and 10, but alleles heterozygous for markers on chromosome 6, was selfed. Analysis of the F_3 progeny confirmed that a partial resistance locus was localized on chromosome 6, may close to CT184. The presence of apartial resistance locus on chromosome 10 was similarly confirmed by analysis of progeny of another F_2 plant chosen on the basis of its marker phenotype. Thoquet *et al.* (1996a) found 7 QTLs using the F_2 and F_3 population derived from *Lycopersicon esculentum* cv. 'Hawaii 7996' (partially resistant) × *pimpinellifolium* 'Wva700' (susceptible), while Carmeille *et al.* (2006a,b) found 4 QTLs with a F_2 and F_3 population of inbred lines (RIL). Balatero *et al.* (2002) found 80 molecular markers, including 72 AFLP markers, 7 resistance gene analogs, 1 SSR marker, during the construction of a tomato linkage map of 12 linkage groups. Yui *et al.* (1999) found 4 RAPD markers, in which RA12–13 and RA12–29 had close linkage to the resistant gene. Most of the loci and markers are on the chromosome 6; the other chromosomes such as chromosome 4, 7, 8, 10, 11 and 12 also contribute to the control of the bacterial wilt resistance (Mangin *et al.*, 1999; Wang *et al.,* 2000).

Several QTLs of resistance have been identified, including a major QTL on chromosome 6 effective towards GMI8217 an isolate of race 1 biovar 1 (Thoquet *et al.*, 1996a,b); of Pss4, an isolate of race 1, biovar 3, phylotype 1 (Wang *et al.*, 1998); and JT516, an isolate representative of race 3-phylotype II (Carmeille *et al.,* 2006a). Wang *et al.* (2000) identified another major QTL of resistance of Hawaii 7996 effective towards Pss4 and located on chromosome 12. Several minor QTLs located on chromosomes 3, 4, 8, some of which having a season dependent expression (Carmeille *et al.*, 2006a) have also been identified. Mejia *et al* (2009) confirmed the QTLs on chromosome 6 and 12 were associated with resistance in Hawaii 7996 observed in Guatemala field evaluation against local phylotype I strains. Work is ongoing at AVRDC for adding markers to the QTLs regions of Hawaii 7996 associated to resistance to bacterial strains belonging to phylotype I, in order to develop tools for marker assisted selection. QTLs of resistance of the resistant line L285 effective towards UW364 an isolate of race 1, biovar 4, have also been located on chromosomes 6, as well as on chromosomes 7 and 10 (Danesh *et al.*, 1994). Pattern of resistance derived from CRA66 has been described as polygenic (Prior *et al.*, 1994) but no molecular data are available for this source.

Miao *et al.* (2009) worked on identification of PCR-based markers amplified fragment length polymorphism (AFLP) linked to the genes that confer resistance to tomato bacterial wilt and reported that thirteen differential fragments were detected using 256 primer combinations, and two AFLP markers were linked to the BW resistance. Subsequently, the AFLP markers were converted to co-dominant SCAR markers, named $TSCAR_{AAT/CGA}$ and $TSCAR_{AAG/CAT}$. Linkage analysis showed that the two markers are on the contralateral side of TRSR-1. Genetic distance between $TSCAR_{AAT/CGA}$ and *TRS-1* was estimated to 4.6 cM, while 8.4 cM between $TSCAR_{AAG/CAT}$ and *TRS-1*. Deberdta *et al.* (1999) reported that the soil-borne bacterial pathogen *R. solanacearum* commonly coexists with polyspecific nematode populations in tropical and subtropical areas. The wounding of roots by nematodes is usually invoked to explain the correlation between nematode infection and bacterial wilt, since this wounding increases the number of sites for bacterial entry. Bacterial wilt development on tomato was investigated in a controlled environment on the susceptible tomato cultivar Floradel and the polygenically wilt resistant cultivar Caraïbo. The bacterial isolate GMI 8217 and two different sedentary plant parasitic nematodes, the endoparasitic root-knot nematode, *Meloidogyne incognita,* and semi-endoparasitic reniform nematode, *Rotylenchulus renifomis,* were cross-infected. At low temperatures (22-27°C), the bacterium GMI 8217 was slightly pathogenic on all tomato lines, except on Floradel coinfected by root-knot nematode. At high temperatures (27-32°C), the root-knot nematode greatly increased wilt severity in susceptible Floradel and resistant Caraïbo, but the reniform nematode had no such effect regardless of temperature × cultivar combination. This showed that infection of tomato roots by root-knot nematodes reduced genetic resistance to bacterial wilt. Only one study clearly demonstrated the specificity of resistance: the tomato cultivar Hawai 7996, which displayed a large-spectrum resistance to several strains from biovars 1 and 3, was submitted to genetic mapping of the resistance. Two QTLs on chromosome 6 and one on chromosome 8 were shown to be involved in resistance to the different strains in different locations, but one QTL on chromosome 12 having a major effect was specifically detected with the Taiwanese strain Pss4 of biovar 3 (Mangin *et al.*, 1999; Wang *et al.*, 2000a). Carmeille *et al.* (2006) reported sources of resistance to tropical race 1 strains of *R. solanacearum* that till now no data are available on *Lycopersicon* spp. source for resistance for race 3, biovar 2 strains. They evaluated eighty-two accessions belongs to *Lycopersicon* spp. in endemic area of Reunion Island against race 3 and *L. esculentum* cv. roma was used as susceptible control. No immunity or complte resistance was observed in any accession tested under above condition. The Hawaii 7996 line represented the best source of partial resistance to race 3 with 52% of wilting. Hence, this was used as starting point for introducing resistance to bacterial wilt caused by *R. solanacearum* race 3 in tomato for genetic control. Champoiseau and Momol (2009) reported that bacterial wilt of tomato is difficult to control, and no single strategy has shown 100% efficiency in control of the disease are popular throughout Asia, Lebeau *et al.* (2011) reported that host resistance is the most effective control strategy against this disease. The wilt resistance is often overcome due to the considerable variation among pathogen strains. They have collected 30 accessions of tomato, eggplant and pepper (Core-TEP), most of which are commonly used as sources of resistance to *R. solanacearum* or for mapping quantitative trait loci. The Core-TEP lines were challenged with a core collection of 12 pathogen strains (Core-Rs2) representing the phylogenetic diversity of *R. solanacearum*. They observed six interaction phenotypes, from highly susceptible to highly resistant. Intermediate phenotypes resulted from the plants' ability to tolerate latent infections (*i.e.*, bacterial colonization of vascular elements with limited or no wilting). The Core-Rs2 strains partitioned into three pathotypes on pepper accessions, five on tomato, and six on eggplant. A "pathoprofile" concept was developed to characterize the strain clusters, which displayed six virulence patterns on the whole set of Core-TEP host accessions. Neither pathotypes nor pathoprofiles were phylotype specific. Pathoprofiles with high aggressiveness were mainly found in strains from phylotypes I, IIB, and III. One pathoprofile included a strain that overcame almost all resistance sources.

Nishi *et al.* (2003) have demonstrated the identification of useful DNA markers for bacterial wilt resistance. The recent advances in DNA marker technology make it possible to characterize complex characters in more detail than ever before. With the use of linkage maps composed of DNA markers, genetic regions containing QTL can now be identified, mapped, and analyzed (Lader and Boststein, 1989). DNA markers have increasingly been used to study complex agronomic traits, indicating fruit characters in tomato (Paterson *et al.* 1988). Danesh *et al.* (1993) confirmed that the existence of a partial resistant gene on chromosome 6, an F_2 individual homozyous for resistant parent's alleles on chromosome 7 and 10, but heterozygous for markers on chromosome 6,was selfed. Analysis of F_3 progeny confirmed that a partial resistance locus was located on chromosome 6, very close to CT184. The presence of a partial resistance locus on chromosome 10 was similarly confirmed by analysis of progeny of another F_2 plant chosen on the basis of its marker phenotype. Carmeille *et al.* (2006b) reported sources of resistance to tropical race 1 strains of *R. solanacearum* and reported that till now, no experimental evidence are available on *Lycopersicon* spp. (now *Solanum* spp.) sources for resistance to temperate race 3, biovar 2 strains. The evaluated eighty-two accessions of *Lycopersicon* spp. in growth chambers for resistance to a race 3 strain in endemic area of Réunion Island using *L. esculentum* cv. roma as susceptible control. No immunity or complete resistance was observed in any accession tested under above conditions. Therefore, race 3 strain was demonstrated to be highly pathogenic on all the *Lycopersicon* accessions tested. Partial resistance was significantly detected in only one accession belonging to the species *L. peruvianum*, and one *L. esculentum* var. *cerasiforme* tomato line. Five other genotypes from the species *L. esculentum*, *L. hirsutum* and *L. peruvianum* were noteworthy. Bacterial wilt-resistant variety Megha has been released for commercial cultivation by crossing NTDR-1 (nematode resistant) with small-fruited CL 1131-0-0-38-40 (AVRDC accession) known for its bacterial wilt resistance.

6.7.2.2 *Brinjal*

Search for host resistance is considered an ecofriendly approach for management of diseases. Babu *et al.* (1998) reported resistant sources from *Solanum melongena* and wild *Solanum* species, for various pests and diseases have been listed and references cited. Several agricultural universities and research institutes have collected rich diversity of brinjal land races and other species of non-tuberiferous *Solanum*. The national germplasm collection and conservation responsibility is with the National Bureau of Plant Genetic Resources (NBPGR) and a total of 2419 accessions have been collected, conserved and evaluated for 44 agro-botanical characters at the Bureau. Sharma *et al.* (2005) have screened in bacterial sick plot against this disease and found five accessions, *viz*, CH243, CH-245, CH-247, CH- 249 and CH309. Several tolerance and resistant lines have been shown in Table 6.9 and 6.10 in world and India.

6.7.2.3 *Potato*

Hermsen (1987) reported that cultivated plants have derived from the wild species and are manipulated by man both agronomically and genetically in order to improve their adaptation to human needs. The potato has its main centre of diversity in the mountainous regions of Latin American countries. The potato was domesticated and has been grown for several millennia. These primitively cultivated potatoes comprise eight *Solanum* species: the diploids *S. phureja,*

S. stenotomum, S. goniocalyx and *S. ajanhuiri,* the triploids *S. chaucha* and *S. juzepczuki,* the important tetraploid *S. tuberosum* ssp. *andigena* and the pentaploid *S. curtilobum.* Not until the latter half of the 16th century was the potato introduced into countries outside Latin America, first into Europe, from there into North America and later on all over the world. The immediate ancestor of our present-day autotetraploid *S. tuberosum* ssp. *tuberosum* cultivars is the auto tetraploid *S. tuberosum* ssp. *andigena.* Although initially only few genotypes *of S. tuberosum* ssp. *andigena* were introduced, a large number of different *S. tuberosum* sp. *tuberosum* cultivar is adapted to various conditions has been derived from that material. Patil *et al.* (2012) had reviewed the global situation with regard to screening of genetic resources and their utilization in resistance breeding for bacterial wilt in potato and also the status and the opportunities that biotechnology offers to combat this disease. Muthoni *et al.* (2012) reported that management of bacterial wilt in potatoes calls for an integrated approach. However, although many potato varieties have been found to have some degree of resistance to the disease, for better results, use of high resistant varieties may be coupled with use of disease-free tubers and clean fields. Because there is a strong host-pathogen-environment interaction effect in the expression of resistance, an essential step in the development of resistant varieties is local screening. The breeders and plant pathologists for a long time have been concerned about reliance on a small gene pool that is limited to two Colombian clones of *S. phureja,* which impart dominant resistance. Hermsen (1987) found valuable bacterial wilt resistance, associated or not with resistance to root-knot nematode. It has been reported by Patil (1989) that incorporation genes from these species into cultivated potatoes is not easy by conventional breeding techniques, but they open an avenue for gene donation through novel methods such as genetic engineering. With current transformation technology, it may be possible to transfer the resistance genes into cultivated potato. Although many potato varieties have been found to have a degree of resistance to bacterial wilt, they still transmit latent infection to their progeny tubers. Laferriere *et al.* (1999) reported the wild potato relative *S. commersonii* is reported to carry resistance to bacterial wilt disease caused by *R. solanacearum.* To overcome sexual incompatibilities due to differences in ploidy and endosperm balance numbers, somatic hybrids were made that combine the *S. tuberosum* and *S. commersonii* genomes. The somatic hybrid plants were vigorous, but their disease resistance level and their fertility was unknown. They tested the *S. commersonii* and *S. tuberosum* as source material cv. Superior, potato cv. Atlantic and six somatic hybrid lines for resistance to a virulent strain of *R. solanacearum* (race 3, biovar 2) at 28°C. *S. commersonii* was significantly more wilt-resistant than the cultivated potatoes. In five of the six somatic hybrid lines, disease resistance levels were similar to that of the resistant *S. commersonii* parent. The resistance level of the sixth somatic hybrid was intermediate, significantly different from both *S. commersonii* and *S. tuberosum.* The somatic hybrids in this study proved both to be male- and female-fertile, were self-compatible and can be crossed with *S. tuberosum* to produce viable seeds. Carputo *et al.* (2009) carried out research on levels of bacterial wilt resistance and genetic diversity of (near) pentaploid sexual hybrids between *S. commersonii* and cultivated *S. tuberosum* following artificial inoculation with *R. solanacearum*, wilting degree was estimated on scale from 0 to 4, and seven genotypes of 26 (27%) displaying a *S. commersonii* like behavior were identified. Latent bacterial colonization was detected in roots of symptomless *S. commersonii* and hybrids, whereas no bacterial population was detected within stem and suggested that the movement and /or growth of bacterium in the aerial parts were strongly inhibited. Bacterial populations were detected within stems. Hybrids displayed good fertility and/or further breeding efforts. The monogenic dominant gene of this accession was easily transferred to the superior cultivar Pusa Kranti (Gopinath, 1983) and selection pressure against bacterial wilt resulted in the commercially acceptable and resistant type Composite 2. Champoiseau *et al.* (2010) reported that normal breeding has achieved for most solanaceous crops is moderate level of host resistance to bacterial wilt on a regional level when conditions are not excessively hot or wet. Some potato cultivars are less susceptible to bacterial wilt at least in some regions. There are active potato resistance breeding programs, some of which are focusing on resistance to latent infection. Seven genotypes from two wild Andean potato species were found to have high levels of resistance to wilt and tuber infection, which provide new resistance sources for developing commercial resistant potato cultivars (CIP, 2004).

Patil (1989) reported that incorporation genes from these species into cultivated potatoes is not easy by conventional breeding techniques, but they open an avenue for gene donation through novel methods such as genetic engineering. With current transformation technology, it might be able to transfer the resistance genes into cultivated potato. Lin *et al.* (2004) reported that transgenic tomato plants expressing the arabidopsis npr1 gene display enhanced resistance to a spectrum of fungal and bacterial diseases.

Belkhindi (1985) reported that incorporation of gene from conventional method by RFLP technique has been a success with brinjal in isolating WCGR 112-8 from the West Coast belt (India) for bacterial wilt resistance in the *S. melongena* group. Several workers attempted to screen varietal resistance for the management of bacterial wilt in brinjal (Hoque *et al.*, 1981; Rahman and Hoque, 1986; Khan, 1974; Hossain *et al.*, 1991; Gao *et al.*, 2000; Makoto, 2003). Resistance sources have been identified in some wild relatives species, such as *S. torvum* and *S. aethiopicum* (Gousset *et al.*, 2004, 2005). Lindhout (2002) reported polygenic resistance has the perspectives in breeding for durable disease resistance. Nirmala Devi and Tikoo (1992) also have studied on genetic resistance to bacterial wilt and root knot nematodes in tomato. The gene governing resistance in disease resistance in solanaceous vegetables have been summarised in Table 6.8.

Table 6.8 Type of gene responsible for resistance to bacterial wilt in solanaceous vegetables

Nature of Gene	*Crop*	*References*
Olegogenic	Tomato	Opena *et al.*, 1988; Domesh *et al.*, 1994, Dinesh *et al.*, 1993
Recessive	Tomato	Opena *et al.*, 1988; Domesh *et al.*, 1994
Small gene	Tomato	Opena *et al.*, 1987; Domesh *et al.*, 1994
Single Dominant	Brinjal	Swaminathan and Srinivasan, 1972; Gopinath and Madslageri, 1986; Zhu *et al.*, 2004; Yang *et al.*, 2006; Ajjappalavora *et al.*, 2008; Cao *et al.*, 2009
Recessive	Brinjal	Singh, 1961; Acosta *et al.*, 1964; Mohamed *et al.*, 1997; Chaudhary, 2000; Gpoalkrishnan *et al.*, 2005; Li *et al.*, 2006; Tian *et al.*, 2007; Sun *et al.*, 2008
Incomplete Dominance	Brinjal	Graham, 1976; Shou *et al.*, 2006

Contd...

Contd...

Dominance	Brinjal	Scott *et al.*, 1988; Grimault *et al.*, 1995; Chanda, 1993; Chaudhary, 2000; Yang *et al.*, 2000; Zhu *et al.*, 2004; Ajjappalavora *et al.*, 2008
Monogenic	Brinjal	Scott *et al.*, 1988; Gowha and Shivasnka, 1990; Grimault *et al.*, 1995; Li *et al.*, 2006
Polygenic	Brinjal	Acosta *et al.*, 1964; Thurston, 1976; Gonzdez and Summers, 1995; Osiru *et at.*, 2001; Lindhout, 2002
Additive	Brinjal	Graham and Yap, 1976; Yu *et al.*, 1995; Hansen *et al.*, 1998; Balatero *et al.*, 2000a,b
Non additive	Brinjal	Mohamed *et al.*, 1997; Balatero *et al.*, 2000a,b
Vertical resistance	Brinjal	Messiaen, 19775; Mochizuki and Yamakawa, 1979b; Che *et al.*, 1997
Dominant polygenetic	Brinjal	Chaudhary, 2000; Li *et al.*, 2002
Recessive polygenic	Brinjal	Feng *et al.*, 2003
Two genes PM687	Chilli	Carporalino *et al.*, 2001

The inheritance of resistance to bacterial wilt caused by *R. solanacearum* in brinjal is complex and different. Various degrees of varietal resistance have been reported (Messiaen, 1975; Mochizuki and Yamakawa, 1979b; Che *et al.* 1997). Cao *et al.* (2009) reported that the inheritance of resistance in brinjal was controlled by a single dominant gene showing Mendelian inheritance model. In addition, a 762 bp molecular marker linked to a bacterial wilt-resistant gene of brinjal was screened by the bulked segregant analysis method and sequence characterized amplified region (SCAR) marker linked to bacterial wilt-resistance gene was also obtained. The genetic distance between this marker and the resistance gene is 3.33 cM. Further, Cao *et al.* (2009) reported that the cytoplasmic factors interfere the bacterial wilt, the resistance was only related to nuclei gene, which was contrary to the result of Gousset *et al.* (2005), which found that bacterial wilt resistance could be interfered by cytoplasmic factors. Gousset *et al.* (2005) have collected accessions of wild relatives of brinjal, *i.e. S. torvum* SW and studied the genetic diversity against *R. solanacearum* by using RAPD and I-SSR markers. Twenty-nine accessions of *S. torvum* and brinjal, cv. Pusa Purple Long were evaluated for resistance to *R. solanacearum* (race 1 biovar 3, strain T926). All inoculated plants of brinjal died within 2 weeks. Bacterial wilt symptoms occurred on lower leaves of *S. torvum* plants tested without causing any plant death. The presence of bacteria, serologically detected in roots of symptomless plants, suggested *S. torvum* to be tolerant to *R. solanacearum*. Sharma *et al.* (2006) reported that out of 7 parental lines and 23 F crosses of brinjal tested in bacterial wilt sick plot. Two parents Swarna Shyamli (CH-249) and Swarna Pratibha (CH-309) were resistant to bacterial wilt. Only one F_1 cross CH 249 × CH-792 showed resistant reaction in all the 3 years of testing. However, one parent, *viz.*, CH 792, was found moderately resistant and 3 F_1 crosses, *viz*; CH 309 × CH381. CH -309 × CH- 249 and CH-381 × Swarna Mani showed moderately resistant reaction to bacterial wilt. Cultivar Swarna Shyamli has been released for cultivation in (Himachal Pradesh, Jammu and Kashmir and Uttarakhand) and Swarna Pratibha has been released for cultivation in Madhya Pradesh. Maharashtra and Jharkhand. Cao Bi-hao *et al.* (2009) reported that the combinations (F_1) were crossed between highly resistant and susceptible to bacterial wilt eggplant parents and its F_2, BC_1 segregation population plants were inoculated with race1of *R. solanacearum* in greenhouse. They reported that the inheritance of bacterial wilt resistance in brinjal was controlled by a single dominant gene showing Mendelian inheritance model. Up till now, many molecular markers related to resistant bacterial wilt have been reported, such as the RAPD marker located at 4.33 cM from the monogenic dominant resistance gene (Zhu *et al.*, 2005), two AFLP markers related to monogenic recessive resistance (Li *et al.*, 2006; Sun *et al.*, 2008) or one 400 bp RAPD marker related to polygenic dominant resistance (Li *et al.*, 2002), but the function of these resistant markers were still not identified. The resistant SCAR marker obtained had been identified, it may be a functional marker linked to the resistant gene and a useful tool for MAS in brinjal breeding in the future.

Resistant cultivars are either location-specific or generally not preferred by growers due to low consumer preference. Goth *et al.* (1991) reported that significance differences in Kaplan-Meier survival curves by the log-rank test for root-dip inoculated parental lines and their corresponding in five of seven eggplant accessions. Resistance in one of these seven F_3 families was significantly greater than that of the parents. The survival curve were plotted for 31 days after inoculation of 3 week old seedlings with a broth culture of *P. solanacearum*, based on time from inoculation to initial wilt symptoms for plants grown in a greenhouse at temperature of 30-37°C in daytime and 24-27°C at night. Significant differences were found in survival curves of parental P_1 lines and their corresponding F_3 families of all four crosses in hybridization program. Ali *et al.* (2003) have reported that to generate amphidiploids hybrids between brinjal (*S. melongena* L. and *S. integrifolium* Poir) attempts were made to obtain easily propagated and disease-resistant rootstocks for eggplant and tomato production. *In vitro* treatment of shoot tip explants of sterile interspecific hybrids, between *S. melongena* and *S. integrifolium*, with 0.05% colchicine induced a high percentage of amphidiploids. The amphidiploids had larger anthers containing approximately 70% viable and larger pollen than those of the interspecific hybrids. The amphidiploid plants produced a profuse number of viable seeded fruits, while the interspecific hybrids produced no fruit or a poor number of parthenocarpic fruits. The seeds of the amphidiploids were bigger than those of the interspecific hybrids. Chromosome number from pollen mother cell and root tips, nucleus volume of the histogenic layers and stomata size assisted in identifying true amphidiploids and chromosomal chimeric types. Seed-grown amphidiploid seedlings were found resistant to two of the most virulent strains of bacteria. Ajjappalavara *et al.* (2008) studied the genetics of bacterial wilt in four F_2 populations which were chosen by 20 F_1 hybrids derived from line × tester crosses of five female × four male parents. The gca effects for parents were noticed that DWD-1 (169.00) and Malapur (410.00) were good general combiners for the fruit yield and DWD-1 (-22.40), DWD-2 (-8.86) and DWD-3 (-15.39) were the best desirable combiners for the bacterial wilt incidence. Among 20 F1's, DWD-1 × Malapur was recorded highest significant desirable sca (-24.93) effects compared to remains. The scoring of bacterial wilt incidence on four F_2 populations noticed that the segregation of population to the inheritance of bacterial wilt resistance was 3 (non-wilted): 1 (wilted) ratio suggested single gene inheritance for bacterial wilt resistance.

6.7.2.4 *Pepper*

Many resistant pepper accessions have been described from local genetic resources, mainly in small-fruited and pungent pepper types (Peter *et al.*, 1984; Matsunaga and Monma, 1995, 1999; Michelmore, 1995). In every accession, resistance was quantitatively expressed, but genetic analyses have never been reported. Race-cultivar specificity was suspected because the resistant accessions displayed different resistance levels in different geographic locations. However, when challenged with known strains of the bacterium rather than heterogeneous populations, most resistant accessions showed a stable resistance compared with the different races or biovars tested, particularly the highly variable biovar1. Quantitative differences were correlated with strain aggressiveness rather than with virulence in potato (Lopes and Quezado-Soares, 1995), Pepper (Quesado-Soares and Lopes, 1995) and tomato (Thoquet *et al.*, 1996). Momol *et al.* (2001) reported from Brazil and Taiwan that the breeding line MC-4 (*C. annuum*), from Malaysia, was found to be very resistant to various isolates of biovars 1 and 3, and is being recommended for breeding programs in Brazil. Champoiseau *et al.* (2009b) reported that both chili and sweet pepper suffer from many diseases and are difficult to grow in hot, humid conditions. The WVC has released 99 chilli pepper and 39 sweet pepper lines with improved resistance to bacterial wilt. The pepper line PM 687 has two genes for nematode resistance on chromosome 9 (Djian-Caporalino *et al.*, 2001) that were anticipated to confer a selective advantage since root-knot nematodes are known to facilitate infection by *R. solanacearum*. PM 687 accession displayed a stable resistance to bacterial wilt in different locations and has been used as a progenitor for breeding programs. Lafortune *et al.* (2005) studied the genetic analysis of resistance of pepper to bacterial wilt was performed in the doubled haploid progeny from a cross between a resistant parental line PM 687 and a susceptible cultivar Yolo Wonder. After artificial inoculation with a local isolate of *R. solanacearum*, the progeny consisting of 90 lines was transplanted into a naturally infested field in Guadeloupe, Lesser Antilles. Accession PM 687 displayed a stable resistance to bacterial wilt in different locations and has been used as a progenitor for breeding programs. Mimura and Yoshikawa (2009) reported that out of eight pepper accessions, the accession LS2341 had the highest level of resistance to all the strains of *R. solanacearum* this accession LS2341 confirmed as an appropriate source for the breeding of resistant cultivars of pepper in Japan and is a candidate for a potential source of resistance in other area. Baruah and Rachid (2000) have identified sources of resistance in chilli germplasm against bacterial wilt from Assam. Pflieger *et al.* (1999) reported that disease resistance gene analogs as candidates for QTLs involved in pepper/pathogen interactions. Pochard *et al.* (1986) reported that use of androgenetic autodiploid lines for the analysis of complex resistance systems in the pepper.

6.7.3 Use of Root Stock

Grafting of susceptible cultivars on resistant root stocks, an alternative to resistant cultivars have been reported by various works. As early as 1969, Gilbert and Chin pointed out that highly resistant tomato lines with poor fruit quality, could be efficiently used as rootstocks on which susceptible scions of good fruit quality could be grafted. These authors reported the bacterial wilt resistance of the root system as being effective evenwhen completely susceptible scions is used. This technique is still used nowadays (Reyes, 1997; Cardoso *et al.*, 2006; Rivard and Louws, 2006; Wang *et al.*, 2009), though the protection provided by the rootstock is sometimes incomplete (Nakaho *et al.*, 2004). Indeed, tomato resistant material harbours the bacteria symptomlessly and the resistance is associated with the ability of the plant to restrict bacteria invasiveness (Grimault *et al.*, 1993). The absence of incompatible interactions in tomato resistant lines (no symptoms, no latent infection) has been confirmed by Lebeau *et al.* (2011) by testing a core collection of bacterial wilt resistant accessions with a core collection of bacterial strains. Brinjal is an alternative rootstock for cultivating susceptible tomatoes in contaminated conditions. It was shown to provide a better protection than tomato rootstocks (AVRDC, 1998). This result was confirmed and extended by Lebeau *et al.* (2011) who found that apart from common cases of latent infection for some brinjal accessions and all tomato accessions, there also exist incompatible interactions between some brinjal resistant lines and some bacterial strains. Further, some brinjal lines control bacterial strains that are not controlled by any of the tomato resistant lines tested so far, as exemplified by Carmeille *et al.* (2006b); Wicker *et al.* (2007) and Lebeau *et al.* (2011). The disease can be managed by grafting tomato scions onto bacterial wilt-resistant eggplant rootstocks. Some tomato varieties show resistance, but the complexities and numerous strains of the disease make it difficult to ensure complete genetic resistance. A grafting technique was described by Paily (1964) who found 90% success in Kerala; the grafted plants branched profusely, started fruiting after 2-3 months and continued fruiting longer than non-grafted plants. The disease can be controlled by grafting on *S. torvum* root stock (Mauritius, 1964). However, Empig *et al.* (1962) screened 14 varieties and 7 strain tomato and reported none was resistant. Obrero (1969) grafted tomato plants to the wilt disease. Plants of tomato line M 52 was grafted on to rootstock of Hawaii selection 5808-2 and grown on soil infested with *P. solanacearum*. The grafted plant grew vigorously and produced higher yields of commercial grade tomatoes as compared to non-grafted plant grown on non-grafted soil. Obrero *et al.* (1971) further reported introducing the bacterium through stem wound, introduction through leaves or the injured resistant rootstock could induce that wilt. Felix (1973) studied methods of wilt disease and nematode control of tomato in greenhouse, and the field trial in which tongue-grafted tomato scion on *S. torvum* rootstock (resistant) was compared with non-grafted highly susceptible plants. The field establishment of grafted seedling was noted more successful than that of non-grafted plants. Tomato may be grafted on KVFV hybrid *(L. esculentum X L. hirsutum*) for protection against corky root (*Pyrenocheeta lycopersici*) and on tomato cv. CRA 66 for protection against bacterial wilt and suggested that grafting should be combined with soil disinfection (Beyries, 1974). Grafting susceptible tomato cultivars onto resistant tomato or other solanaceous rootstocks is effective against Asian strains of *R. solanacearum* and is used on the commercial scale in different locations worldwide. The effectiveness of grafting for use against R3b2 has not been tested (Saddler, 2005). Cheng and Chua (1976) used grafted tomato-brinjal plants for tomato prodution. Scions of the tomato cv. Farmer no. 1 were grafted on to rootstock of *S. melongena* (a local brinjal), *S. wrightii* (a perennial tree) or a local tomato cultivar, all resistant to bacterial wilt. The grafting was successful and no graft died from wilt. Tomato- *S. melongena* grew well but produced smaller fruit, on tomato/tomato, while tomato/*S. wrightii* was slightly drawfted and gave a low yield of small fruit. Lum and Wang (1976) reported that the problem of bacterial wilt can be successfully overcome when tomato scions are grafted onto selected resistant brinjal rootstocks. They recommended four weeks old scion should be grafted on 6 weeks old rootstockon on local eggplants cv. Hitam Bulat or Sabah Common as rootstocks and below 10% wilt incidence was consistently recorded in field whereas ungrafted tomato cvs suffered total loss. Tikoo *et al.* (1979) have used scion of tomato cv. Dingra's Multiple Purple (DMP) on tomato root stock successfully. The grafts were tested for their survival, height at flowering, average fruit weight and yields and were compared with non-grafted plants of Pusa Ruby and CR 66-Sel A with non-grafted plants Pusa Ruby. The highest yield was obtained from Pusa Ruby on

DMP, followed by Pusa Ruby on CRA 66-Sel A. Drought tolerance was also maximum with Pusa Ruby on DMP in bacterial wilt sick soil. Resistant rootstocks have been tried at several places to study the compatibility and the usefulness. Some of such rootstocks are *S. torvum, S. diversiforium, L. esculentum* (own root stock), *S. melongena* and *S. wrightii.* The tolerance of graft combination of *S. diversifolium* and *S. lycopercicum* to bacterial wilt was studied by Reyes (1997). He reported that resistance was transmitted to tomato scions and the production was prolonged. A second crop was also obtained by pruning. The rootstock was resistant to rootstock nematode (*Meloidogyne* sp.). The wild brinjal *S. melongena* var. *insanum* (Srinivasan and Gopimony, 1969) and *S. torvum* (Goussetic *et al.,* 2005) are resistant against *R. solanacearum,* but till today resistant to bacterial wilt in many local germplasm is not known. Several workers reported the mechanism of resistance due to stem colonization (Grimault *et al.* 1994b,d, 2007; Gopalakrishnan *et al.*, 2005) which is a useful to explore bacterial wilt resistant marker in brinjal. Reboul (1981) grafted scions of tomato cv. susceptible to *P. solanacearum* onto brinjal. Graft on brinjal rootstock yielded 54.4 t/ha whereas ungrafted plants yield only 16.5t/ha. Peregrine and Ahmad (1982) reported that of the three *Solanum* spp. tested, *S. torvum* proved to be the most resistant to *P. solanacearum*. Plants of Roma and Floradel grafted onto *S. torvum* rootstock were planted once or twice a month for one year, giving 18 planting dates. Plant survival was 61.6% and 50.5% for two cvs, respectively and yields (average of 18 crops) were equivalent to 23544 and 18678 kg/ha, respectively. Iizuka *et al.* (1988) and Wang (1998) recommendation of grafting susceptible tomato cultivars onto resistant tomato or other solanaceous rootstocks is an effective measure for use in commercial scale in Japan, Bangladesh, and the Philippines. Grafting has not been tested for use against R3bv2 and some reports suggest that bacterial wilt-resistant tomato cultivars and breeding lines may have poor resistance to R3bv2. Hanson *et al.* (1988) studied the genetic basis of bacterial wilt in tomato and Grimault and Prior (1994a) have investigated the mechanism of resistance to *R. solanacearum* and reported that resistant scion grafted on susceptible rootstock wilted, indicating that vascular tissue of resistance were not tolerant to higher bacterial population. Colonization frequencies and higher densities observed in plants grafted on resistant or susceptible was correlated to the limitation to bacterial spread in the lower part of the stem. Lu *et al.* (1992) grafted local tomato cultivars onto CH-2-21, CH-2-25 and CH-2-26 tomato lines resistant to wilt as rootstock. Grafts on CH-2-26 gave 100% control of bacterial wilt, while grafts on Ch-2-25 and Ch-2-21 controlled 86.7 and 80%, respectively and delayed the time of disease development by 25 and 40 days, respectively, compared with non-grafted control. However, Nakaho *et al.* (1996) reported that the bacterial wilt of susceptible scion grafted on resistant rootstock wilted and this was due to movement of the pathogen through rootstock. Lin (2008) reported that resistant brinjal (EG203) and tomato (Hawaii 7996) rootstocks were selected based on stable resistance against representative strains of *R. solanacearum* at the seedling stage. Integrated use of Hawaii 7996 as the rootstock and mixture of urea and slaked lime as the soil amendment provided significantly greater control of wilt than use of Hawaii 7996 as root stock alone in only one of the four locations, whereas the soil amendment did not provide significant control effect when EG203 was used as the rootstock. Guan *et al.* (2012) reported that grafting technology is considered an important and innovative practice of integrated pest management and a promising alternative for soil fumigants in vegetable production. Inherent resistance within rootstocks and improved plant nutrient uptake are generally suggested as the main reasons for improved disease control in grafted vegetables.

However, increasing evidence indicated that systemic defense mechanisms may also play an important role in plant defense as a result of grafting. Nakaho *et al.* (1996) discussed the condition that increase latent infection of grafted on non-grafted tomatoes with *P. solanacearum.* Nakaho (1999) has studied on mechanism of resistance in tomato rootstock cultivars to bacterial caused by *R. solanacearum.*

McAvoy *et al.* (2012) have used seven hybrid tomato rootstocks resistant cultivar to *R. solanacearum* were tested as grafting rootstocks to impart resistance to a bacterial wilt-susceptible cultivar, BHN 602. Greenhouse studies showed resistance of all the rootstocks to bacterial wilt. The disease incidence and yield of 'BHN 602' grafted to these rootstocks were evaluated in open-field tomato production in Florida and Virginia over four seasons. Significant differences in bacterial wilt incidence were observed between grafted entries in three of the four trials. In these three trials, grafted entries consistently exhibited the least bacterial wilt incidence compared with the controls; the self-graft, and non-grafted entries. Over all the trials, tomato plants grafted onto 'Cheong Gang', 'BHN 1054', and 'BHN 998' displayed the least bacterial wilt incidence. Rootstocks had a significant effect on total marketable yield in all the trials with certain grafted entries yielding significantly greater than non-grafted 'BHN 602'. Field studies show that grafting holds promise for decreasing the impact of bacterial wilt on tomato cultivars as well as increasing the overall productivity of tomato cultivars.

An attempt was done by Singh *et al.*, (2013) to avoid incidence of bacterial wilt. The rootstock replacement through tube- grafting has been considered effective in controlling this disease, Swarna Baibhav, a promising F1 hybrid of tomato, was on the bacterial wilt resistant rootstocks of brinjal, Swarna Shyamali, Swarna Pratibha, HAB 900, HAB 901, and Swarna Lalima of tomato resulted in complete compatibility between the rootstock and the scion. The grafts grew vigorously and their cropping season also lengthened by a month as compared with the conventional methods. Swarna Baibhav on brinjal HAB 900 rootstock recorded the maximum yield (67.66 tonnes/ha) and higher number of fruits per plant (40) than other rootstocks. There was a huge loss in yield due to high mortality of non-grafted Swarna Baibhav (23.71 tons/ha). Percentage survival of rootstocks against bacterial wilt was 100% in Swarna Pratibha, followed by HAB 900 and HAB 901 (98.24%).

6.7.4 Bioluminescence from Lux-Marked *R. solanacearum*

A new simplified method of selection of tomato plant resistant to bacterial wilt (*R. solanacearum*) has been reported by Mikichi *et al.* (1999). The roots of 3-weeks-old tomato seedling were dipped in a solution containing *R. solanacearum* YN5 (Lucifer erase, Lux marked) for 30 min.Wilting was observed 5-7 days after inoculation. Images of bacterial bioluminescence were obtained in tomato seedling after 5 and 7 days of inoculation. A video intensified microtomy (VIM) camera equipped with ARGUS 50 was used to detect bioluminescence. Bioluminescence was observed in the root and collar regions of all 25 cultivars tested, whereas in some of the cultivars bioluminescence was also observed in upper part of the stem only. Quantitative differences were observed in the bioluminescence of susceptible and resistant culitivars. Percentage of wilted seedlings positively correlated with the incidence of bacterial bioluminenscece in upper part of stems. Zuluage Cruz *et al.* (2014) described a new method to study the interaction between

R. solanacearum and potato germplasm using a luminescent *R. solanacearum* reporter strain. The *R. solancearum* UY031 was genetically modified to constitutively generate light from a synthetic *LUXCDABE* operon stably inserted in its chromosome. Colonization of this reporter strain on different potato lines was followed using life image.

6.7.5 Tissue Culture

Toyoda *et al.* (1989) reported that leaf explants-derived callus tissue which were resistant to toxic substances, derived from *P. solanacearum* in culture filtrate were selected and regenerated into plants. These plants expressed resistance to *P. solanacearum* at the early infection stage, by suppressing or delaying the growth of the inoculated bacteria. Complete resistance was obtained in self-pollinated progeny of regenerates from non-selected callus tissue. These plants showed high resistance when inoculated with the virulent strain and were also resistant when planted in fields with different strains of the pathogen in tomato.Toyoda *et al.* (1998) suggested that the resistance to bacterial wilt was maintained in self-pollinated progenies of callus-derived regenerates.

Table 6.9 Resistant and tolerant lines to bacterial wilt of solanaceous vegetables in world

Place /Country	*Resistant lines/varieties*	*Reaction*	*References*
Tomato			
Bangladesh	Tusti, Bikash	R	Hosain *et al.*, 1991
Brazil	65 - S2, 68 –S4, 65 –S52	R	Akiba *et al.*, 1972
	Caraibo, CL 113 -0-0-13-0-6, 72 TR 4-4	R	Quezado-Soares and Lopes, 1994
	Hawaii 7998, Rotam 4	R	Gonzalez and Summers, 1995, Oliverira *et al.*, 1999
	Hawaii 7998	R	Thoquet, 1993
	Yashimatsu 4-11	R	Lopes *et al.*, 1994
	Rdade, Venus	R	Oliverira *et al.*, 1999
	Saturn	R	Lopes, 1994
	CL5915-93	R	Silveira *et al.*, 1999
China	Zao Kang 1	R	Zhang Zan Hong and Liu, 1998
Hermansen	CL 1131-0-0-38-40	R	Hermsen, 1987
France	CRA 66 CRA 66, Hawaii 7996, Caraibo	R R	Beyries, 1974 Grimault and Prior, 1993
Georgia	Hawaii 7996, L 285	R	McGarvey *et al.*, 1999
Hawaii	N-52	R	Obrero, 1969
	CRA 66	R	Lasso, 1974
	NC 1953-63 N, NC 1953-64N	R	Suzuki *et al.*, 1964
	Hawaii 7997, Hawaii 7998 & CRA 66	R	Chellem *et al.*, 1994

Contd...

Contd...

Korea Republic	Kwangyyong	MR	In Moo Seong *et al.*, 1996
	MST 21-23	MR	
	Naebyong Changsu, SeoJin	MR	
Lembang (Indonesia)	Ratna,Intan	R	Gunwan, 1988a,b
	LV-209 (Gelatik Hijau)	R	Hanudin and Gaos, 1993
Mauritius	MST 32-1	MR	Girad *et al.*, 1989
North Carolina	L285. cln65-349		AVRDC, 1986
	L285, CLN 65-349		Opena and Tschanz, 1987
New york	CL 1131-0-0-38-40	R	Hermsen, 1987
Nepal	Mallajh local	R	Panthee and Bhttrai, 1999
Queensland	Australian variety	R	Aberdee, 1946,
	Scorpio	R	Peterson *et al.*, 1983
	Redland summer etaste	R	Herringglon and Saeanah, 1985
	Oxheart, TM 003, TM 008, Raton	MR	Hosain *et al.*, 1991
	BT-12, BT	R	Mahanta *et al.*, 1998
Reunion Island	Hawaii7996	R	Carmeille *et al.*, 2006b
South Africa	Rodade (Flora Dade X Floradade x Ld 2048A resistant strain	R	Bosch *et al.*, 1985; Anderson, 1947
Taiwan	Vc/Nova	R	Bissonauth, 1980
	VC 8-1-2-1	R	Mew and Ho, 1976
	TM 103	R	Taiwan, 1984
	H 7997, CRA 84, L 285	R	Hanson *et al.*, 1988a,b
	FMTT 277, FMTT 285, FMTT 269, FMTT 267, FMTT 301, FMTT 304	R	Lin and Horng, 1995
Thialand	PBC195, CA364		Watcharachaivakup *et al.*, 2005
	LA 1421, Cascade	R	Mohamed *et al.*, 1997
USA	Venus, Saturn	R	Henderson and Jenkin, 1971
	IRAT L, OTB 2	R	Daly, 1973, 1976
	Hawaii 7981, Hawaii 7997	HT	Sonoda and Augustine, 1978
	Hawaii 7997, Hawaii 7998, CRA 66	R	Chellemi *et al.*, 1994
	CRA 66 (SELA, D&E), P1 212441, P1 263722, P1 263722, P1 365930, Venus, Saturn	HT	Sonoda and Augustine, 1979
Western Samoa, Malaysia	F8 line FP-1	T	Anonymous, 1976
	F8 line FP-2, F8 line FP-5	T	Anonymous, 1977
Brinjal			
Argentina	PI 337597	R	Hermsen, 1987

Contd...

Contd...

Dhanka (Bangla-desh)	Kata Begun	R	Rahman *et al.*, 2011
	Marich Begun and Dundul	MR	
	Kazla and Luffa	MR	
	Solanum torvum	R	Hermsen, 1987; Gousset *et al.*, 2005
New york	*PI* 196300	R	Hermsen, 1987
Chilli			
California (USA)	PM 687 (USDA PI 322719 accession)	R	Lafortune *et al.* 2005
Malaysia	MC-4	R	Quezado-Soares and Lopes, 1995; Momol *et al.*, 2001; Lopes and Boitbux, 2004
	PM 687	R	Djian-Caporalino *et al.*, 2001; Lafortune *et al.*, 2005.
	MC5	R	Matos *et al.*, 1990
	LS2341	R	Mimura and Yoshikawa, 2009
Japan	Manganji	R	Mimura *et al.*, 2000
	Fushimi-amanaga	R	Tsuro *et al.*, 2007
	Fushimi-amanaga	R	Hashimoto *et al.*, 2001
	Mie-midori, Ishii-midori, Akashi	R	Matsunaga and Monma, 1999
Potato			
Brazil	Achat	R	Priou *et al.*, 1999
East Africa/Peru	Cruza 148	R	Laferriereet *et al.*, 1999
	Solanum commonersonii	R	Laferriereet *et al.*, 1999
Nicaragua	PI 196300	R	Hermsen, 1987

Rashid *et al.* (2004) used a scale for reaction of bacterial wilt as resistant (30% and below wilted plant); moderately resistant (31%-50% wilted plant); moderately susceptible (51%-70% and wilted plant); Susceptible (71% and above wilted plant) in brinjal in Bangladesh.

Table 6.10 Resistant and tolerant lines to bacterial wilt of solanaceous vegetables in India

Place/Country	*Resistant Lines/Varieties*	*Reaction*	*References*
Tomato			
Andaman & Nicobar Island	BT118-4-1-1, BT-116-8-1-1 and Tomato-415	R	Swaroop and Suryanarayan, 2005
Assam	EC 178823, EC 179924, EC 179926, EC 179931	R	Kapoor *et al.*, 1991
	Sonali (Sel-2)	R	Patil *et al.*, 1990
	DPT 38	R	Kumar *et al.*, 1997
	VC 48	R	Roy *et al.*, 1992
	BWR 5 (Arka Alok)	MR	Kapoor *et al.*, 1991
	VC 9-1 UG	MR	Bedekar, 1977

Contd...

Contd...

	VC 11-1 UG	MR	Bedekar, 1977
	Sonali (Sel-2)	R	Patil *et al.*, 1990
	DPT 4, DPT 14	R	Banerjee and Singh, 2002
	DPT 4, DPT 14	R	Roy *et al.*, 1992
	VC 48	R	Banerjee and Singh, 2002
Himachal Pradesh	EC386019, IC214633, EC386023	R	Aggarwal *et al.*, 2008
	BL-342-1, EC-191536	R	Thakur *et al.*, 2004
	EC191536, Hawaii 7998	R	Gautam and Prakash, 2006
	Hawai 7996	R	Mangin, *et al.*, 1999; Wang *et al.*, 2000
Jharkhand	BT-1, BT-10,BT 17, BT-18,BT-21, LE-79-5,BRH-1, BRH-2, Arka Alok, Arka Abha	R	Sharma and Kumar, 1994, 1995
	BWR1, BWR5	R	Sharma and Kumar, 1997
	EC339074 (Swarna Lalima), EC369060 A (Swarna Naveen)	R	Sharma *et al.*, 1999, 2001
	EC –386021 x BT-17, Sonali xx BT-17, EC339074 x EC-386021 released as Swarna Sampada)	R	Sharma *et al.*, 2006
	LE 79 (Shakthi)	MR	Sharma and Kumar, 1997
Karnataka	CRA 66(Sel-A)	R	Rao *et al.*, 1975
	Hawaii 7996	R	Gopalakrishnan *et al.*, 2005
	BWR 1 (sel.from VC 8-1-2-1)	R	Tikoo *et al.*, 1988
	Hybrids Arka Alok × SP-2-2 and Arka × L-101, Sonali × SP-2-2	R	Dharmati *et al.*, 2009
	Arka Abha, Sonali, DPT-38 and Arka Alok	R	Anonymous, 2008
	L-15	R	Madalageri and Dharmatti, 1991
Kerala	LE 79 (Shakthi)	R	Seshadri, 1998
	CL32-d-0-1-19 GS (Shakthi)	R	Peter *et al.*, 1992
	LE 415		Banerjee and Singh, 2002
	FL7514, BHN 466, Hawaii 7996	R	Gopalakrishnan *et al.*, 2005
New Delhi	LA 2639A, LA 2691	R	Seshadri, 1998
	BWR 1(Arka Abha), BWR 5(Arka Alok)	R	Seshadri, 1998
	88 BWR, 83-211, 84 BWR, CR15955-223, D4 -22-0	R	Sharma *et al.*, 1997

Contd...

Contd...

Orissa	BT 1(Utkal Pallavi), BT 2(Utkal Deepti)	R	Seshadri, 1998
	BT12, BT 14, BT 18, Pusa Early Dwarf, Navodaya, Selection 7	R	Mishra *et al.*, 1999
	Pusa Sheetal	R	Mahanta *et al.*, 1998
	BWR 1(Arka Abha), BWR 5(Arka Alok)	R	Seshadri, 1998
	BT 1(Utkal Pallavi), BT 2(Utkal Deepti)	R	Seshadri, 1998, Mishra *et al.*, 1999
	BT 2	T	Mishra *et a1.*, 1995
	BT3,BT17	MR	Mishra *et a1.*, 1995
	BT12, BT14, BT18	R	Mishra *et a1.*, 1995
	Pusa Sheetal	R	Mahanta *et al.*, 1998
	Pusa Early Dwarf, Navodaya, Selection 7	R	Mishra *et al.*, 1999
	GA1565-2-413 WT, GA 219-1-2 BWT, GA 1095-1-4 BWT, GA 1405-1-2 BWT	T	Jaworski *et al.*, 1987
	BT102-1-2-2	R	Banerjee and Singh, 2002,
Tripura	Arka Alok, BT-1, BT- 10, CKVT-17 and Sikkim local	R	Biswas and Singh, 2008
	Phlippines 1169, Hawaii 7580,	R	Krausz and Thurston, 1975
	Arka Abha, Sonali, T-38 and Arka Alok	R	Mallikarjun *et al.*, 2008
Brinjal			
Andaman & Nicobar Island	BB 60C, CHES 309	R	Kishan-Swaroop *et al.*, 2000 a, b
	95-17, BB 46, CH 243, BB 44, Arka Keshav	R	
Jharkhand	CH243, CH-245, CH-247, CH-249 and CH-309	R	Sharma *et al.*, 2005
	CH-309	R	Banerjee and Singh, 2002
	CH-157	MR	Sharma *et al.*, 2005
	SM-6-6, BB-44, BB60C Arka Keshav, CH-249	R	Sharma and Kumar, 1994
	BB44, sm6-6, BB13-1, BB12, BB60C, JC-1, BWR-12	R	Sharma and Kumar, 1995
	CH-249, CH -309 SM-141, BB-60C, BB64	R	Sharma and Kumar, 1996
	Swarna Shyamli, Swarna Pratibha, BB 64, JC 8, Arka Keshav, Arka Nidhi,	R	Sharma and Kumar, 2007

Contd...

Contd...

Karnataka	IHR-171, IHR-180, IHR-181, IHR-182, IHR-85	R	Sadashiva *et al.*, 1994
	SM6-6	R	Sadashiva and Reddy, 1995
	FL7514 , BHN 466	R	Gopalakrishnan *et al.*, 2005
	Arka Keshav	R	Ponnuswami, 1997
	Arka Nidhi, Arka Keshav, Arka Neelkantha, BB1, BB44, BB49, EP143, Surya, BB 13-1, BB2, Kerala local, SM6-6	R	Ponnuswami, 1999
Kerala	CA 219, *i.e.* CA 219-1-19-9 and CA 219-1-19-6, CA 33, *i.e.* CA 33-1-4-6 and CA 33-1-12-3	R	Gopalakrishanan and Peter, 1991
	SM132, SM6-2, SM6-6, SM6-7	R	Geetha and Peter, 1993
	SM6-7 (Released as Surya)		Peter *et al.*, 1992
	Swetha (White fruit), Haritha (long, light grrenfrut), Neelima(F1)		Gopalakrishanan *et al.*, 2005
	SM6-1, SM 6 and SM6-9	T	Peter *et al.*, 1992
	SM121, SM126, SM133, SM135 and SM140	R	
Nagaland	Arka Keshav	R	Singh *et al.*, 1990
	Arka Keshav, Arka Neelkanth, Arka Nidhi, SM6-6	R	Chaudhary and Sharma, 2000
	EG 203	R	Hussain *et al.*, 2005
Orissa	BB40, BB64	R	Banerjee and Singh, 2002; Biswas and Singh, 2007
Tripura	Singnath, BB 40, BB 64 , Green round	R	Biswas and Singh, 2007
UP	EG 203	R	Hussain *et al.*, 2005
Chilli and Sweet paper			
Karnataka	Kandhari	R	Rahim and Samraj, 1974
	CA 219, *i.e.* CA 219-1-19-9 and CA 219-1-19-6	R	Gopalkrishanan and Peter, 1991
	Ujwala	R	Gopalakrishanan, 1996
	Manjeri	R	Jyothi *et al.*, 1993
	Kulai	R	Rahman and Abdulah, 1997
	Manjeri and Kandari, White Kandari, Chna, KAUcluster (Released as Manjeri)	R	Peter *et al.*, 1992

R = Resistant, MR = Moderately Resistant, T = Tolerant, HT = Highly Tolerant

6.8 INDUCING SYSTEMIC RESISTANCE

Induced resistance inplants opens new horizons in plant protection, being a promising tool for ecologically-friendly disease control and sustainable agriculture. It remains a challenge for both fundamental and applied research (Edreva, 2004; Hacisalihoglu *et al.*, 2007). Most plant protection methods currently applied use toxic chemicals noxious to the environment for pathogen and pest control. Induced resistance exploiting natural defense machinery of plants could be proposed as an alternative, non-conventional and ecologically-friendly approach for plant protection. Its introduction into agricultural practice could minimize the scope of chemical control, thus contributing to the development of sustainable agriculture. Induced resistance can be defined as an increased expression of natural defence mechanisms of plants against various types of pathogens, provoked by a range of factors: pathogens causing hypersensitive necrotic reaction; avirulent or attenuated pathogenic strains; elicitors of pathogenic origin (glucans, proteins, lipids, etc.); abiotic elicitors, including synthetic harmless chemical products, such as 2,6-dichloroisonicotinic acid (INA), b-aminobutyric acid (BABA), benzothiadiazole (BTH), etc. Induced resistance, being based on the expression of latent genetic information present in plants, is not underlied by genome alterations (mutations, introgression of foreign genetic material), thus enhancing its biological safety. Molecular bases of induced resistance, involving receptor-elicitor interactions, signal transducing pathways and SAR gene expression. Presently, disease control is largely based on the use of fungicides, bactericides and insecticides – chemical compounds toxic to plant invaders, causative agents or vectors of plant diseases. However, the hazardous effect of these chemicals or their degradation products on the environment and human health strongly necessitates the search for new, harmless means of disease control. Since the late 1950s increasing body of evidence on the natural phenomenon of induced resistance has been accumulated, culminating in its successful practical application in the last decade (Kuc, 2001). When a plant is inoculated with a pathogen ("primary inoculation"), and after a time interval is subjected to a secondary ("challenge") inoculation, reduced disease symptoms develop, *i.e.,* the induced plant becomes more resistant than the normal, non-induced plant. Later, stimuli other than pathogens, such as some nontoxic chemicals, were found to be effective at inducing resistance. Thus, the induced resistance can be defined as an increased expression of natural defense mechanisms of plants against different pathogens provoked by external factors of various type and manifested upon subsequent inoculation. Hence, the low specificity is an inherent character of induced resistance. Ramamoorthy *et al.* (2001) reported that plant growth promoting rhizobacteria (PGPR) belonging to *Pseudomonas* spp. are being exploited commercially for plant protection to induce systemic resistance against various pests and diseases. Mixtures of different PGPR strains have resulted in increased efficacy by inducing systemic resistance against several pathogens attacking the same crop. Seed-treatment with PGPR causes cell wall structural modifications and biochemical/ physiological changes leading to the synthesis of proteins and chemicals involved in plant defense mechanisms. Lipopolysaccharides, siderophores and salicylic acid are the major determinants of PGPR-mediated ISR. The performance of PGPR has been successful against certain pathogens, insect and nematode pests under field conditions. Liu *et al.* (1995) reported that plant growth-promoting rhizobacteria produced induction of systemic resistance duration of protection and effect of host resistance on protection and root colonization in cucumber.

6.8.1 Inducers of Resistance

A multitude of factors are reported to induce resistanc in plants: pathogens (fungi, bacteria, viruses) causing hypersensitive necrotic reaction (HR); avirulent and attenuated pathogenic strains; pests (insects, nematodes); elicitors of biotic origin; abiotic elicitors, *i.e.* chemical products, such as benzothiadiazole (BTH), β-aminobutyric acid (BABA), 2, 6-dichloroisonicotinic acid (INA), salicylic acid, inorganic salts, etc. (Kessman *et al.*, 1994; Lyon *et al.*, 1995; Schneider *et al.*, 1996; Benhamouand Picard, 1999; Cohen, 2001; Kuc, 2001). Plant growth-promoting rhizobacteria can suppress diseases through antagonism between the bacteria and soil-borne pathogens, as well as by inducing a systemic resistance in the plant against both root and foliar pathogens. The generally non-specific character of induced resistance constitutes an increase in the level of basal resistance to several pathogens simultaneously, which is of benefit under natural conditions where multiple pathogens may be present. Specific *Pseudomonas* strains induce systemic resistance in, e.g., carnation, cucumber, radish, tobacco and *Arabidopsis*, as evidenced by an enhanced defensive capacity upon challenge inoculation. Although some bacterial strains are equally effective in inducing resistance in different plant species, others show specificity, indicating specific recognition between bacteria and plants at the root surface. In carnation, radish and *Arabidopsis*, the O-antigenic side chain of the bacterial outer membrane lipopolysaccharide acts as an inducing determinant, but other bacterial traits are also involved. Pseudobactin siderophores have been implicated in the induction of resistance in tobacco and *Arabidopsis*, and another siderophore, pseudomonine, may explain induction of resistance associated with salicylic acid (SA) in radish. Although SA induces phenotypically similar systemic acquired resistance (SAR), it is not necessary for the systemic resistance induced by most rhizobacterial strains. Instead, rhizobacteria-mediated induced systemic resistance (ISR) is dependent on jasmonic acid (JA) and ethylene signaling in the plant. Upon challenge inoculation of induced *Arabidopsis* plants with a pathogen, leaves expressing SAR exhibit a primed expression of SA-, but not JA/ethylene-responsive defense-related genes, whereas leaves expressing ISR are primed to express JA/ethylene-, but not SA-responsive genes. Combination of ISR and SAR can increase protection against pathogens that are resisted through both pathways, as well as extend protection to a broader spectrum of pathogens than ISR or SAR alone. Chaudhary *et al.* (2007) stated that induced resistance is a state of enhanced defence capacity developed by a plant when approtriately stimulated. SAR and ISR are twin forms of induced existence wherein plant defences are preconditioned by prior infection or treatment that results in resistance against subsequent challenge by a pathogen or parasite. Rhizhobacteria mediated ISR resembles that of pathogen induced SAR in that both types of induced resistance render uninfected plant parts more resistance towards broad spectrum of plant pathogens. Several rhizobacteria trigger the SA-dependent SAR pathway by producing SA at the root surface whereas other rhizobacteria trigger different signaling pathway by independent of SA. The exixtence of SA-independent ISR pathway has been studied in *Arabidopsis thiaiana*, which is dependent on jasmonic acid (JA) and ethyle signaling. Combination of ISR and SAR can increase protection against pathogens that are resisted through both pathways besides protection to a broader spectrum of pathogens than isr/sar alone. Besides *Pseudomonas* strains, ISR is conducted by *Bacillus* spp. wherein published results show that several specific strains of species *B. amyloliquifaciencs, B. subtilis, B. pasteurii, B. cerus, B. pumilus, B. mycoides, and B. sphaericus* elicit significant reduction in the incidence or severity of various diseases on diversity of hosts. Zehnder *et al.* (2001) have developed a biological control system in vegetables based on ISR mediated by plant growth-promoting rhizobacteria (PGPR) to bacterial wilt disease in cucumber caused by *Erwinia tracheiphila* suggesting that PGPR-ISR protects cucumber against bacterial wilt through the induction of other plant defense mechanisms after the pathogen has been introduced into the plant. Kyungseok *et al.* (2007) have studied on biocontrol activity of five strains of selected rhizobacteria in tomato against bacterial wilt caused by *R. solanacearum* by challenge inoculation with the pathogen, all of the rhizobacterial strains efficiently suppressed the bacterial wilt in tomato. The strain had been proven to produce an efficient elicitor for inducing systemic resistance in many crops and that confirmed that EXTN-1 has the ability for reducing the pathogen spread in tomato, the strain could be effectively used as a potential biocontrol agent against bacterial wilt. Several reports have been reported on *B. vallismortis* strain EXTN-1 had induced systemic resistance properties against various pathogens (Park *et al.* 2001). Ahn *et al.* (2002) reported that the major mechanisms by which EXTN-1 bring about disease suppression in crops is by the induction of systemic resistance to the host plant. Grimault *et al.* (1994) reported that the host defence modifications could slow down the rate of bacterial (*R. solanacearum*) spread in tomato. Induced systemic resistance by *B. vallismortis* EXTN-1 to suppress tomato bacterial wilt. The hypersensitive reaction (HR), oxidative burst, lignifications and cell structural defence modifications in host plants were observed with EXTN-1 strain (Park *et al.* 2001; Jeun *et al.* 2001; Ahn *et al.*, 2001). Sarowar *et al.* (2005) showed that the pathogenesis related genes responsive to ethylene pathway in transgenic tomato enhanced its resistance to bacterial wilt. Chern *et al.* (2008) reported that SAR in plants confers lasting broad-spectrum resistance to pathogens and requires the phytohormone salicylic acid. Arabidopsis NPR1/NIM1 is a key regulator of the SAR response. Studies attempting to reveal the function of NPR1 and how it mediates SA signaling have led to isolation of two classes of proteins that interact with NPR1: the first class includes rice NRR, *Arabidopsis* NIMIN1, NIMIN2, and NIMIN3, and tobacco NIMIN2-like proteins; the second class belongs to TGA transcription factors. Overexpression of NRR in rice suppresses both basal and Xa21-mediated resistance.

Yin (2013) reported that the induction of defense responses in tomato against bacterial wilt by stress ethylene level reduction mediated by the ACC deaminase activity of *Methylobacterium* strains. Under greenhouse conditions, the disease index value in *Methylobacterium* sp. inoculated tomato plants was lower than control plants. Plants treated with *Methylobacterium* sp. challenge inoculated with *R. solanacearum* showed significantly reduced disease symptoms and lowered ethylene emission under greenhouse condition. The ACC and ACO (1-aminocyclopropane-1-carboxylate oxidase) accumulation in tomato leaves were significantly reduced with *Methylobacterium* strains inoculation. While ACC oxidase gene expression was found higher in plants treated with *R. solanacearum* than *Methylobacterium* sp. treatment, PR proteins related to induced systemic resistance like b-1, 3-glucanase, PAL, PO and PPO were increased in *Methylobacterium* sp. inoculate plants. A significant increase in b-1, 3-glucanase and PAL gene expression was found in all the *Methylobacterium* spp. treatments compared to the *R. solanacearum* treatment. This study confirms the activity of *Methylobacterium* sp. in increasing the defense enzymes by modulating the ethylene biosynthesis pathway and suggest the use of methylotrophic bacteria as potential biocontrol agents in tomato cultivation.

Table 6.11 Bacterial determinants and types of host resistance incduced by bacterial antagonists.

Antagonist	*Host*	*Bacterial determinant*	*Type*	*Reference*
B. amyloliquefaciens	Tomato	Polyphenol oxidase, PAL, peroxidase	ISR	Tan *et al.*, 2013
Pseudomonas fluorescens WCS374	Tomato	Siderophore		Ran *et al.*, 2005
P. fluorescens WCS417	Tomato	Lipopolysaccharide	ISR	Duijff et al., 1997
Methylobacterium spp.	Tomato	Polyphenol oxidase, PAL, peroxidase	ISR	Yim, 2013

6.8.2 Endophytic Bacteria

The endophytic bacteria live in plant tissues without doing substantive harm or gaining benefit other than residency (Kobayashi and Palumbo, 2000). Hallmann *et al.* (1997) defined an endophyte as any microorganism that resides inside the plant without regard to the specific tissue colonized and these bacterial endophytes can be isolated from surface-disinfected plant tissue or extracted from internal plant tissue. Endophytes enter plant tissue primarily through the root zone; however, aerial portions of the plants such as flowers, stems, cotyledons may also be used for entry (Kobayashi and Palumbo, 2000). Once inside the plant tissue, endophytic bacteria may either remain localized at the point of entry or spread throughout the plant (Hallmann *et al.* 1997), these microorganisms can reside within the cells (Jacobs *et al.,* 1985), in the intercellular spaces or in the vascular system (Bell *et al*., 1995). Moreover, Ciampi-Panno *et al*., (1989) have proved the use of antagonistic microbes in the control of *R. solanacearum* under field conditions. Ramamoorthy *et al.* (2001) reported that endophytic bacteria colonize an ecological niche similar to that of plant pathogens, especially vascular wilt pathogens, which might favour them as potential candidates for biocontrol agents and plant growth-promoting agents. A total of 144 endophytic bacteria were isolated from three crops, more endophytic bacteria were isolated from stem tissue than from root tissue in case of brinjal. More than 60% endophytes from these crops were Gram-negative. Prious *et al.* (1999) reported that the bacterium affects more than 30 plant species, the most susceptible crops being potato, tomato, eggplant, pepper, banana and groundnut. Spread of the pathogen worldwide has been associated with its dissemination in latently infected planting material. Jetiyanon and Kloepper (2002) have studies of induced systemic resistance using strains of plant growth-promoting rhizobacteria (PGPR) have concentrated on the use of individual PGPR as inducers against multiple diseases of a single crop. Twenty-one combinations of PGPR and seven individual PGPRs were for induced resistance activity in which results indicated that four mixtures of PGPR significantly reduced the severity compared to the non-bacterized control. The use of individual PGPR strains, mixtures of several strains resulted more stable rhizosphere community, provide several mechanisms of biological control, and suppress a broader range of pathogens. Agriculture in many parts of the world involves intercropping or mixtures of different crops in a confined geographical area, in contrast to large-acreage monoculture which typifies agriculture in North America. Under conditions of intercropping or small fields with multiple crops, induced resistance should be operable against several diseases of different crops. It has been reported by various workers (Viswanathan and Samiyappan, 2002; Bashan and de Bashan, 2002) that the endophytic bacteria have ability to inhibit the growth of *R. solanacearum* which is the first step towards the selection of potential biocontrol agents. The antagonistic ability of the endophytic bacteria towards *R. solanacearum* was assayed *in vitro* by the agar diffusion method and also by the streaking method. Numerous plant growth promoting rhizobacteria (PGPR) have been reported which promote plant growth and also control plant diseases. Diffusible metabolites are produced from cell-free culture filtrates isolates inhibited *R. solanacearum* growth indicated the role of secondary metabolites in the pathogen suppression *in vitro*. Xiao and Kisaalita (1998) have reported that pseudomonads exert their beneficial effects in part by producing yellow-green, fluorescent siderophores called pyoverdins, under iron-deficient conditions. Plant growth-promoting bacteria have frequently found success with bacterial genera *Pseudomonas* and *Bacillus* (Merriman *et al.*, 1974). Reiter *et al.* (2002) have reported isolation of fluorescent *Pseudomonas* from potato plants. Biochemical characterization and identification fairly yielded a correct identification as reported by Zinniel *et al.* (2002). These bacteria move upward and downward from the point of application and by colonizing the internal root tissues, can exclude the entry of pathogen in the vascular stele. The endophytic bacteria can out-compete the vascular plant pathogens. So establishing beneficial bacterial populations in the rhizosphere seems to be a key for obtaining a healthy microfloral balance within the plants. Under field conditions, the effectiveness of the biocontrol agent depends on various factors like time of application, age of the plant, strength of the biocontrol agent formulation, mode of application and many other factors (Botelho and Mendonca-Hagler, 2006).

6.8.3 Production of Secondary Metabolites

One of the mechanisms by which *P. fluorescens* suppresses plant pathogens is the production of antibiotics such as 2, 4-diacetylphloroglucinol (DAPG), pyrrolnitrin pyoluteorin, and phenazine-1-carboxylate. Raaijmakers and Weller (2001) reported the existence of wide genotypic diversity of DAPG-producing *Pseudomonas* species which could be exploited for the identification of superior biocontrol agents against plant pathogens. Production of siderophores is another mechanism by which endophytic biocontrol agents suppress the pathogens indirectly by stimulating the biosynthesis of other antimicrobial compounds by increasing availability of minerals to the biocontrol agent in addition to iron chelation (Duffy and Defago, 1999).

6.8.4 Antibacterial Agents from Animals and Mammals

Dobesh and Latham (1998) reported about the platelet aggregation and thrombus formation, resulting from disruption of a coronary artery plaque, play a critical role in the pathology of acute coronary syndromes. Currently, aspirin and heparin are administered to decrease platelet aggregation. The discovery of the platelet integrin receptor alpha (IIb) beta 3, also known as the platelet glycoprotein (GP) IIb-IIIa receptor, is a breakthrough in antiplatelet therapy. The GP IIb-IIIa receptor is responsible for the crucial binding of fibrinogen to platelets, leading to crosslinks between platelets and further platelet aggregation. Since the introduction of abciximab, the first GP IIb-IIIa-receptor antagonist, several other non-antibody agents have been studied for use in percutaneous transluminal coronary angioplasty and also in stent placement, treatment of unstable angina, and myocardial infarction. Ohshima *et al.* (1999) reported that enhanced resistance to bacterial diseases of transgenic tobacco plants overexpressing sarcotoxin IA, A bactericidal peptide of insect. The expression in plants of antibacterial proteins from higher animals has scarcely been investigated. The best known peptides are given in Table 6.12.

Table 6.12 Antibacterial protein produced from higher animals trigering in Induced System Resistance

Name of animal	*Antibacterial Protein*	*References*
Frog skin		Barra and Simmaco, 1995
	Peptides pore-forming Magainins	Bevins and Zasloff, 1990
	Bombinins	Simmaco *et al.*, 1991
	Brevinins and Esculentin	Simmaco *et al.*, 1993, 1994
	Rugosins	Suzuki *et al.*, 1995
	Temporins	Simmaco *et al.*, 1996
Amphibian skin	Esculentin-1	Ponti *et al.*, 1999
Cytoplasmic granules of bovine blood neutrophils, peptides	Indolicidin	Selsted *et al.*, 1992
Chiken egg	White lysozyme, Cecropin,Attacin	Dodds, 1989
Bactericidal peptide of insect	Sarcotoxin IA	Ohshima *et al.* 1999 Okamoto *et al.* 1998

The phytoalexins are secondary metabolites, antibiotics, low molecular weight produced by plants in response to physical stress, chemical, or biological. They are able to prevent or reduce the activity of pathogens, the rate of production is dependent on the genotypes of host and/or pathogen (Daniel and Purkayastha, 1995).

6.8.5 Antibodies

An interesting strategy for control of bacteria may be the expression in transgenic plants of antibodies that target specific bacterial pathogenicity factors (Düring, 1996). Specific binding of antibodies to one or more bacterial factors, such as secreted lytic enzymes or extracellular proteins, is expected to compromise bacterial pathogenicity. Remenant *et al.* (2011) reported that the *R. solanacearum* species complex includes *R. solanacearum*, *R. syzygii*, and the Blood Disease Bacterium (BDB). All colonize plant xylem vessels and cause wilt diseases, but with significant biological differences. *R. solanacearum* is a soil-borne bacterium that infects the roots of a broad range of plants. *R. syzygii* causes Sumatra disease of clove trees and is actively transmitted by cercopoid insects. BDB is also pathogenic to a single host, banana, and is transmitted by pollinating insects. Sequencing and DNA-DNA hybridization studies indicated that despite their phenotypic differences, these three plant pathogens are actually very closely related, falling into the Phylotype IV subgroup of the *R. solanacearum* species complex. To better understand the relationships among these bacteria, we sequenced and annotated the genomes of *R. syzygii* strain R24 and BDB strain R229. These genomes were compared to strain PSI07, a closely related Phylotype IV tomato isolate of *R. solanacearum*, and to five additional *R. solanacearum* genomes. Whole-genome comparisons confirmed previous phylogenetic results: the three phylotype IV strains share more and larger syntenic regions with each other than with other *R. solanacearum* strains. Furthermore, the genetic distances between strains, assessed by an *in-silico* equivalent of DNA-DNA hybridization, unambiguously showed that phylotype IV strains of BDB, *R. syzygii* and *R. solanacearum* form one genomic species. Based on these comprehensive data we propose a revision of the taxonomy of the *R. solanacearum* species complex. The BDB and *R. syzygii* genomes encoded no obvious unique metabolic capacities and contained no evidence of horizontal gene transfer from bacteria occupying similar niches. Genes specific to *R. syzygii* and BDB were almost all of unknown function or extrachromosomal origin. Thus, the pathogenic life-styles of these organisms are more probably due to ecological adaptation and genomic convergence during vertical evolution than to the acquisition of DNA by horizontal transfer.

6.8.6 Bacterial Avirulence Genes (Gene–for-Gene Interactions)

The first avirulence genes were cloned from bacteria (Staskawicz *et al.,* 1984) and the very first plant resistance genes defined and cloned in tomato (Martin *et al.,* 1993; Salmeron *et al.,* 1996) and *Arabidopsis* (Bent *et al.,* 1994; Mindrinos *et al.,* 1994; Grant *et al.,* 1995) conferred resistance specifically to strains of *P. syringae*. Plant-bacterial interactions have provided one of the best systems for studying plant-pathogen relationships and the molecular basis of disease resistance in plants. The gene-for-gene hypothesis (Flor, 1971; Keen, 1990) explains host-pathogen recognition events on the assumption that single genes at particular loci in the plant and single genes at particular loci in the pathogen together determine whether the plant-pathogen interaction will be compatible (resulting in infection and disease development) or incompatible resulting in the initiation of hypersensitive defense response (HR) and disease resistance. Biological races of pathogens that contain a set of genes with alleles for virulence and avirulence will induce a specific pattern of compatible and incompatible reactions when inoculated on a collection of host plant cultivars that contain (or lack) a set of genes with alleles for resistance and susceptibility. Incompatibility results only when the resistance allele at a particular gene locus in the plant complements the corresponding avirulence allele in the pathogen. This concept presupposes that plants with a particular resistance gene (R gene) recognize a pathogen-produced elicitor, a direct or indirect product of the corresponding avirulence (effector) gene (Gabriel and Rolfe, 1990). It is likely that when the interaction is compatible, the direct or indirect product of most avirulence genes may contribute to the pathogen's virulence (Dangl, 1994). It has been found that a gene cluster called hrp (hypersensitive response and pathogenicity) determines the expression of avirulence genes (Huynh *et al.,* 1989) in a wide range of plant pathogenic bacteria. The gene cluster controls secretion and transport of effector proteins outside of bacterial cells since at least eight of the known 20 hrp genes show remarkable homology to proteins involved either in protein secretion pathways (Van Gijsegem *et al.,* 1993; Bogdanove *et al.,* 1996) or in flagellum biogenesis (Rosqvist *et al.,* 1995) of animal pathogenic bacteria. The hrp gene cluster itself encodes and controls the secretion of a glycine-rich protein called harpin (HrpZ from *P. syringae*, HrpN from *E. amylovora*, and PopA from *R. solanacearum*), which is also able to elicit the HR when infiltrated into plant leaves (He *et al.,* 1993). Though the precise role of harpin is not well understood, some results indicate that it may have potential for some pathogen control in transgenic plants. When PopA was expressed in tobacco under the control of a pathogen-inducible promoter, it rapidly accumulated in the infection sites, which resulted in a localized HR (Belbahri *et al.,* 2001). This result has now been confirmed by constitutive expression of PopA in *R. solanacearum*, which rendered the recombinant bacteria avirulent on tobacco leaves and roots but not in xylems (Kanda *et al.,* 2003). Taken together these observations indicate that an early recognition of PopA by tobacco plants may result in effective protection in the apoplast. However, potential negative effects of the induced HR on plant health should be first addressed by extensive field testings. During the past two decades, about 50 bacterial effector genes, mainly from species of *Pseudomonas* and *Xanthomonas*, have been cloned. More recently, several plant disease-resistance genes corresponding to effector genes have also been cloned and characterized at a molecular level. However, little is known about the gene products, their function and the mechanism by which they interact during incompatible plant-bacterial interactions. So far only one effector gene product of *P. syringae* has been shown to function as an enzyme that is involved in the synthesis of syringolides, which are able to elicit the hypersensitive response (Keen *et al.,* 1990). The indirect evidence that effector proteins may be delivered into plant cells via the Hrp system indicates that they could be involved in bacterial pathogenesis rather than merely controlling genotype-specific elicitation of the HR (Collmer, 1996; Gabriel, 1999). Indeed, two effectors, AvrBs3 from *X. campestris* pv. *vesicatoria* (Szurek *et al.,* 2002) and PopP2 from *R. solanacearum* (Deslandes *et al.,* 2003) have recently been shown to be localized in plant nuclei by immunolocalization and monitoring GFP fusions, respectively. In the case of PopP2, the corresponding RRS1 resistance gene was colocalized in the nucleus only in the presence of PopP2, which is consistent with the model that R genes might act as a guard to interact with the targets of effector proteins rather than directly interacting with the effectors (van der Biezen and Jones, 1998). Natori (1994) reported about the function of antimicrobial proteins in insects.

6.8.7 Plant Resistance Genes

At least 11 plant resistance genes conferring resistance to bacterial diseases have been cloned to date (Martin *et al.,* 1993; Bent *et al.,* 1994; Mindrinos *et al.,* 1994; Grant *et al.,* 1995; Song *et al.,* 1995; Salmeron *et al.,* 1996; Yoshimura *et al.,* 1998; Warren *et al.,* 1998; Gassmann *et al.,* 1999; Tai *et al.,* 1999; Swiderski and Innes, 2001; Deslandes *et al.,* 2002; Lee *et al.,* 2002). Huang *et al.* (2007) reported the expression of a foreign gene to enhance plant disease resistance to bacterial pathogens is favourable strategy. Some of the examples of foreign genes are mentioned in Table 6.13.

Table 6.13 Stuctural protein derived from foreign gene responsible for production of Induced Systemic Resistance against bacterial wilt in tomato

Foreign Gene	*Structure Protein*	*References*
A leucine-rich of *Arabidopsis thaliana*	RPS2	Bent *et al.*, 1994
Sweet Pepper	Ferre Doxin-I Protein (PFLP)	Huang *et al.*, 2007
Tomato	Ferredoxin ferredoxin-NADP reductase	Green *et al.*, 1991
Antimicrobial Protein	Thionins	Chan *et al.*, 2005
Tomato Stress-Responsive Factor	TSRF1	Zhang *et al.*, 2004
Transcription factor binding protein	NPR1 Expression of 2-1,3-glucanase genes	Chern *et al.*, 2005a,b, Fitzgerald *et al.*, 2004; Lin *et al.*, 2004 Ishihara *et al.*, 2012
Magainin Analog Peptide	Myp30	Lin *et al.*, 2004.

Contd...

Contd...

Sweet pepper	Fd-I protein	Dayakar *et al.*, 2003
Active Oxygen Species (AOS)		Dayakar *et al.*, 2003
Hypersensitive Reaction (HR)		Dayakar *et al.*, 2003

Ger *et al.* (2002) reported that hypersensitive response-assisting protein (HRAP) has been previously reported as an amphipathic plant protein isolated from sweet pepper that intensifies the harpin (Pss)-mediated hypersensitive response (HR). The hrap gene has no appreciable similarity to any other known sequences, and its activity can be rapidly induced by incompatible pathogen infection. Li *et al.* (2002) also reported that a tridecapeptide possesses both anti-microbial and protease-inhibitory activities. Lin *et al.* (2003) reported that acyl-homoserine lactone acylase from *Ralstonia* strain XJ12B represents a novel and potent class of quorum-quenching enzymes. Mäe *et al.* (2001) reported that transgenic plants produced the bacterial pheromone N-acyl-homoserine lactone exhibit enhanced resistance to the bacterial phytopathogen *Erwinia carotovora*. Lottmann *et al.* (1999) reported about the influence of transgenic T4- lysozyme-producing potato plants on potentially beneficial plant-associated bacteria. Further Lottmann *et al.* (2000) have reported about establishment of introduced antagonistic bacteria in the rhizosphere of transgenic potatoes and their effect on the bacterial community. Lottmann and Berg (2001) reported about phenotypic and genotypic characterization of antagonistic bacteria associated with roots of transgenic and non-transgenic potato plants. Mc Nellis *et al.* (1998) reported glucocorticoid-inducible expression of a bacterial avirulence gene in transgenic arabidopsis induces hypersensitive cell death. The bacterial wilt caused by *R. solanacearum* can be alleviated by foreign proteins such as anti-microbial protein shown in Table 6.12. Ferredoxin-I (Fd-I) is an important member of photosynthesis proteins that transfer electrons from photosystem I (PSI) to the enzyme Fd: NADP+ reductase for photoreduction of NADP+ via the linear electron flow (Green *et al.*, 1991). This reaction is involved in several fundamental metabolic processes (Meyer *et al.*, 1999). Recently, an Fd-I protein isolated from sweet pepper was identified that shares 84% identity in nucleotide sequence and 89.58% similarity in protein sequence compared with Fd-I of tomato (Dayakar *et al.*, 2003). Overexpression of Fd-I in transgenic plants led to intensified production of active oxygen species (AOS) and activation of the hypersensitive reaction, when plants were challenged with bacterial pathogens or elicitors of defense responses (Dayakar *et al.*, 2003, Huang, *et al.*, 2004). The bacterium infects root tissue and accumulates in stem tissue of hosts to cause serious wilting symptoms (Genin and Boucher, 2004). To create a tomato variety resistant to *R. solanacearum* using PFLP, the protein will need to be expressed in root and stem tissues of transgenic tomato plants. However, PFLP is a member of Fd-I protein that only exist in the green tissue of plants. This tissue-specific expression of Fd-I may be due to two reasons. First, the original promoter of Fd-I is light inducible. Second, a certain specific sequence located in the 5'UTR is sensitive to degradation of protease that might cause instability of *fd-I* gene mRNA under dark conductions. To circumvent these problems, the original promoter of the transgene was replaced with CaMV 35S promoter and its 5′UTR was deleted. This modification may be beneficial for expression of PFLP in the root and stem tissue of transgenic tomato. The selection of transgenic lines with highly expressed levels of PFLP. Challenge of these lines with *R. solanacearum* resulted in a decrease in disease symptom production, indicating their potential for resistance to disease caused by these bacterial pathogens. Huang *et al.* (2007) reported expression of a foreign gene to enhance plant disease resistance to bacterial pathogens is a favorable strategy. It has been demonstrated that expressing sweet pepper ferredoxin-I protein (PFLP) in transgenic plants can enhance disease resistance to bacterial pathogens that infect leaf tissue. In this study, PFLP was applied to protect tomato (*L. esculentum* cv. cherry Cln1558a) from the root-infecting pathogen, *R. solanacearum*. Independent *R. solanacearum* resistant T_1 lines were selected and bred to produce homozygous T_2 generations. Selected T2 transgenic lines 24-18-7 and 26-2-1a, which showed high expression levels of PFLP in root tissue, were resistant to disease caused by *R. solanacearum*. In contrast, the transgenic line 23-17-1b and non-transgenic tomato, which showed low expression levels of PFLP in root tissue, were not resistant to *R. solanacearum* infection. The expansion of *R. solanacearum* populations in stem tissue of transgenic tomato line 24-18-7 was limited compared with the nontransgenic tomato Cln1558a. Using a detached leaf assay, transgenic line 24-18-7 was also resistant to maceration caused by *E. carotovora* subsp. *carotovora*; however, resistance to *E. carotovora* subsp. *carotovora* was less apparent in transgenic lines 26- 2-1a and 23-17-1b. These results demonstrate that PFLP is able to enhance disease resistance at different levels to bacterial pathogens in individual tissues of transgenic tomato. Pieterse *et al.* (1996) reported that systemic resistance in arabidopsis induced by biocontrol bacteria is independent of salicylic acid accumulation and pathogenesis-related gene expression.

6.9 CHEMICAL METHODS

The control of bacterial wilt has proved to be very difficult. Chemical control is nearly impossible to apply. Chemical control means any natural and synthetic compounds that could be employed to combat the threats. This method works by directly eliminating disease-causing organisms or curbing their spread; however, it has been shown to have too broad an effect, typically, to be good for the local ecosystem. From an economic standpoint, all but the simplest natural additives may disqualify a product from "organic" status; potentially reducing the value of the yield. Champoiseau *et al.* (2010) reported that direct control of bacterial wilt diseases caused by *R. solanacearum* in the field is difficult due to wide host range, strain variation, and lack of adequate chemical treatments. Bacterial wilt control through chemical has been summarized in Table 6.13. Several chemicals *viz*; Actigard (acibenzolar-*S*-ethyl) and phosphorous acid (Anith *et al.*, 2004, Pradhanang *et al.*, 2005; Norman *et al.*, 2006; Ji *et al.*, 2007) have been used for controls; however it required validation on a larger scale and against R3b2. Low doses of chlorine or per acetic acid have been used in irrigation or effluent water which effectively reduced the pathogen. However, the bacterial wilt of solanaceous vegetable can also be controlled by various means as follows:

6.9.1 Fumigants

Soil fumigants showed either slight or no effects (Murakoshi and Takahashi, 1984). Soil fumigation with vapam, methyl bromide, or chloropicrin and formalin (Fig. 6.17c) is of limited efficacy and utility. Chemical control by soil fumigation (chloropicrin) or application of phosphorous acid is expensive and soil fumigation has been reported to achieve limited success if combined with other control methods (Anonymous, 2008). Several reports are available on use of soil fumigats (Anon, 1960; Domica, 1964; Patil *et al.*, 1973) and reported that the bacterial wilt of tomato was successfully prevented by soil fumigation with DD at 200 l/ha. Whereas Cox (1960) reported that soil treatment three weeks before planting with chlopicrin at 160 or 320 lb/acre, or Mylone 85 W at 250lb/acre or ammonium thiocynate at 250lb/acre have given good control. Chlopicrin gave excellent control (Cox, 1961) and yielded 15% higher than control of those treated with sulphur and were slightly higher than Mylone treated plot. Ishikawa *et al.* (1996) reported that Nemex was effective for control of bacterial wilt in tomato. Mallikarjun *et al.* (2008) reported soil treatment with chloropicrin, methyl bromide or mixture of both was found effective in retarding the wilt development. Iizuka *et al.* (1988) have recommended that the most effective soil fumigants fungicide such as chloropicrin and use of resistant rootstock to bacterial wilt. Enfinger and McCarter (1976) stated that application of Chlopicrin either covered or scaled with water was most effective in reducing the disease development; however, all Vapam treatments were less effective than chlopicrin, vorlex and methyl bromide. Chemical control by soil fumigation (chloropicrin) or application of phosphorous acid is also expensive to apply. Soil fumigation has been reported to achieve limited success if combined with other control methods. Chemical control should be integrated with other methods to reduce selection pressure for pathogen resistance (Grimault and Prior, 1994; Champoiseau and Momol, 2009) in potato.

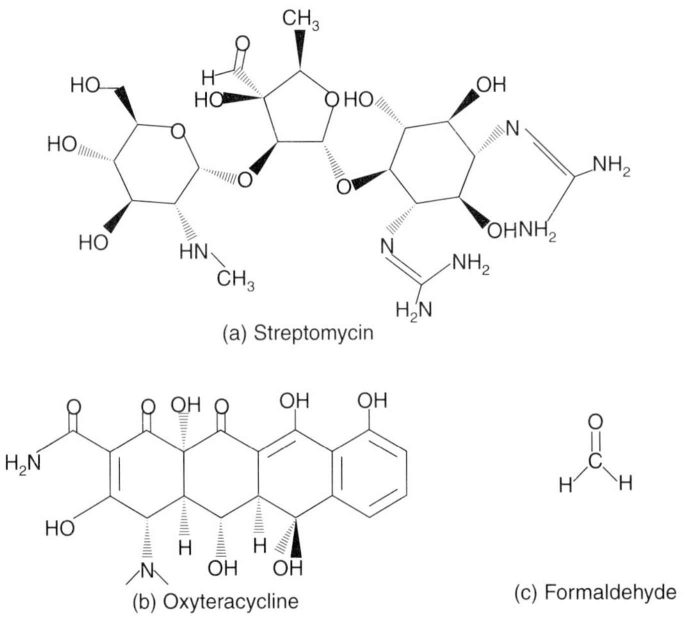

Fig. 6.17 Structural formula of antibiotics (a) Streptomycin, (b) Oxyteracycline and (c) Formaldehyde

6.9.2 Antibiotics

Nearly 40 antibiotics were screened for plant disease control (Goodman, 1959) stated that of the screened compounds, fewer than 10 were used commercially and only streptomycin and oxytetracycline have significant usage. Falla *et al.* (1996) reported that peptides have many exciting properties that would seem to recommend them as novel antibiotics. It is some clinical studies (with α-helical peptides). Habashy *et al.* (1993) studied the sensitivity of the pathogen to antibiotics and also see the pathogenesis by streptomycin resistant mutant in potato plants to cause wilt diasease.

6.9.2.1 *Streptomycin*

Streptomycin application increased the incidence of bacterial wilt in potato in Egypt (Farag *et al.,* 1982, 1986). Streptomycin (Fig. 6.17a) is an aminoglycoside antibiotic. The primary mechanism of action of streptomycin is binding irreversibly to bacterial ribosomes and thereby inhibiting protein synthesis. In the U.S., streptomycin use is registered on 12 plant species, but the primary uses are on apple, pear, and related ornamental trees for the control of fire blight caused by *Erwinia amylovora.* Minor uses include floriculture, seed treatment, and on certain vegetable seedlings in the greenhouse and/or field. At high concentrations, streptomycin can be phytotoxic to plants; thus it is applied to the surface of plants and not injected.

6.9.2.2 *Oxytetracycline*

Oxytetracycline (Fig. 6.17b) is a member of the tetracycline antibiotics. Tetracyclines inhibit protein synthesis by binding reversibly to bacterial ribosomes. This antibiotic is used primarily on peach, nectarine, and pear. Streptomycin and Oxytetracycline have been used on crop plants for the past 45 years and 25 years, respectively, without reports of adverse effects on humans. The efficacy of these antibiotics for control of plant diseases has been diminished in some regions due to the emergence of antibiotic-resistant strains of pathogens. However, until effective and economic alternatives become available, antibiotics will remain important tools for the management of some of the most devastating plant diseases. The genes *strA* and *strB* usually reside on plasmid-borne transposons, which are mobile pieces of DNA that are generally not required for basic bacterial functions but that frequently carry antibiotic resistance genes. Transposon Tn*5393* carries *strA* and *strB* in streptomycin-resistant strains of *Erwinia, Pseudomonas*, and *Xanthomonas* on plants (Jones *et al.*, 1991; Burr *et al*., 1993; Huang and Burr, 1999; Jones and Schnabel, 2000). *StrA* and *strB* also have been identified in at least 17 environmental and clinical bacteria populating diverse niches (Sundin and Bender1996a, b), and in most cases, the genes are on plasmids. Champoiseau and Momol (2009) reported that reported that bacterial wilt of tomato is difficult to control, and no single strategy has shown 100% efficiency in control of the disease so far. Bactericides (copper) and antibiotics (streptomycin, ampicillin, tetracycline and penicillin) have shown little efficiency on suppression of *R. solanacearum* these methods should be used in combination for best management of bacterial wilt of tomato. Most of following recommendations are applicable to race 1 strain of *R. solanacearum* only. Seedling dip with streptocycline avoids early invasion and infection by the pathogen through wounds formed during transplanting. Bacterinol-100 can be used as dry seed dresser, for nursery spray and field application. Streptomycin sulphate or oxytetracycline when sprayed at 200 ppm at 7 days interval provides good control. Sharma and Kumar (2009b) reported that at initial stage the *R. solancearum* population was (85.0×10^4 cfu/g soil) but the population decreased later on to 60.66×10^4 cfu/g soil at 90 days in seedling dip treatment whereas in spray treatment it was 77.49×10^4 cfu/g soil at initial stage and it reduced to 55.83 × 10^4 cfu/g soil at 90 days. These data indicate reduction in *Ralstonia* population by 28.64 and 27.95% respectively. Seedling dip treatment of Streptocycline @500 ppm resulted reduction in the apparent infection rate of – .0497 unit/day and spraying @500 ppm resulted in reduction of apparent infection rate to –0.1776 unit/day). The trend was found to be linear in reduction in both the doses (Fig. 6.18) whereas the infection rate in control was higher (–0.0409 unit/day). Rao (1990) reported that streptocycline was most effective in reducing the *P. solanacearum* in soil.

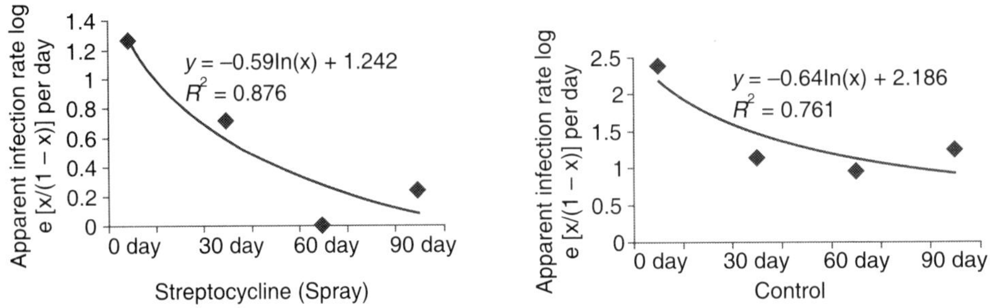

Fig. 6.18 Logarithmic reduction of *R. solanacearum* population under (a) Streptocycline and (b) Control

Partyka *et al.* (1973) have reported that the disease was controlled by soaking plant in Cu as Kocide 101 for 60 min. Pegg (1975) and Pegg Crenshaw (1976) used ethylene for the control of *P. solanacearum*. Hanudin and Machmud (1994) used Terlai, an antibiotics and *P. fluorescens* resulted in some control either singly or in combination. Ishikawa *et al.* (1996) reported that application of validamycin A (VM-A) — the active ingredient of validacin @250 mg/ml delayed the appearance of disease symptom in tomato when injected into plant. At higher concentration *i.e.* @500 mg/ml the bacterial population reduced to (2.13 × 10^9 cfu/g fresh weight) against non-treated (3.84 × 10^{10} cfu/g fresh weight). Rao and Kishun (1986) reported that pre-planting soil treatment with streptocycline resulted wilt disease 70% against control (83%).

6.9.3 Amino Acids

The effect of four amino acids, unutilizable by *R. solanacearum*, on tomato bacterial wilt was tested in three Japanese soils and reported (Posas *et al.*, 2010) that 2.5 mg g^{-1} soil of lysine and serine suppressed tobacco bacterial wilt, but that of tyrosine and valine did not. The number of the pathogen in non-rhizosphere soil, rhizosphere soil and the rhizoplane was markedly lower in the lysine and serine treatments than in the tyrosine and valine treatments, while the opposite result was obtained for the total bacterial population. Substrate-induced respiration analysis revealed that bacteria became more predominant in the amino acid treatments, especially in lysine. There were no apparent correlations between the microbial activities in soil and the disease index. PCR-DGGE targeting the 16S rRNA gene of the rhizoplane samples from lysine-added Nagoya, Fuchu and Iwate soils revealed 10, 1 and 2 unique bands, respectively, indicating shifts in bacterial community. DGGE patterns in the suppressive treatments belonged to the same cluster while the not suppressive treatments clustered differently. The lower disease incidence of TBW following lysine treatment is likely related to a specific bacterial community in the rhizoplane that developed on the addition of lysine.

Table 6.14 Chemicals used in management of bacterial wilt of solanaceous crops

Chemical Group	*Crop/Chemicals*	*Rate of application*	*Mode of application*	*Varieties*	*Reduction (%)*	*References*
Organic/ Inorganic	**Tomato**					
	Calcium chloride	0.1%	Spray		32.5	Gopal Krishnan and Krishna, 2010
		0.25%	Spray		33.0	
	Lime (Calcium Carbonate)		Soil appl.	Pongalgreen	38.0	Ramesh *et al.*, 2009
		25 q/ha	Soil appl.	Pusa Ruby	18.7	Sharma and Kumar 2000, 2009b
	Gypsum (Calcium Sulfate)	125, 187.5, 250.0, 375.0, 500kg/ha	Soil appl.	Malamar	26	Ajjappalara *et al.*, 2008
		10g and 5g/litre	Spray			Mazumder, 1998
		500 ppm	Seedlig dip			Sharma and Kumar, 2009b
		2500ppm for 60 minute	Seed soak			Partyka *et al.*, 1973
	Ammonium thiocynate	250 lb/acre	Soil appl.			Cox, 1960, 1961
	Urea	1000 lb/acre	Soil appl.			
	Urea + Ammonium thiocynate	1000lb/acre + 125 lb/acre	Soil appl.			
	Urea + Lime	200 + 5000kg/ha	Soil appl.			Michel *et al.*, 1997
	Sulphur		Soil appl.			Cox, 1960, 1961
Antibiotics/ Nematicides	Carbamate-Nematicide		Soil appl.			Fogoonee, 1978
	Mylone	250 lb/Acre	Soil appl.			Cox, 1961
	Chloropicrin	160 or320 lb/acre	Soil appl.			
		326 L/ha	Soil appl.			Enfinger *et al.*, 1979

Contd...

Contd...

	DD	200 L/ha	Soil appl.			Anon, 1960; Domika, 1964
	Dichloropropane		Soil appl.			Enfinger and Mc Carter, 1976; Enfinger *et al.*, 1979
	Methyl isothiocynate	944 ml/100 cft	Soil appl.	102-1, 103-1-4, 135-1-2		Sonoda, 1979
	Nemex (Chloropicrin + DD)		Spray			Ishikawa *et al.*, 1996
	Vapam		Soil appl.			Enfinger and Mc Carter, 1976
	Vapum		Soil appl.		66.0	Rao and Kishun, 1986
	Agrimycine	200 ppm	Spray			
	Bactrinol-100	2000 ppm	Spray			Rao, 1990
		200 ppm	Seed & spray			Champoiseau and Momol, 2009
	Bacterocin		Spray			Ren *et al.*, 1988
	CCC (Chlormequate)	2000 ppm	Spray			Power, 1983
	Kocide 101	2500 ppm	Seed soak			Patyka *et al.*, 1973
	Oxytetracycline or Streptomycin	200 ppm at 4 to 7 days intervals	Spray	30.0		Gunawan, 1988d
	Streptomycin	2500 ppm	Seedilng soak			Patyka *et al.*, 1973; Perez, 1961
	Streptocycline		Soil appl.		70.0	Rao and Kishun, 1986; Rao, 1990; Hussain and Bora, 2008
		2000 ppm at 7 days interval	Spray			Champoiseau *et al.*, 2009
	Validamycin A	250 mg/ml	Plant Injection	Solar Set	12.0	Ishikawa *et al.*, 1996

Contd...

Contd...

	Bleaching powder		5 or 10 gr/L	Spray		62%	Mazumder, 1998
			0.025%	Spray		30.0	GopalKrishnan and Krishna, 2010-11
			30 and 45 kg/ha	Soil appl.	Pusa Ruby		Sharma and Kumar, 2000, 2009b
			15 kg/ha	Soil appl.			Mallikarjun *et al.*, 2008; Champoiseau and Momol, 2009
	Formalin		10 ml /Litre drench	Soil appl.			Rao and Kishun, 1986; Sharma and Kumar, 2000
Sugar	Sugar	YURif43	5 mg/g	Spray			Posas *et al.*, 2007
Amino Acid	Anino Acids	YURif43	0.5 to 5 mg/g + inoculation	Soil appl.			
	Lycine		2.5 mg/g soil				Posas *et al.*, 2010
Hormones	IPA (3-indol, Propionic Acid)		5-10 mg/ml Spray	Spray			Dogo *et al.*, 1997; Matsuda *et al.*, 1998
Potato							
Organic/ Inorganic	Bleaching Powder		50 g/m^2	Soil appl.		68.4%	Verma and Shekhawat, 1991
			12 kg + fertilizer followed by earthing up.	Soil appl.			Arora, 2012

6.9.4 Use of 3-Indole Propionic Acid (IPA)

Growth-hormone such as 3-indole propionic acid (5-10 mg/ml) controlled the disease (Dogo *et al.*, 1997) and no adverse effect of plant was also reported (Matsuda *et al.,* 1998). Matsuda *et al.* (1998) reported that 3-indole propionic acid (IPA)-related compounds having a benzo [*b*] thiophene or an indazole ring and derivatives having various substituents in the propionic acid moieties were tested for their antibacterial activity against *R. solanacearum.* Substitution of the indole ring for other aromatic rings resulted in lowered activity, whereas addition of a methyl or a trifluoromethyl group to the propionic acid moiety had little effect. Of the derivatives, 3-(3-indolyl) butanoic acid (3-IBA) was as active as IPA, exhibiting a 10-fold higher activity with the *S* configuration than with the *R* configuration. In contrast with the strong phytotoxicity of IPA, 3-IBA was able to suppress bacterial wilt without affecting the growth of tomato plants.

7
Conclusion and Future Research

- The bacterium *Ralstonia solanacearum* is a soil-borne bacterium, which belongs to the betaproteobacteria under phylum protobacteria. It is responsible for bacterial wilt on more than 450 plant species from 54 botanical families, including important crops such as potato, tomato, eggplant, pepper, tobacco, banana, ginger and mulberry. In fact, bacterial wilt is considered the single most destructive bacterial plant disease because of its extreme aggressiveness, wide geographic distribution, and unusually broad host range.
- *Ralstonia solanacearum* is a Gram-negative motile rod. The organism grows aerobically and does not form endospores. Cells are 0.5-0.7 × 1.5-2.0 μm and are non-encapsulated. It is a catalase positive, oxidase positive, and reduces nitrates. The pathogen does not hydrolyze starch and does not readily degrade gelatin. In broth culture, the organism is inhibited by concentrations of NaCl greater than 2%.
- *Ralstonia solanacearum* inhabits the vascular tissue of its hosts. The bacterium normally invades plant roots from the soil through wounds or natural openings where secondary roots emerge, colonizes the intercellular space of the root cortex and vascular parenchyma, and eventually enters the xylem vessels and spreads up into the stem and leaves, where the pathogen cell density commonly surpasses 10^9 cfu/g of host tissue. After *R. solanacearum* has colonized the xylem, large numbers of bacterial cells are shed from roots, providing a pathway for bacteria to return to the soil and initiate new infections.
- Affected plants suffer chlorosis, stunting, wilting, and usually die rapidly. In potato, rapid wilting of leaves and stems occurs and later stages plants fail to recover, become yellow and brown necrotic and die. Tubers show bacterial ooze on the eyes and stolon end attachment and browning of vascular ring of tuber is important symptoms. In tomatoes, brinjal and chillies, leaves wilt, usually at the warmest time of day and whole plant wilting is observed, whereas less favourable conditions, the disease develop less rapidly, stunting of plants and large number of adventitious roots on the stem are produced. "Water streaming test" is presumptive diagnostic value in the field.
- Isolation from symptomatic material can easily be performed using YPGA non-selective medium or Kelman's tetrazolium medium. In some cases when secondary infections are present, isolation on selective media (TTC medium, SMSA medium) for isolation from latently infected material or difficult substrates like soil, waste, or surface water are used.

- Detection of latent infection is by performing an immuno-fluorescence test and/or selective plating on SMSA medium eventually combined with optional PCR assays, ELISA or fluorescent *in situ* hybridization tests which can be performed following enrichment procedures for added sensitivity. Quick diagnosis of bacterial wilt, lateral flow device and immuno strips are commercially available.
- RFLP and PCR based techniques using a specific primer are developed to detect *R. solanacearum* from seeds, tubers and asymptomatic planting materials. Various genes (16S r RNA, hrc, hrpB) of bacterial are used to developed specific primers. Molecular characterization and fingerprinting of bacteria are also done by using AFLP, EFLP, RAPD, Rep- PCR and multi-locus sequence typing.
- *Ralstonia solanacearum* can spread through soil, irrigation water, seeds, infected seed potato and other vegetative propagating material. The bacteria from tomato and brinjal survive up to 2- 3 years in fallow soil and up to unlimited period in cultivated land. Seed-borne nature of the pathogen in eggplant has also been reported. Once inside the host, the bacterium has an affinity for the vascular system where it multiplies rapidly filling the xylem vessels with bacterial cells and slime. Wilt occurs within 2-5 days after infection depending upon the host susceptibility, temperature and virulence of the pathogen. As the host wilts, the pathogen invades parenchyma cells in the pith and cortex and pockets filled with masses of bacteria develop around the vascular bundles. Under favourable conditions the bacterium may move through the cortex and exudes on the surface of the stem. It is released into the soil from the roots of the infected plants and from decaying host material.
- *Ralstonia solanacearum* is considered a species complex-α heterogeneous group of related but genetically distinct strains. DNA-DNA hybridization studies have revealed that the identity between *R. solanacearum* genomes is often less than the 70% threshold level commonly expected within a bacterial species. Gillings and Fahy first used the term "species complex" to describe the high genetic variation between isolates, and Taghavi *et al.* expanded the concept of the *R. solanacearum* species complex by including two closely related species from Indonesia, *R. syzygii* (a pathogen of clove trees) and the agent of blood disease of banana, known as the BDB. *R. solanacearum* species complex strains are highly competent for genetic exchange *in planta*. Two different systems, race and biovar, have been widely used to differentiate *R. solanacearum* strains. *R. solanacearum is* subdivided into five "races" based loosely on host range. Race 1 strains attack tobacco, many other solanaceous crops and many hosts in other plant families; race 2 strains are limited to musaceous species including *Heliconia* spp. and triploid banana; race 3 strains primarily attack potato (called potato race) and has lower temperature optimum (27°C), biovar 2A or RFLP group 26 with a worldwide distribution; race 4 strains are particularly virulent on ginger; race 5 strains infect and cause disease on mulberry tree, and are only found in China.
- Based on their different abilities to utilize and oxidize several disaccharides (cellobiose, lactose, and maltose) and hexose alcohols (dulcitol, mannitol, and sorbitol), *R. solanacearum* strains were originally divided into five biovars. Biovar 1 strains metabolize none of them; biovar 2 strains only metabolize disaccharides; biovar 3

strains metabolize all of them; biovar 4 strains metabolize only hexose alcohols; biovar 5 strains metabolize all of them except dulcitol and sorbitol. Later, a new group of *R. solanacearum* isolates is differentiated from biovar 2 using ribose and trehalose and named biovar 2-T or biovar N2 and the original biovar 2 strains are now referred to as 2-A. Except for biovar 2-A, which almost always corresponds to race 3, and biovar 5, which is identical to race 5, there is no correlation between biovars and races.

- A new phylogenetic classification system consisting of four phylotypes is formed based on phylogenetic analysis of sequence data from the 16S-26S internal transcribed spacer region, the *egl* gene, the *hrp*B gene, and the *mutS* gene. The phylotypes correlate with the geographical origin of the strains as phylotype I includes strains originating primarily from Asia, phylotype II from America, phylotype III from Africa and surrounding islands in the Indian Ocean, and phylotype IV from Indonesia. Each phylotype is further divided into sequevars, which are defined as isolates with less than 1% nucleotide variation in the endoglucanase *(egl)* locus.
- The extensive genetic diversity of *R. solanacearum* strains responsible for the various bacterial wilt diseases has in recent years led to the concept of an *R. solanacearum* species complex. Genome sequencing of more than 10 strains representative of the main phylogenetic groups has broadened our knowledge of the evolution and speciation of this pathogen and led to the identification of novel virulence-associated functions.
- Many factors contribute to virulence and pathogenicity of *R. solanacearum* strains, including a heterogeneous polymer of N-acetylated extracellular polysaccharide I (EPS I), the type III secretion system (T3SS), flagella-driven swimming motility and type IV pili-driven twitching motility, cell-wall-degrading enzymes and type II secretion sytem (T2SS). The functions have been hypothesized for EPS is physically blocking the vascular system and thereby altering water movement and is protecting *R. solanacearum* from plant antimicrobial defenses by cloaking bacterial surface features that could be recognized by hosts. In *R. solanacearum*, the T3SS is encoded by a 23-kb *hrp* gene cluster on the mega plasmid and is essential for pathogenicity. Both flagella-driven swimming motility and type IV pili-driven twitching motility contribute to the ecological fitness and virulence of *R. solanacearum*. While swimming motility contributes to virulence only in the early stage of host colonization and invasion and the type IV pili and twitching motility are important for several stages of wilt disease development. *R. solanacearum* secretes several plant CWDEs via the T2SS that contribute to the pathogen's ability to cause wilt, although none is essential for disease.
- The key virulence and pathogenicity factors of *R. solanacearum* are controlled by a complex regulatory network that responds to multiple signals. The central player of this regulatory network is the transcriptional regulator PhcA, involved in the Phc cell density sensing system. PhcA positively regulates the production of EPS I, Pme and Eglexoproteins, and an acyl-homoserine lactone (AHL) quorum sensing system called SolI/R. PhcA also represses the expression of genes involved in motility, T3SS, polygalacturonase and siderophore production. The levels of active PhcA protein are controlled by a diffusible endogenous signal molecule, 3-hydroxypalmitic acid methyl

ester (3-OH PAME). When extracellular 3-OH PAME accumulates above threshold concentrations (*i.e.*, at high cell density in a confined space, such as the plant vascular system), PhcA is expressed, resulting in activation or repression of its various target genes.

- The disease is more severe in sandy loam soil than red loam soil. Infection always occurs through wounds caused by transplanting, cultural operations or nematode infection. Root knot nematodes may predispose tomato plants to bacterial wilt by interfering with nutrition of plants and root wounding to provide entry to the pathogen. *Meloidogyne incognita* and *R. solanacearum* complex increase the severity bacterial wilt in brinjal. Optimum temperature for the growth of the bacterium is 35-37°C. Disease does not develop at 15°C and most favourable temperature for rapid development of the disease takes place at 37°C. The disease mostly develops when mean temperature is above 20°C. No infection occurs at or below 50 per cent soil moisture. The pathogen can grow over a wide range of pH (6.2 – 7.4). However, alkaline soils are generally unfavorable for disease development. Organic manures promote the activity of bacterium, while inorganic fertilizers decrease its activity. The pathogen infects over 450 plant species under 54 families including grasses, herbs, shrubs and trees and serve as a reservoir for the pathogen.
- Soil microfloras grown in the rhizosphere are ideal for use as biocontrol agents, which provide the frontline defense for roots against any attack by the pathogens. Several biological control agents like Actinomyces, *Bacillus* spp., *B. subtilis*, *B. licheniformis B. cereus*, *Enterobacter* sp., *Stenotrophomonas maltophilia*, *Pseudomonas fluorescens*, *P. aeruginosa* ,*Pseudomonas* sp., a Hrp-mutant of *R. solanacearum*, Bacteriophages ($PT2_1$), vesicular arbuscular mycorrhiza, *T. viride*, *T. harzianum*, *B. amyloliquefaciens*, *T. pseudokoningii* and *Candida ethanolica* have been found effective to delay the wilt development and reduce disease incidence in solanaceous crops.
- Two types of resistance exist in tomato; one is polygenic, originating from *L. esculentum* var. *ceresiformae* and needs to be sustained by additional crop management measures. The second type is a dominant resistance present in cv. Hawaii, which is of simpler heredity, like the monogenic resistance described in Hawaii 7998. Most of the resistant sources against *P. solanacearum* were found in *L. pimpinellifolium* or ecotypes of *L. esculentum* var. *ceresiformae*, which are wild tomato germplasm with small fruits and resistant to elevated temperatures. Graft of susceptible cultivars of tomato on resistant rootstocks belonging to *Solanum* spp. has been reported to be effective in management of this disease. The disease was controlled by grafting susceptible tomato cultivars on to resistant eggplant (Pusa Purple Cluster) rootstocks.
- Bio-efficacy of microbial antagonists is further improved to control bacterial wilt diseases of solanaceous crops by using in combination of chemicals. Application of actigaurd (Acibenzolar-s- methyl) in combination with *P. putida* 89B61 or Bioyield reduced bacterial wilt incidence compared with the untreated control of tomato.
- Karanj cake along with bleaching powder was more effective in reducing the bacterial growth at flowering stage thereby increasing the tomato yield. The combination of seed treatment with streptocycline sulphate (1 g/ kg seed), bleaching powder application in

the soil (20 kg / ha), root dip treatment (streptocycline sulphate (1 g) + copper sulphate (1.75 g) in liter of water) on Arka Keshav (resistant variety) recorded the highest yield with minimum disease indigence in brinjal.

- The best strategy for controlling bacterial wilt in the field consists primarily of phytosanitation and cultural practices, where bacterial wilt of potato is endemic or in locations where *R. solanacearum* is present but not yet established, these methods can be effective under some conditions. The cultural practices can reduce disease, by planting healthy (tested) seed potatoes, early detection and accurate identification of the pathogen, quarantine measures on infected fields and farms, sufficient crop rotation, use of cover crops and other measures to reduce the impact of weed hosts and volunteer plants (and in some cases of nematodes), avoidance or testing and treatment of surface water for irrigation. Integrated application of these strategies will help for successful management of diseases caused by *R. solanacearum*. Recently, the rootstock replacement through tube-grafting has been considered effective in controlling this disease.

Future Research

- *Ralstonia solanacearum* has been genome completely sequenced, and this milestone enhanced research progress in many direct and indirect ways. The present need to understand the complexity and diversity within the *R. solanacearum* species complex based on gene-specific and genomic sequences available for multiple strains from around the world.
- Rapid, reliable and more sensitive advanced diagnostic tool either serological or nucleic acid based and microarray technique to detect bacteria from asymptomatic plants, seeds, soil and irrigation water is required to develop.
- Develop diagnostic kit to differentiate races of *R. solanacearum*. Utilization of isothermal amplification techniques may be used to identification and diagnosis of the disease. Quantitative PCR is more expensive than the PCR, but it can be adopted for high consequences particularly in regulatory circumstances, because of their high sensitivity and specificity.
- To understand mechanism, how the morphological changes take place during storage of *R. solanacearum* in low temperature and how they can survive in the soil and water without presence of host.
- The major phylogenetic groups of the *R. solanacearum* species complex are now well established, there is need to make proposal to divide the *R. solanacearum* species complex into new species and subspecies.
- High-through put, next-generation sequencing methods will definitely expand the known genetic repertoire of *R. solanacearum* species complex strains. Most notably, we expect that examining environmental isolates (e.g., from soil or water sources) will promote understanding of the evolutionary dynamics of these strains in relation to survival and pathogenicity.
- Genome sequencing of multiple closely related strains with different host ranges (such as phylotype IIB strains that differ in their ability to wilt banana) will hopefully identify

consistent differences in genes encoding T3E associated with host preferences and other candidate host-specificity factors. These results will then provide testable hypotheses for future functional and mechanistic studies.

- Complete genome sequencing also opens the way to develop experimental evolution methods to unravel determinants of bacterial adaptation. For example, serial inoculation of legume plants with a *R. solanacearum* strain carrying a symbiotic plasmid from a nitrogen-fixing symbiont resulted in selection of variants able to reproduce and multiply intracellularly into root nodules.
- Identify molecular traits associated with the pathogen's adaptation to different environments and thereby lead to significant advances in our knowledge of the genes conferring selective advantages or adaptability in these conditions.
- The genetic resources now available will also provide the foundation upon which advanced transcriptomic studies can be pursued.
- It will be greatly modified once, we determine how extracellular and intracellular signals affect the true regulons of key regulators (PhcA, VsrAD, HrpG, HrpB, etc.) and how gene expression varies with time and location outside and within the plants. Knowing when and where many presumed pathogenicity factors (such as most T3E or T2SS substrates) are produced will help to decipher how they help to make *R. solanacearum* such a successful pathogen omits diverse hosts.
- Identify of resistance genes against this pathogen in the plants and incorporation of these genes in susceptible high yielding cultivars/variety through marker assisted selection and other high throughput biotechnological tool.
- There is a case of continuous application of antibiotics and copper fungicides lose their utility because of pathogen become resistant, often through acquisition of broad host range transmissible plasmids from other bacteria. It is required to overcome this problem by searching alternative source to use for control of the disease.
- The field now seems prepared to move forward with greater assurance toward fulfilling our dream that basic research on the pathogen can help to provide innovative tools and strategies to control *R. solanacearum* wilt diseases.
- More emphasis should be done on resistant rootstock replacement through tube-grafting, which proved effective in controlling this disease.

References

Aarons, S.R., Danesh, D. and Young, N.D. (1993). DNA genetic marker mapping of genes for bacterial wilt resistance in tomato In: Bacterial wilt ACIAR Proc. 45 G.L. Hartman and A.C. Hayward, eds. Watson Ferguson Co, Brisbane Pages 170-175.

Abdalla, M.Y., AL-Mihanna, A.A., AL-Rokibah, A. A. and Ibrahim, G.H. (1999). Tomato bacterial wilt in Saudi Arabia and the use of antagonistic bacteria for its control. *Ann. Agri. Sci. (Cairo),* **44**(2): 511-521.

Abdullah, H. (1980a). A disease of winged bean (*Psophocarpus tetragonolobus*) to *Pseudomonas solanacearum* in Malaysia. *Plant Dis.* **64**: 798-799.

Abdullah, H. (1980b). *Pseudomonas solanacearum* isolated from two weed hosts. *FAO Plant Protec. Bull.* **28**: 79-81.

Abdullah, H. (1982). Resistance of winged (*Psophocarpus tetragonolobus*) to *Pseudomonas solanacearum* in Malaysia. *Appl. Biol.* **11**: 35-39.

Abdullah, H. (1983). Record of additional new host of bacterial wilt pathogen (*Pseudomonas solanacearum*) in Malaysia. *Malaysian Appl. Biol.* **12**: 59-60.

Abdullah, H., Maene, L.M.J., and Naib, H. (1983). The effects of soil types and moisture levels on bacterialwilt disease of groundnut *(Arachishypogaea). Pertanika*, **6**: 26-31.

Abdullahi, M.S. and Bitrus Choji, K (2009). The origin and benefits of tomatoes as a home garden and commercial vegetable. *EJEAFChe*, **8** (11): 1156-1169.

Aberdeen, J.E.C. (1946). Experiments in the control of tomatoes in Southeastern Queensland. *Qd. J. Agric. Sci.* **3**: 87-95.

Acosta, C.J., Jilbert, J.C, and Quinon, V.L. (1964). Heritability of bacterial wilt in tomato. *Proc. Am. Soc. Hortic. Sci.* **84**: 455-462.

Adhikari, T.B. and Basnyat, R.C. (1998). Effect of crop rotation and cultivars resistance on bacterial wilt of tomato in Nepal. *Can. J. Plant Pathol.* **20**(3): 283-287.

Adsuar, J. (1961). Use of root dipping inoculation method in testing tomatoes for wilt resistance. *J. Agric. Univ. P.R.* **45**: 16-18.

Aeschbach, R., Loliger, J., Scott, B. C., Murcia, A., Butler, J., Halliwell, B., and Aruoma, O. I. (1994). Antioxidant actions of thymol, carvacrol, 6-gingerol, zingerone and hydroxytyrosol. *Food Chem. Toxicol.* **32:** 31-36.

Agati, J.A. (1949). Brown rot of solanaceous plant. *Plant Indust. Digest,* **12**: 31-34.

Aggarwal, P., Sood, A.K. and Kumar, P. (2008). Performance of tomato (*Lycopersicon esculentum*) cultivars against bacterial wilt. *Indian J. Agric. Sci.* **78**(4): 379-381.

Agrios, G.N., (2005). *Plant Pathology.* 5th edn., Academic Press, New York, USA. Pages 922.

Agrios, G.N. (2008). *Plant Pathology*. Amsterdam [u.a.], Elsevier Academic Press, pp 647-649

Agrios, G.N., (2005). *Plant Pathology*. 5th edn., Academic Press, New York, USA. Pages 922.

Ahmed, H. U. and Hossain, M. (1985). Crop disease survey and establishment of herbarium at BARI. final report of the project (1982-85), Plant Pathology Div. BARI, Joydebpur Gazipur. 107 pp.

Ahn, I. P., Park, K. S. and Kim, C. H. (2002). Rhizobacteria induced resistance perturbs viral disease progress and triggers defense-related gene expression. *Mol. Cells* **13:** 302-308.

Ajjappalavara, P.S., Dharmatti, P.R., Salimath, P.M., Patil, R.V., Patil, M.S. and Krishnaraj, P.U. (2008). Genetics of bacterial wilt resistance in brinjal karnataka. *J. Agric. Sci.* **21**: 424-427.

Akiba, F., Ribeiro, R., De, L.D., Sudo, S., Robbs, C.F. and Kimura, O. (1972). The elevation of introduced tomato strain for susceptibility to Brazilian isolates of *Pseudomonas solanacearum* wilt. *Arquives do Instituto Biologic,* **9** (4): 243-247.

Aldon, D., Gueneron, M., Brito, B. *et al.* (2000). Molecular dialogue between *Ralstonia solanacearum* and plant cells. *Mol. Plant-Microbe Interact.* **2**: 297-302.

Aldrick, S.J. (1971). Bacterial wilt (*Pseudomonas solanacearum*) of *Stylosanthes humilis* in Northern Territories. *Tropical Grassland,* **5**: 23-61.

Algam S.A., Xie, G.L., Li, B., Coosemans, J. (2004). Comparative performance of *Bacillus* spp. in growth promotion and suppression of tomato bacterial wilts caused by *Ralstonia solanacearum*. *J. Zhejiang Univ. (Agriculture and Life Science)* **30**: 603-610.

Algam, S.A., Xie, E. G. , Li,B. Yu ,S. , Su ,T. and Larsen, J. (2010). Effects of *Paenibacillus* strains and chitosan on plant growth promotion and control of *Ralstonia* wilt in tomato. *J. Plant Pathol.* **92** (3): 593-600.

Ali M.S. (1993). Bacterial wilt of potato in Bangladesh. In: Integrated Management of Bacterial Wilt: Proc. Int. Workshop, held in New Delhi, India, October 11 -16, 1993/edited by Bill Hardy, Edward R.French. -Lima, Peru: International Potato Center, 1995. P1-10.

Ali, A.M., Ahmed, O.M., Misaka, B.C., Latif, A.H., Elsiddig, K. and Babiker, E. A. (2003). Characterisation of cotton germplasm and its utilization in breeding for major production constraints in Sudan. Proc. World Cotton Research Conf. Cape Town, South Africa. pp 16-21.

Ali, M., Alam. M.Z. and Akanda, M.A.M. (1994). Grafting: A technique of control soil borne diseases of tomato and eggplant. IPSAJICA publication No. 4. Institute of Post graduate Studies in Agriculture (IPSA), Gazipur-1703. Bangladesh. 10 pp.

Aliye N., Fininsa C., Hiskias Y., (2008). Evaluation of rhizosphere bacterial antagonists for their potential to bioprotect potato (*Solanum tuberosum*) against bacterial wilt (*Ralstonia solanacearum*). *Biol. Contr.* **47**: 282-288.

Allen, C., Kelman, A. and French, E.R. (2001). Brown rot. *In:* Stevenson WR, Loria R, Franc G.D., Weingartner D.P., *eds. Compendium of Potato Diseases,* 2nd edn. St Paul, MN, USA: APS Press, 11–13.

Allen, C., Prior, P. and Hayward, A. C. (2005). Bacterial wilt disease and the *Ralstonia* species complex *St. Paul, Minnesota*, USA: APS Press, pp. 205-224.

Alvarez, A. M., Adams, P. D., (1999). Flow cytometry: a promising tool for seed health testing. Ames, Iowa, USA: International Seed Testing Association, 93-97.

Alvarez, A.M., Berestecky, J., Stiles, J.I., Ferreira, S.J. and Benedict, A.A. (1992). Serological and molecular approaches to identification of *Pseudomonas solanacearum* strains from Heliconia. Presented at Bact. Wilt. Int. Conf., Kaohsiung, Taiwan.

Alvarez, A.M. (2001) Differentiation of bacterial populations in seed extracts by flow cytometry. In: *Plant Pathogenic Bacteria* (De Boer, S.H., ed.), pp. 393–396. Dordrecht: Kluwer Academic Publishers.

Alvarez, A.M. and Adams, P.D., (1999). Flow cytometry: a promising tool for seed health testing. Ames, Iowa, USA: International Seed Testing Association, 93-97.

Alvarez, B. (2009). *Biology of Ralstonia solanacearum phylotype II in host and non-host environments. PhD Thesis*, Universidad. Politecnica de Valencia.

Alvarez, B., Biosca, E. G., and López, M. M. (2010a). On the life of *Ralstonia solanacearum*, a destructive bacterial plant pathogen. *Current Research, Technology and Education Topics in Applied Microbial Biotechnology* (A. Mendez-Vilas (Eds): pp. 287-279

Alvarez, B., Lopez, M.M. and Biosca E.G. (2008b). Survival strategies and pathogenicity of *Ralstonia solanacearum* phylotype II subjected to prolongedstarvation in environmental water microcosms. *Microbiology,* **154:** 3590-3598.

Alvarez, B., Lopez, M.M. and Biosca, E.G. (2007). Influence of native microbiota on survival of *Ralstonia solanacearum* phylotype II in river water microcosms. *Appl. Environ. Microbiol.* **73:** 7210–7217.

Alvarez, B., Vasse, J., Le-Courtois, V., Trigalet-Demery, D., Lopez, M.M., Trigalet, A. (2008a). Comparative behaviour of *Ralstonia solanacearum* biovar 2 in diverse plant species. *Phytopathology,* **98**: 59-68

Alvarez, B., Lopez, M.M. and Biosca E.G. (2008b). Survival strategies and pathogenicity of *Ralstonia solanacearum* phylotype II subjected to prolongedstarvation in environmental water microcosms. *Microbiology,* **154**: 3590-3598.

Alvarez, B., Lopez, M.M. and Biosca, E.G. (2010b). Ability of *Ralstonia solanacearum* phylotype II to adapt to environmental prevailing factors in water. In: Mendez-Vilas A., ed. *Microorganisms in Industry and Environmentfrom Scientific and Industrial Research to Consumer Products. World Scientific Publishing Co, in press.*

Alvarez, B., Vasse, J., Le-Courtois, V., Trigalet-Demery, D., Lopez, M.M., Trigalet, A. (2008a). Comparative behaviour of *Ralstonia solanacearum b*iovar 2 in diverse plant species. *Phytopathology,* **98**: 59-68.

Ambardar, V.K. (2011). Potential of endomycorrhizae and bacterial antagonist on the growth performance of tomato cultivars against bacterial wilt. *J. Res. Dev.* **11**: 1-12.

Anahosur, K.H. (1999). Management of plant disease through antagonist: An overview. In: *Recent advances in plant pathology* (eds. Vaidya, J. G. and S.Y. Kamble.) Dept. of Botany, Univ. of Pune pp. 38-56.

Anandaraj, M. (2009). Challenges in managing soil borne diseases with reference to *Phytophthora, Fusarium* and *Ralstonia. In 5th International Conference on Plant Pathology in the Globalized Era,* Nov. 10-13, 2009, New Delhi, India. 033.

Anderson, F.G. (1947). Horticulture Problem: Research and Control Work (in South Africa). *Fmg S. Abr.* **22**: 1073-81.

Andersona R. C. and Gardner, D. E. (1999). An evaluation of the wilt-causing bacterium *Ralstonia solanacearum* as a potential biological control agent for the alien kahili ginger (*Hedychium gardnerianum*) in Hawaiian forests. *Biol. Contr.* **15** (2): 89-96.

Anith, K.N., Momol, M.T., Kloepper, J.W., Marois, J.J., Olson, S.M. and Jones, J.B. (2004). Efficacy of plant growth-promoting rhizobacteria, acibenzolar-s-methyl, and soil amendment for integrated management of bacterial wilt on tomato. *Plant Dis.* **88**: 669-673.

Anonymous (1912). *The J. Amer. Pharmacists Assoc.* **1**: 453-4.

Anonymous (1954). Tomato diseases and their control. *J. Dep. Agric. Viet.* **52**: 187-90.

Anonymous (1960). Precautions against wilt disease. *Not. Agric. Serv. Shell. Agric.* Venezuela **2**: 90-1.

Anonymous (1969). Annual Report of the ICAR Scheme on bacterial of solanauous crops.

Anonymous (1976). Possible bacterial wilt control. *AlafuaAgri. Bull.***1**(1):12.

Anonymous (1977). Wilt tolerant tomatoes for low-level Malaysia. *Penorbitan Fakulti*, Faculty of Agriculture, University of Malaya No. 8: 10.

Anonymous (1989). Office of International Affairs, *Lost Crops of the Incas: Little-Known Plants of the Andes with Promise for Worldwide Cultivation* (1989) online.

Anonymous (2008). Tomato dishes may protect Skin. *BBC News* 2008-04-28.http://news.bbc.co.uk/2/hi/health/7370759.stm. Retrieved 2010-01-06

Anonymous (2012). Global facilitation unit for underutilized species. *Andean Native Potatoes* (*Solanum tuberosum subsp. andigena*). p. 5. Retrieved 28 February 2012

Anonymous Health24.com – Aubergine

Ansari, M.M. (1990). *Cosmos sulphureus*- a new host of *Pseudomonas* race-1 from Andaman. *Indian Phytopath.* **43**(3): 438-9.

Ansari, M.M., Ramesh, C.R. and Bandyopadhay, A.K. (1996). Multiplication of *Pseudomonas solanacearum in vivo* as affected by host and growth stage. *Trop. Horti.***73**(1): 74-76.

Anzai, Y., Kim, H., Park, J.-Y., Wakabayashi, H. and Oyaizu, H. (2000). Phylogenetic affiliation of the pseudomonads based on 16S rRNA sequence. *Int. J. Syst. Evol. Microbiol.* **50**: 1563–1589.

Araud-Razou, I., Vasse, J., Montrozier, H., Etchebar, C. and Trigalet, A. (1998). Detection and visualization of the major acidic exopolysaccharide of *Ralstonia solanacearum* and its role in tomato root infection and vascular colonization. *Eur. J. Plant Pathol.* **104**(8): 795-809.

Arlat, M., Van Gijsegem, F., Huet, J.C., Pernollet, J.C. and Boucher, C.A. (1994). PopA1, a protein which induces a hypersensitivity-like response on specific *Petunia* genotypes, is secreted via the Hrp pathway of *Pseudomonas solanacearum. EMBO J.* **13**: 543-553.

Arneson, P.A. (2001). Plant Disease Epidemiology: Temporal Aspects. *Plant Health Instructor* (American Phytopathological Society). DOI:10.1094/PHI-A-2001-0524-01. http://www.apsnet.org/education/Advanced Plant Path/Topics/Epidemiology/Epidemiology.htm.

Arora, R.K. (2012). Eco-friendly management of soil and tuber borne diseases of potato. *Indian Phytopath.* **65** (2): 116-121.

Arora, R.K., Sunaina, V. and Somani, A.K. (2008). Eco-friendly technologies to manage soil and tuber borne diseases of potato. In: *Twenty Steps Towards Hidden Treasure – Technologies that Triggered Potato Revolution in India* (eds. S.K. Pandey and S.K. Chakrabarti). Central Potato Research Institute, Shimla, pp. 137-150.

Arwiyanto, T., Sakato, K., Goto, M., Tsuyumu, S. and Takikawa, Y. (1994a). Introduction of tomatine in tomato plant by an avirulent strain of *Pseudomonas solanacearum. Ann. Phytopathol. Soc. Japan*, **60**(4): 281-287.

Arwiyanto, T., Goto, M., Tsuyumu, S. and Takikawa, Y. (1994b). Biological control of bacterial wilt of tomato by avirulent strain of *Pseudomonas solanacearum* isolated from *Strelitzia reginae. Ann Phytopathol. Soc. Japan,* **60**(4): 421-430.

Arwiyanto, T., Sakato, K., Goto, M., Tsuyumu, S. and Takikawa, Y. (1994a). Introduction of tomatine in tomato plant by an avirulent strain of *Pseudomonas solanacearum. Ann. Phytopathol. Soc. Japan,* **60**(4): 281-287.

Asati, B.S. and Yadav, D.S. (2004). Diversity of Horticultural Crops in North Eastern Region. *ENVIS Bull. Himalayan Ecol.* **12**(1): 1-11.

Ashrafuzzaman, H. and Islam, T. (1975). Bacterial wilts in tomato-A review. *Bangladesh Hort.* **3**(2): 37-44.

Assis, S.M.P., Silveira, E.B., Mariano, R.L.R. and Menezes, D. (1998). Bacte´riasendofı´ticas-Me´todo deisolamentoe potencial antago^nico nocontrole da podrida˜o negra do repolho. *Summa Phytopathol.* **24**: 216–220.

AVRDC. (1988). Progress Report 1986. *Asian Vegetable Research and Development Center*. Shanhua, Taiwan. 540 pp.

Ayana G., Fininsa, C., Ahmed, S. and Wydra, K. (2011). Effects of soil amendment on bacterial wilt caused by *Ralstonia solanacearum* and tomato yields in Ethiopia. *J. Plant Protec. Res.* **51**(1): 72-76.

Babu, A.G.C. and Kelaniyangoda, D.B. (1993). Bacterial wilt in Sri Lanka. p 55-58.In: Integrated Management of Bacterial Wilt: *Proc.Int.Workshop held in New Delhi*, India, October 11-16, 1993/edited by Bill Hardy, Edward R. French-Lima, Peru: International Potato Center, 1995. 196 p.

Babu, B. Sarath, Varaprasad, K.S., Chakrabarty, S.K., and Sivaraj, N.(1998). Status of resistant germplasm in eggplant and wild *Solanum* species. *Indian J. Plant Genetic Reso.* **11**(1): 49-61.

Bagyaraj, D.J. (2002). VA Mycorrhiza: Utility in the control of plant pathogens. In ICAR Sponsored Winter School on "*Frontiers in ecofriendly management of crop diseases and pests and utility of biofertilizers in improving soil fertility*" (Aug. 6 to Sept. 4). Department of Plant Pathology and Agriculture Microbiology. Mahatma Phule Krishi Vidyapeeth, Rahuri. pp. 37-41.

Bahal, V.K. and Shekhawat, G.S. (1993). Bacterial wilt of potatoes in the Northeastern Hills of India. P32-39. In: Integrated management of bacterial wilt: *Proc. Int. Workshop held in New Delhi,* India, October 11-16, 1993/edited by Bill Hardy, Edward R.French-Lima, Peru: International Potato Center, 1995. 196 p.

Baker, C.J., Neilson, M.J., Sequeira, L.and Keegstra, K.G. (1984). Chemical characterization of the lipopolysaccharide of *Pseudomonas solanacearum. Appl. Environ. Microbiol.* **47**: 1096-1100.

Baker, K.F. and Cook, R.L. (1994). Biological scontrol of plant pathogen. *W.H. Freeman and Company, San Francisco*, 433.

Balatero, C.H., Hautea, D.M., Hanson, P.M. and Narciso, J.O. (2002b). Development of molecular markers for marker-assisted breeding for bacterial wilt resistance in tomato. *Philipp Agric. Sci.* **85**(2): 170-181.

Balatero, C.H., Hautea, D.M. and Narciso, J.O. (2000a). Genetic of resistance and host pathogen interaction in tomato *P. solanacearum* system: Implications in breeding for tomato. *Philipp. J. Crop Sci.* **25**: 8.

Banerjee, M.K. and Singh, B. (2002). Proceedings of XXth Group Meeting held at Kerala Agricultural University, Thrissur on 9th-12th April, 2002 Session-XI. pp 72-73.

Banerjee, N. and Chatterjee, P. ( ). Control of bacterial wilt of tomato by *Lantana Camara* L. *International Seminar on "Multidisciplinary Approaches in Angiosperm Systematics.* 812-818.

Bang, S.K. and Wiles, G.C. (1995). Control of bacterial wilt (*Pseudomonas solanacearum*) in potato by crop rotation, SAPPRAD on the third year of phase 3: selected research papers; v. 1: potato July 1993-June 1994, Rasco, E.T. Jr. Aromin, F.B. (eds.). - Manila (Philippines), May 1995. pp. 103-106

Bareisis, R., Sniauka, P. and Viselga, G. (2002). Investigation of ecological potato cultivation possibilities. *Proc. Int. Conf.,* Raudondvaris, Lithuania. pp. 57-66.

Barksdala, T.H., Jaworski, C.A. and Mc Carter, S.M. (1971). Early tomato transplant escape wilt. *Agri. Res.*, USA,**19** (11): 14.

Baruah, S.J.N. and Deka, P.C. (1995). *In vitro* selection of tomato plants resistant to *Pseudomonas solanacearum. Acta Hort.* **392**: 115-121.

Baruah, S.J.N., Binoy, Mudiar, Rachid, H.A. and Mudiar, B. (2000a). Yield potentiality of some brinjal cultivars in severely bacterial wilt infected condition. *Vege. Sci.* **27**(1): 76-77.

Baruah, S.J.N. and Rachid, H.A. (2000). Sources of resistance in chilli germplasm against bacterial wilt and anthracnose. *Indian J. Plant Genetic Res.* **13**(1): 93-98.

Bashan, Y. and Bashan, L.E.L. (2002). Protection of tomato seedlings against infection by *Pseudomonas syringae* pv. *tomato* by using the plant growth-promoting bacterium *Azospirillum brasilense. Appl. Environ. Microbiol.* **68**(6): 2637-2643.

Bassler, B.L. (1999). How bacteria talk to each other: Regulation of gene expression by quorum sensing. *Curr. Opin. Microbiol.* **2**: 582–587.

Bauske, E.L. Zehnder, G.M., Sikora, E.J. and Kemble, J. (1998). Southeastern tomato growers adopt integrated pest management. *Hort Technol.* **8**(1):40-44.

BBC News Online. (2007). Chilies heated ancient cuisine. Friday, 16 February. Available from: http://news.bbc.co.uk/2/hi/americas/6367299.stm. Retrieved 16 February 2007.

Bedekar, H.S. (1977). Reaction of four tomato cultivars of different mixtures of isolates of *Pseudomonas solanacearum.Veg. Sci.* **4**:1-5.

Begum, N., Haque, M.I., Mukhtar, J., Naqvi, S.M. and Wang, J.F. (2012). Status of bacterial wilt caused by *Ralstonia solanacearum*in Pakistan. *Pak. J. Phytopath.* **24**(1):11-20.

Belbahri, L., Boucher, C., Candresse, T., Nicole, M., Ricci, P. and Keller, H. (2001). A local accumulation of the *Ralstonia solanacearum* PopA protein in transgenic tobacco renders a compatible plant-pathogen interaction incompatible. *Plant J.* **28**: 419-430.

Belkhindi, G. (1985). Studies on resistance to bacterial wilt *(Pseudomonas solanacearum* E.E. Smith) in brinjal *(Solanum melongena* L.). *M.Sc. Thesis*. University of Agricultural Sciences, Bangalore, India.

Bell, C. R., Dickie, G. A., Harvey, W.L.G. and Chan, J.W.Y.F. (1995). Endophytic bacteria in grapevine. *Can. J. Microbiol.* **41**:46–53.

Beneduzi, A., D. Peres, P. Beschoren da, Costa, M.H.B. Zanettini and Passaglia, L.M.P. (2008). Genetic and phenotypic diversity of plant-growth-promoting bacilli isolated from wheat fields in Southern Brazil. *Res. Microbiol.* **159**: 244-250.

Benhamou, N. and Picard, K. (1999). La résistance induite: unenouvellestratégie de défense des plantescontre lesagents pathogènes. P*hytoprotection,* **80**: 137-168.

Benhamou, N., Kloepper, J.W. and Tuzun, S. (1998). Induction of resistance against fusarium wilt of tomato by combination of chitosan with an endophytic bacterial strain: ultrastructure and cytochemistry of the host response. *Planta,* **204:** 153-168.

Bent, A.F., Kunkel, B.N., Dahlbeck, D., Brown, K.L., Schmidt, R., Giraudat, J., Leung, J. and Staskawicz, B.J. (1994). RPS2 of *Arabidopsis thaliana*: A leucine-richrepeat class of plant disease resistance genes. *Science,* **265:** 1856-1860.

Bent, E., Tuzun, S., Chanway, C.P. and Enebak, S. (2001). Alterations in plant growth and in root hormone levels of lodge pole pines inoculated with rhizobacteria. *Can. J. Microbiol.* **47**: 793-800.

Berg, G., Marten, P and Ballin, G. (1996). *Stenotrophomonas maltophilia* in the rhizosphere of oilseed rape – occurrence, characterization and interaction with phytopathogenic fungi. *Microbiol. Res.* **151**: 19-27.

Berniac, M. and Coleno, A. (1974). The Use of immunofluorescence in detecting phytopathogenic bacteria. *Symp. Crop Protec. in the Caribbean. St. Augustine, Trinidad.*

Bertolini, E., Olmos, A., Lo´ pez, M.M. and Cambra, M. (2003a). Multiplex nested reverse-transcription polymerase chain reaction in a single tube for sensitive and simultaneous detection of four RNA viruses *and Pseudomonas savastanoi*pv. *savastanoi* in olive trees. *Phytopathology,* **93:** 286-292.

Bertolini, E., Penyalver, R., Garcı´a, A., Olmos, A., Quesada, J.M., Cambra, M. and Lo´ pez, M.M. (2003 b). Highly sensitive detection of Pseudomonas savastanoipv. savastanoi in asymptomatic olive plants by nested-PCR in a single closed tube. *J. Microbiol. Methods,* **52**: 261-266.

Besnard, V., Federighi, M. and Cappelier, J. M. (2000). Evidence of viable but non-culturable state in *Listeria monocytogenes*by directviable count and CTC-DAPI double staining. *Food Microbiol.* **17**: 697-704.

Bevins, C.L. and Zasloff, M. (1990). Peptides from frog skin. *Ann. Rev. Biochem.* **59**: 395-414.

Beyries, A. (1974). Grafting market garden solanaceous crops. *Pepinierstes Horticultures Maraichers No.* **152:** 27-32.

Bhattarai, S.P. Sharma, S. and Subedi, P.P. (1998). Heat and bacterial wilt tolerant tomato varietal evaluation for rainy season at river basin Dorwin. *Working paper – Lumple Agricultural Research Centre No.* 98/21 (iii) +IIPP.

Bigat, B. and Escudie, A. (1967). A Survey of bacterial wilt of solanaceous plants in the French Antilles. *Phytiat-Phytopharm*, Paris, **16**: 187-97.

Bingham, F. T., Chapman, H. D. and Pugh, A. L. (1954). Solution-culture studies of nitrite toxicity to plants. *SoU Sci. Soc. Amer. Proc.* **18**: 305-308.

Bissonauth, O. (1980). Observations on some new varieties of tomato from Taiwan. *Technical Bulletin Ministry of Agriculture and Natural Resources and Environment, Mauritius No.* 2, 16-17.

Biswas, S. and Singh, N.P. (2007). Effect of host genotypes and cultural practices for the management of bacterial wilt in brinjal (*Solanum melongena* L.). *Indian Phytopath.* **60**(4): 438-441

Biswas, S. and Singh, N.P. (2008). Integrated management of wilt of tomato caused by *Ralstonia solanacearum. Indian J. Mycol. Plant Pathol.* **38** (2): 182-184.

Blat, Y. and Eisenbach, M. (1995). Tar-dependent and –independent pattern formation by *Salmonella typhimurium. J. Bacteriol.***177**: 163– 169.

Bogdanove, A.J., Beer, S.V., Bonas, U., Boucher, C.A., Collmer, A., Coplin, D.L., Cornells, G.R., Huang, H.C., Hutcheson, S.W., Panopoulos, N.J., and Van Gijsegem, F. (1996). Unified nomenclature for broadly conserved *Hrp* genes of phytopathogenic bacteria. *MOI. Microbiol.* **20**: 681-683.

Bonants, P., De Weerdt, M., Van Beckhoven, J., Hilhorst, R., Chan, A., Boender, P., Zijlstra, C., and Schoen, C. (2002). Multiplex detection of plant pathogens by microarrays: an innovative tool for plant health management. In Abstracts Agricultural Biomarkers for Array Technology, Management Committee Meeting, Wadenswil, 20.

Bonants, P.J.M., Schoen, C.D., Szemes, M., Speksnijder, A., Klerks, M.M., van den Boogert, P.H.J.F., Waalwijk, C., van der Wolf, J.M., and Zijlstra, C. (2005). From single to multiplex detection of plant pathogens: pUMA, a new concept of multiplex detection using microarrays. *Phytopathol. Polonica,* **35**: 29-47.

Boonham N, Tomlinson J, Mumford R. (2007). Microarrays for rapid identification of plant viruses. *Annu.Rev. Phytopathol.* **45**: 307–328.

Bora, L.C. (1995). Management of bacterial wilt of tomato by organic formulation. *Plant Hith.* **1**: 74-6.

Bora, L.C. and Addy S.K. (1982). Quantitative aspects of assessing virulence of *Pseudomonas solanacearum* in tomato plants. *Indian Phytopath.* **35**: 285-290.

Bora, L.C. and Bora, P. (2008). Biological control strategies for management of bacterial wilt of brinjal (*Solanum melongena*). *J. Mycol. Plant Pathol.* **38**(3): 542-545.

Bora, L.C. and Deka, S.N. (2007). Wilt disease suppression and yield enhancement in tomato (*Lycopersicon esculentum*) by application of *Pseudomonas fluorescens* based biopesticide (Biofor-Pf) in Assam. *Indian J. Agric. Sci.* **77**(8): 490-494.

Bora, L.C. and Deka, S.N. (2008). Bioformulation of *Pseudomonas aeruginosa* in organic substrates and its role in management of bacterial wilt of chilli (*Capsicum annum*). *J. Mycol. Plant Pathol.* **38**(3): 80-83.

Bora, L.C., Gogoi, P.K. and Das, B.C. (1996). Bacterial wilt of tomato in relation to few environmental parameters in Assam. *J. Agri. Sci. Soc. North East India,* **9** (2): 185-188.

Bora, L.C., Gogoi, P.K. and Samuel, Jaya (1997). Use of botanicals in management of bacterial wilt of tomato caused by *Pseudomonas solanacearum* (Abstr.) *In Symposium on Major Diseases of Crop Plants in Eastern India and Their Management.* Dec.10-11. *Assam Agril Univ*. pp. 11 (Mimeograph).

Bora, L.C. and Deka, S.N. (2007). Wilt disease suppression and yield enhancement in tomato (*Lycopersicon esculentum*) by application of *Pseudomonas fluorescens* based biopesticide (Biofor-Pf) in Assam. *Indian J. Agric. Sci.* **77**(8): 490-494.

Bosch, S.E., Louw, A.J. and Aucamp, E. (1985). Rodade, bacterial wilt resistant tomato. *Hort.Sci.* **20**(3): 458-459.

Bose, T.K.and Som, M.G. (1986). Vegetable crops in India. Nayaprashan, 206, Bidhan Sarani, Calcutta, 700006, India, pp. 775.

Boshou, L. (2005). A broad review and perspective on breeding for resistance to bacterial wilt, In: Bacrerial wilt disease and the *Ralstonia solanacearum* species complex eds C. Allen, P. Prio end A.C. Hayward, APS, USA, pp. 225-238.

Bosland, P.W. (1996). Capsicums: Innovative uses of an ancient crop. pp. 479–487. In: J. Janick (ed.), Progress in new crops. ASHS Press, Arlington, VA. Hort.purdue.edu. 1997-08-22. http://www.hort.purdue.edu/newcrop/proceedings1996/V3-479.html. Retrieved 2010-12-23

Botelho, G. R. and Mendonca-Hagler, L. C. (2006). Fluorescent *Pseudomonas* associated with the rhizosphere of crops—An overview. *Braz. J. Microbiol.* **37**: 401-416.

Boucher, C., Barberis, P., Trigalet, A. and Demery, D. (1985). Transposon mutagenesis of *Pseudomonas solanacearum*: Isolation of Tn5-induced avirulent mutants. *J. Gen. Microbiol.* **131**: 2449-2457.

Boucher, C., Genin, S., Arlat, M. (2001). Current concepts of pathogenicity in plant pathogenic bacteria. *Comptes Rendus de l' Academie des Sciences Serie III: Sciences de la Vie-Life Sciences,* **324**: 915-922.

Boucher, C.A., Gough, C.L. and Arlat, M. (1992). Molecular genetics of pathogenicity determinants of *Pseudomonas solanacearum* with special emphasis on *Hrp* genes. *Annu. Rev. Phytopathol.* **30:** 443-461.

Boudazin G., Le Roux A.C., Josi K., Labarre P., Jouan B., 1999. Design of specific primers of *Ralstonia solanacearum*and application to identification of European isolates. *Eur. J. Plant Pathol.* **105**: 373-380.

Brito, B., Aldon, D., Barberis, P., Boucher, C., Genin, S.A. (2002). Signal transfer system through three compartments transduces the plant cell contact-dependent signal controlling *Ralstonia solanacearum Hrp* genes. *Mol. Plant-Microbe Interact.* **15**:109-119.

Brown, D. (2009). *Ralstonia solanacearum* and bacterial wilt in the post-genomics era. In: Jackson RW, Ed. Plant Pathogenic Bacteria. *Genomics and molecular biology*. London: Caister *Academic Press*, 175.

Brown, P.O., and Botstein, D. (1999). Exploring the new world of the genome with DNA microarrays.*Nat. Genet.* **21**: 33-37.

Brumbley, S.M. and Denny, T.P. (1990). Cloning of wild-type *Pseudomonas solanacearum phca*, a gene that when mutated alters expression of multiple traits that contribute to virulence. *J. Bacteriol.* **172**: 5677-5685.

Brumbley, S.M., Carney, B.F. and Denny, T.P. (1993). Phenotype conversion in *Pseudomonas solanacearum* due to spontaneous inactivation of *phca*, a putative lysr transcriptional regulator. *J. Bacteriol.* **175:** 5477-5487.

Buddenhagen, I.W. (1960). Strains of *Pseudomonas solanacearum* in indigenous hosts in banana plantations of costa rica, and their relationship to bacterial wilt of bananas. *Phytopathology,* **50**: 660-664.

Buddenhagen, I. W., Sequeira, L. and Kelman, A. (1962). Designation of races in *Pseudomonas solanacearum. Phytopathology,* **52**: 726.

Buddenhagen, I.W. and Kelman, A. (1964). Biological and physiological aspects of bacterial wilt caused by *Pseudomonas solanacearum. Annu. Rev. Phytopathol.* **2**: 203-230.

Burkhead, K.D., Schisler, D.A. and Slininger, P.J. (1994). Pyrrolnitrin production by biological control agent *Pseudomonas cepacia* B37w in culture and in colonized wounds of potatoes. *Appl. Environ. Microbiol.* **60**: 2031-2039.

Burkholder, W.H. (1948). Bacteria as plant pathogen. *Annu. Rev. Microbiol.* **2**: 389-412.

Burr, T.J., Norelli, J.L. and Reid, C.L. (1993). Streptomycin-resistant bacteria associated with fire blight infections. *Plant Dis.* **77**: 63-66

CABI (2003). *Crop protection compendium* United Kingsom Wallingford On-line CABI International http//www.cabicompendium.org/cpc. Accessed Nov. 4, 2003

CABI. (2006). *Crop protection compendium*. United Kingdom: Wallingford: CAB International. Online database. http://www.cabicompendium.org/cpc. Accessed July 2007

Call, D.R., Borucki, M.K., and Loge, F.J. (2003). Detection of bacterial pathogens in environmental samples using DNA microarrays. *J. Microbiol. Methods,* 53: 235-243.

Cao Bi-hao, Lei Jian-jun, Wang Yong and Chen Guo-ju (2009). Inheritance and identification of SCAR marker linked tobacterial wilt-resistance in eggplant. *Afri. J. Biotechnol.* **8** (20): 5201-5207.

Cappel, E.L. (1892). A note on a potato disease prevalent in Poona district and elsewhere. *Rep. Dept. Ld. Rec. Agric*. Bombay. p. 16.

Cardoso, S.C., Soares, A.C.F., Brito, A., dos S., Laranjeira, F.F., Ledo, C.A.S., and Santos, A.P. (2006). Control of tomato bacterial wilt through the incorporation of aerial part of pigeon pea and crotalaria to soil. *Summa Phytopathol.* **32**: 27-33

Carmeille, A. Prior, P., Kodja, H. and Chiroleu, F. (2006b). Evaluation of resistance to race 3, biovar 2 of *Ralstonia solanacearum* in tomato germplasm. *J. Phytopathol.* **154**(7-8): 398–402.

Carmeille, A., Caranta, C., Dintinger, J., Prior, P., Luisetti, J., Besse, P. (2006a). Identification of QTLs for *Ralstonia solanacearum* race 3-phylotype II resistance in tomato. *Theor. Appl. Genet.* **113**(1): 110-121

Carputo, D., Aversano, R., Barone, A., Di Matteo, A., Iorizzo, M., Sigillo, L., Zoina, A., and Frusciante, L. (2009). Resistance to *Ralstonia solanacearum* of sexual hybrids between *Solanum commersonii* and *S. tuberosum. Am. J. Pot. Res.* **86**: 196–202.

Cartin, J.M. and Wang, A. (1997). Isolation of suppressing agent of *Pseudomonas solanacearum* E.F. Smith in tomato (*Lycopersicon esculentum*). *Agron. Mesoamericana,* **8**(2): 54-58.

Caruso, P., Bertolini, E., Cambra, M. and Lopez, M. M. (2003). A new and sensitive co-operational polymerase chain reaction for rapid detection of *Ralstonia solanacearum* in water. *J. Microbiol. Methods,* **55**: 257 -272.

Caruso, P., Palomo, J. L., Bertolini, E., A lvarez, B., Lo pez, M. M. and Biosca, E. G. (2005). Seasonal variation of *Ralstonia Solanacearum* biovar 2 populations in a spanish river: recovery of stressed cells at low temperatures. *Appl. Environ. Microbiol.* **71**: 140–148.

Cavalcantle, E.B., Mariano, R.L.R, Leit, J.P. and Coelho, R.S.B. (1995). Influence of mineral nutrition on the reaction of tomato cv. Yashimatsu and Santa Cruz to *Pseudomonas solanacearum. Bacterial wilt News Lett. No.* **12**: 3-8.

Cellier, G. and Prior, P. (2010). Deciphering phenotypic diversity of *Ralstonia solanacearum* strains pathogenic to potato. *Phytopathology*, **100:** 1250-1261.

Chakrabarti, S.K, Slkhawat, G.S. and and Gadewar, A.V. (1993b). Bacterial wilt of potato in India. pp. 23-31. In: Integrated management of bacterial wilt: Proc. Int. Workshop held in New Delhi, India, October 11-16, 1993/edited by Bill Hardy, Edward R. French.-Lima, Peru: International Potato Center, 1995. 196 p.

Chakrabarti, S.K., Singh, M.N. and Shekhawat, G.S. (1993a). ELISA for *Pseudomonas solanacearum* from potato stem and tubers. *Proc. Nati. Symp. Potato: Present and Future.*

Chakraborty, M., Chatterjee, B., Habib, A.K.M.A., Sengupta, C., Som, M.G. and Samaddar, K.R. (1994). Occurrence of *Pseudomonas solanacearum* biovar 3 on eggplant, tomato and potato in West Bengal. *Bacterial wilt News Lett. No.* **11**: 3-8.

Chakravarty, G. and Kalita, M. C. (2011a). Management of bacterial wilt of brinjal by *P. fluorescens* based bioformulation. *ARPN J. Agric. and Biol. Sci.* **6**(3): 1-11.

Chakravarty, G. and Kalita, M. C. (2011b). Comparative evaluation of organic formulations of *Pseudomonas fluorescens* based biopesticides and their application in the management of bacterial wilt of brinjal (*Solanum melongena* L.). *Afri. J. Biotechnol.* **10**(37): 7174-7182.

Champoiseau, P. G. (2010). *Ralstonia solanacearum*race 3 biovar 2. E-learning Module. Published online, March 2010. NPDN Training and Educational Site: http://cbc.at.ufl.edu/.

Champoiseau, P.G. Jones, J.B., Allen, C. and Momol, T. (2009a). Description and strategies for best anagementof *Ralstonia solanacearum* race 3 biovar 2 as a potential incitant of bacterial wilt of tomato. *24th Annual Tomato Disease Workshop,* November 3-5, 2009 State College, Pennsylvania.

Champoiseau, P. G., Jones, J.B., and Allen, C. (2009 c). *Ralstonia solanacearum* race 3 biovar 2 causes tropical losses and temperate anxieties. online. *Plant Health Progress*. DOI: 10.1094/PHP-2009-0313-01-RV.

Champoiseau, Patrice G. and Timur Momol, M. (2009b). Bacterial wilt of tomato. *R. solanacearum* race 3 biovar 2 USDA-NRI Project. P. 11.

Chan, Y.L., Prasad, V., Sanjaya, Chen, K.H., Liu, P.C., Chan, M.T., and Cheng, C.P. (2005). Transgenic tomato plants expressing an arabidopsis *thionin* (Thi2.1) driven by fruit-inactive promoter battle against phytopathogenic attack. *Planta,* **221**: 386-393.

Chand, J.N. and Mishra, B. (1995). Studies on bacterial wilt of solanacearum crops caused by *Pseudomonas solanacearum* in Chotanagpur Plateau of Bihar. *Indian Phytopath. –Suppl. Issue* **45**.C XXIV pp.

Chandrashekhara, K.N. and Prashannakumar, M.K. (2010). New host plant for *Ralstonia solanacearum* from India. *Plant Pathol,* **59**(6): 11-78.

Chandrashekhara, K.N., Prashannakumar, M.K., Deepa, M., Vani, A. and Khan, A.N.A. (2012). Prevalance of races and biotypes of *Ralstonia solancearum* in India. *J. Plant Protec.* **52**(1): 53-58.

Chapman, H. D. and Liebig, G. F. Jr. (1952). Field and laboratory studies of nitrite accumulation in soils. *Soil Sci. Soc. Am. Proc.* **16**: 276-282.

Chatterjee, S., Mukherjee, N., Khatua, D.C. (1997). Status of bacterial wilt in West Bengal. *Bengal J. Interacademicia,* **1** (1): 97-99.

Chatterjee, B., Chakraborty, M., Habib, A.K.M.A. and Samaddar, K.R. (1994). Survival of Pseudomonas solanacearum biovar 3 on seeds of eggplant. *Bacterial Wilt Newslet.* No. 11: 11

Chattopadhyay, S.B. and Mukhopadhyay, A.N. (1968). Outbreaks and new records. *Pi. Prot. Bull.* FA.O. **16**:52-53.

Chaudhary, B. (1976). Vegetables (4th edn.), National Book Trust, New Delhi, pp. 50-58.

Chaudhary, D. R. (2000). Inheritance of resistance to bacterial wilt (*Ralstonia solanacearum* E. F. Smith) in eggplant. *Haryana J. Horti. Sci.* **29** (½): 89-90.

Chaudhary, D. R. and Sharma, S. D. (2000). Screening of brinjal cultivars against bacterial wilt and fruit borer. Agri. *Sci. Dig.* **20** (2): 129-130.

Che, N.C., Li, H.M. and Wang, J.F. (1997). Bacterial wilt resistance sources in eggplant, *Solanum melongena. Capscicum and Eggplant News*, **16**: 111-114.

Chellemi, D.O., Dankers, H.A. Olson, S.M., Hodge, N.C. and Scott, J.W. (1994a). Evaluating bacterial wilt resistant tomato genotypes using a regional approach. *J. Amer. Soc. Horti. Sci.***119**(2): 325-329.

Chellemi, D.O., Mitchell, O.J. and Barkdol, A.W. (1992). Effect of composted organic amendments on the incidence of bacterial wilt of tomato. *Proc. Florida Hort. Soc.* **105**: 364-66.

Chellemi, D.O., Dankers, H.A. Olson, S.M., Hodge, N.C. and Scott, J.W. (1994a). Evaluating bacterial wilt resistant tomato genotypes using a regional approach. *J. Amer. Soc. Horti. Sci.* **119**(2): 325-329.

Chellemi, D.O., Olson, S.M. and Mitchell, O.J. (1994b). Effects of soil solarization and fumigation on survival of soil borne pathogens of tomato in northern florida. *Plant Dis.* **78**(12): 1167-71.

Chellemi, D.O., Olson, S.M. and Scott, J.W. (1995). Field evaluation of tomato genotypes for resistance of bacterial wilt. *Proc. Florida Soc. No.* **107**:151-153.

Chellemi, D.O., Olson, S.M., Mitchell, D.J., Secker, I. and McSorley, R. (1997). Adaptation of soil solarization to the integrated management of soil-borne pests of tomato under humid conditions. *Phytopathology,* **87**: 250–258.

Cheng, Y.W. and Chua, S.E. (1976). The possible use of grated tomato in brinjal plants for tomato production in low land tropics. *Singapore J. Primary Indus.* **3**(2): 94-102.

Chern, M., Canlas, P.E. and Ronald, P.C. (2008). Strong suppression of systemic acquiredresistance in arabidopsis by nrr is dependent onits ability to interact with npr1and its putativerepression domain. *Molecular Plant.* **1** (3): 552–559

Chern, M., Canlas, P.E., Fitzgerald, H.A., and Ronald, P.C. (2005a). Rice NRR, a negative regulator of disease resistance, interactswith Arabidopsis NPR1 and Rice NH1. *Plant J.* **43:** 623–635.

Chern, M., Fitzgerald, H.A., Canlas, P.E., Navarre, D.A., and Ronald, P.C. (2005b). Over-expression of a rice NPR1 homologueleads to constitutive activation of defense response and hypersensitivity to light. *Mol. Plant–Microbe Interact* **18**: 511–520.

Chester (2008). "Curiosities of I-5, facts about King and the benefits of volunteers". *Chester Progressive*. 2008-01-16.

Choudhary, D.K., Prakash, A. and Johri, N. (2007). Induced systemic resistance in plants: Mechanism of action. *Indian J. Life Sci.* **47**(4): 289-297.

Ciampi-Panno, L., Fernandez, C., Bustamante, P., Andrade, N., Ojeda, S. and Conteras, A. (1989). Biological control of bacterial wilt of potatoes caused by *Pseudomonas solanacearum. Am. Potato J.* **66**:315–332.

CIP. (2004). Breakthrough in bacterial wilt resistance. International Potato Center Annual Report 2004. http://www.cipotato.org/publications/annual_reports/2004/ar2004_10.asp [accessed November 2009].

Clough, S.J., Flavier, A.B., Schell, M.A. and Denny, T.P. (1997). Differential expression of virulence genes and motility in *Ralstonia* (*Pseudomonas*) *solanacearum* during exponential growth. *Appl. Environ. Microbiol.* **63**: 844-850.

Cockerill, F. R. & Smith, T. F. (2002). Rapid-cycle real-time PCR: a revolution for clinical microbiology. *ASM News*, **68**: 77–83.

Coenye, T., Goris, J., de Vos, P., Vandamme, P. and Li Puma, J.J. (2003). Classification of *Ralstonia pickettii*-like isolates from the environment and clinical samples *as Ralstonia insidiosa* sp. nov. *Inter. J. Sys. Bacteriol.* **53**: 1075-1080.

Cohen Y. (2001). The BABA story of induced resistance. *Phytoparasitica.* **29**: 375-378.

Coleman, L.C. (1909). The ring disease of potato. *Bull. Mycol. Ser. Dept. Agric.* Mysore. **1**:15.

Coleno, A., Hingard, L. and Rat, B. (1976). Some aspects of the serology of *P. solanacearum. Conf. Bacterial Wilt*. Raleigh, N. C.

Collmer, A. (1996). Bacterial avirulence proteins: Where's the action? *Trends in Plant Sci.* **1:** 209-210.

Colwell, R. R. and Huq, A. (1994). *Vibrios* in the environment: Viable but nonculturable *Vibrio cholerae*. In: *Vibrio cholerae* and cholera: Molecular to Global Perspectives, pp. 117-133. Edited by I. K. Wachsmuth, P. A. Blake & Ø. Olsvik. Washington DC.

Cook, D. and Sequeira. L. (1991). Genetic and biochemical characterization, of a *Pseudomonas solanacearum* gene cluster required for extracellular polysaccharide production and for virulence. *J. Bacteriol.* **173**: 1654–1662

Cook D. and Sequeira, L. (1994**).** Strain Differentiation of *Pseudomonas solanacearum* by molecular genetic methods. Pages 77-93. In: Bacterial Wilt: The Disease and Its Causative Agent, *Pseudomonas solanacearum.* A.C. Hayward and G.L. Hartman, eds. CAB International, Wallingford, U.K.

Cook, D. and Sequeira. L. (1991). Genetic and biochemical characterization, of a *Pseudomonas solanacearum* gene cluster required for extracellular polysaccharide production and for virulence. *J. Bacteriol.* **173**: 1654–1662

Cook, D., Barlow, E. and Sequeira, L. (1989b). Genetic Diversity of *Pseudomonas solanacearum*: Disease and the *Ralstonia solanacearum* species Complex, pp. 205–224. Edited by C. Allen, P. Prior & A.C. Hayward. St Paul, MN: *APS Press*

Cook, D., Barlow, E., and Sequeira,. L. (1989a). Genetic diversity of *Pseudomonas solanacearum*: detection of restriction fragment length polymorphisms with DNA probes that specify virulence and the hypersensitive response. *Mol. Plant-Microbe Intera.* **2**: 113-121.

Cook, D., Barlow, E. and Sequeira, L. (1989b). Genetic Diversity of *Pseudomonas solanacearum*: Disease and the *Ralstonia solanacearum* species Complex, pp. 205–224. Edited by C. Allen, P. Prior & A.C. Hayward. St Paul, MN: *APS Press*

Coons, A.H. and Kaplan, M.H. (1950). Location of antigen in tissue cells II. Improvements in a method for the detection of antigens by means of fluorescent antibody *J. Exp. Med.* **91**: 1.

Cox, R.S. (1960). Studies on the control of southern bacterial wilt on trellis-grown tomato in South Florida. *Proc. Fla. Sta. Hort. Soc.* **73**: 120-123.

Cox, R.S. (1961). Progress on the control of southern bacterial wilt on trellis-grown tomato in South Florida. *Proc. Fla. Sta. Hort. Soc.* 1962, **74**:148-151.

Cox, S (2000). *I Say Tomayto, You Say Tomahto.* www.landscapeimagery.com/tomato.html

Crocetti, G.R., Hugenholtz, P., Bond, P.L., Schuler, A., Keller, J., Jenkins, D., and Blackall, L.L. (2000). Identification of polyphosphate-accumulating organisms and design of 16S rRNA-directed probesfor their detection and quantitation. *Appl. Environ. Microbiol.* **66:** 1175-1182.

Crowder, D.W., Northfield, T.D., Strand, M.R. and Snyder, W.E. (2010). Organic agriculture promotes evenness and natural pest control. *Nature* (London), **466**: 109-112.

Cummings, J.H., Beatty, E.R., Kingman, S.M., Bingham, S.A. and Englyst, H. N. (1996). Digestion and physiological properties of resistant starch in the human large bowel. *Br. J. Nutr.* **75** (5): 733–747.

Cunnac, S., Boucher, C. and Genin, S. (2004a). Characterization of the cis-acting regulatory element controlling Hrpb-mediated activation of the type III secretion system and effector genes in *Ralstonia solanacearum. J. Bacteriol.* **186**: 2309-2318.

Cunnac, S., Occhialini, A., Barberis, P., Boucher, C. and Genin, S.(2004b). Inventory and functional analysis of the large hrp regulon in *Ralstonia solanacearum*: Identification of novel effector proteins translocated to plant host cells through the type III secretion system. *Mol. Microbiol.* **53**: 115-128.

Daly, P. (1973). Study of three tomato varieties tolerant to *Pseudomonas solanacearum. Agronomie Tropical,* **28**(1): 28-33.

Daly, P. (1976). IRAT L3, a new tomato variety combining resistance to several diseases. *Agronomie Tropical* **31**(4): 398-401.

Danesh, D, Aarons, S., McGill, G. E. and Young, N.D. (1994). Genetic dissection of oligogenic resistance to bacterial wilt in tomato. *Mol Plant Microbe Interact,* **7:** 464–471.

Danesh, D. and Young, N.D. (1994). Partial resistance loci for tomato bacterial wilt show differential race specificity. *Tomato Genet Coop Rep.* **44**: 12-13

Dangl, J.L. (1994). The enigmatic avirulence genes of phytopathogenic bacteria. *Curr. Topics in Microbiol. and Immunol.* **192**: 99-118.

Danial, J.R. C., McHugh, G.S. and Saddler, G. (2006). Molecular characterization of the potato brown rot pathogen *R. solanacearum* race 3/biovar 2A. *Proceedings of 4th International Bacterial Wilt Symposium.* The Lakeside Conference Centre, United Kingdom.

Daniel, M. and Purkayastha, R.P. (1995). Handbook of Phytoalexin Metabolism and Action. Marcel Dekker, Inc., New York.

Danks, C. and Barker, I. (2000). On-site detection of plant pathogens using lateral flow devices. OEPP/EPPO Bulletin **30:** 421-426 See comment in Pub Med Commons below

Das, C.R. and Chattopadhyay, S.B. (1955). Bacterial wilt of eggplant. *Indian Phytopath.* **8**: 130-135.

Das, G. P.; Ramaswamy, S. and Bari, M. A. (2000). Integrated crop management practices for the control of the brinjal shoot and fruit borer in Bangladesh. *DAE-DANIDA Strengthening Plant Protection Services (SPPS) Project.* Dept. of Agril. Extension. Khamarbari, Dhaka. 12pp.

Daughtrey, M. (2003a). New and re-emerging diseases in 2003. Department of Plant Pathology, Cornell University, Long Island Horticultural Research & Extension Center.

Daughtrey, M. (2003b). Southern bacterial wilt, caused by *Ralstonia solanacearum.* In: Summary of Presentation at Society of American Florists' 19th Annual Conference on Insect and Disease Management on Ornamentals. http://www.greenhouse.cornell.edu/pestsdis/MD.SAFtalk.pdf.

Dauny, M.C. (2008). *Vegetable II.Hand book of Plant Breeding* **2**(3): 163-220.

Dayakar, B.V., Lin, H.J., Chen, C.H., Ger, M.J., Lee, B.H, Pai, C.H., Chow, D., Huang, H.E., Hwang, S.Y., Chung, M.C., and Feng, T.Y. (2003). Ferredoxin from sweet pepper (*Capsicum Annuum* L.) Intensifyingharpin (Pss)-mediated hypersensitive response shows an enhanced production of active oxygen species (AOS). *Plant Mol. Biol.* **51**: 913-924.

De Baere, T., Steyaert, S., Wauters, G., De Vos, P., Goris, J., Coenye, T., Suyama, T., Verschraegen, G. and Vaneechoutte, M. (2001). Classification of *Ralstonia pickettii*biovar 3}'*thomasii'* strains (Pickett 1994) and of new isolates related to nosocomial recurrent meningitis as *Ralstonia mannitolytica* sp. nov. *Int J. Syst. Evol. Microbiol.* **51:** 547-558.

De Meyer, G., Audenaert, K., and Höfte, M. (1999). *Pseudomonas aeruginosa* 7NSK2-induced systemic resistance in tobacco depends on *in planta* salicylic acid accumulation but is not associated with *pr1a* expression. *Eur. J. Plant Pathol.* **105**: 513-517.

Deberdt, P., Perrin, B. and Coranson-Beaudu, R. (2012). Effect of *Allium fistulosum* extract on *Ralstonia solanacearum* populations and tomato bacterial wilt. *Plant Dis.* **96**(5): 687-692.

Deberdta, P., Queneherveb, P., Darrassea, A. and Prior, P. (1999). Increased susceptibility to bacterial wilt in tomatoes by nematode galling and the role of the mi gene in resistance to nematodes and bacterial wilt. *Plant Pathol.* **48**: 408-414.

Delaspre, F., Penalver, C.G. N. and Saurel, O. *et al.* (2007). The *Ralstonia solanacearum* pathogenicity regulator HrpB induces 3-hydroxyoxindole synthesis. *Proc. Natl. Acad. Sci. USA,* **104**: 15870-15875.

Denny T.P. (2006a). Plant pathogenic *Ralstonia* species in Gnanamanickam, S. S. Plant-Associated Bacteria. Dordrecht, Springer. pp. 1-62

Denny, T.P. (1999). Auto regulator- dependent control of extracellular polysaccharide production in phytopathogenic bacteria. *Eur. J. Plant Pathol.* **105:** 417- 430.

Denny, T.P., Brumbley, S.M., Carney, B.F., Clough, S.J. and Schell, M.A. (1994). Phenotypic conversion of *Pseudomonas solanacearum*: its molecular basis and potential function. In: A. C. Hayward and G.L. Hartman, Bacterial wilt: The Disease and itas causative agent, *Pseudomonas solanacearum*. pp. 137-143. Wallingford, U.K.: CAB International.

Denny, T.P. (1995). Involvement of bacterial polysaccharides in plant pathogenesis. *Annu. Rev. Phytopathol.* **33:** 173-197

Denny, T.P. (2000). *Ralstonia solanacearum*, a plant pathology in tough with its host. *Trinda. Microbiol.* **8**(11): 486-489.

Denny T.P. (2006a). Plant pathogenic *Ralstonia* species in Gnanamanickam, S.S. Plant-Associated Bacteria. Dordrecht, Springer. pp. 1-62

Denny, T.P. (2006b). Plant pathogenic *Ralstonia* species In: *Plant-associated bacteria*. in Gnanamanickam, S.S. (ed). *Springer Publishing*, Dordrecht, Springer. pp: 573-644.

Denny, T.P. and Beak, S.R. (1991). Genetic Evidence that Extra Cellular Polysaccharide is a Virulence Factor of *Pseudomonas solanacearum*. *Mol. Plant-Microbe Intera.* **4**(2): 198-206.

Denny, T.P., Carney, B.F. and Schell, M.A. (1990). Inactivation of multiple virulence genes reduces the ability of *Pseudomonas solanacearum* to cause wilt symptoms. *Mol. Plant-Microbe Interact.* **3**: 293-300.

Denny, T.P. and Hayward, A.C. (2001). Gram-negative bacteria: *Ralstonia*. Pages 151-174 In: Schaad, N.W., Jones, J.B., and Chun, W., eds. Laboratory Guide for Identification of PlantPathogenic Bacteria. 3rd edn. American Phytopathological Society, St. Paul, MN. 373pp.

Denny, T.P., Carney, B.F. and Schell, M.A. (1990). Inactivation of multiple virulence genes reduces the ability of *pseudomonas solanacearum* to cause wilt symptoms. *Mol. Plant-Microbe Interact.* **3**: 293-300.

Department for Environment, Food and Rural Affairs of United Kingdom (DEFRA). (2003). Plant Pathology. www.defra.gov.uk/planth/phnews/ openday/brown.pdf. Accessed March 2003.

Deslandes, L., Olivier, J., Peeters, N., Feng, D.X., Khounlotham, M., Boucher, C., Somssich, I., Genin, S. and Marco, Y. (2003). Physical interaction between RRS1-R, a protein conferring resistance to bacterial wilt, and Popp2, a type III effector targeted to the plant nucleus. *Proc. Nat. Acad. Sci.* USA **100:** 8024-8029.

Deslandes, L., Olivier, J., Theulieres, F., Hirsch, J., Feng, D.X., Bittner-Eddy, P., Beynon, J. and Marco, Y. (2002). Resistance to *Ralstonia solanacearum* in *Arabidopsis thaliana* is conferred by the recessive RRS1-R gene, a member of anovel family of resistance genes. *Proc. Nati. Acad. Sci.* USA, **99**: 2404-2409.

Devi, L.R. and Raj, J.S. (1966). The Rhizosphere microflora of tomato as influenced by the age of the plants. *Agric. Res. J, Kerala,* **4**: 60-66.

Devi, R.L. and Menon, M.R. (1980). Seasonal incidence of bacterial wilt of tomato. *Indian J. Microbiol.* **20:** 13-15.

Dey, P. K.(1947). Plant Pathology. United Provinces Agric. Dept. Report. 1944-45. pp. 38-40.

Dharmatti, P.R., Patil, R.V., Revenappa and Mannikeri, I.M. (2009). High yielding and bacterial wilt resistant tomato hybrids. *Karnataka J. Agric. Sci.* **22**(1): 158-160.

Digat, B. (1968). Why and how to distinguish the *Pseudomonas solanacearum* strains, causal agent of the bacterial cells of solanaceous and musaceous crops in the Caribbean Zone. *In: Proc. Caribbean Food Crops Soc.,* 6th Annual Meeting (Trinidad) **6**: 89-91.

Digat, B. and Cambra, M. (1976). Specificity of antigens in *Pseudomonas solanacearum* E.F. Smith and application of serology for studying bacterial wilt. *In: Proc. First Int. Planning Conf. and Workshop on the Ecol. and Control of bacterial Wilt caused by Pseudomonas solanacearum* (ed. Sequeira, L. and Kelman, A.), Raleigh North Carolina, July 18-24.

Dinesh, D., Aarons, A., McGill, G.E. and Yong, D. (1993). Genetic dessection of oligogeneic resistance to bacterial wilt in tomato. *Mol. Plant –Microbe Intera.* **7**(4): 464-471.

Dinesh, D., Aarons, A., McGill, G.E. and Yong, D. (1993). Genetic dessection of oligogeneic resistance to bacterial wilt in tomato. *Mol. Plant –Microbe Intera.* **7**(4): 464-471.

Dittapongpitch, V. and Surat, S. (2003). Detection of Ralstonia solanacearum in soil and weeds from commercial tomato fields using immunocapture and the polymerase chain reaction. *J. Phytopathol.* **151**: 239- 246.

Djian-Caporalino, C., Pijarovsky, L., Fazari, A., Samson, M., Gaveau, L., Lefebvre, V., Caranta, A., Palloix, A., and Abad, P. (2001). High resolution genetic mapping of the pepper (*Capsicum annuum* L.) resistance loci Me3 and Me4 conferring heat-stable resistance to root knot nematodes (*Meloidogyne* spp.). *Theor. Appl. Genet.* **103**: 592-600.

Dodds, J.H. (1989). Collaborative research at the international potato center on genetic manipulation for potato improvement. In: A. Mujeeb-Kazi and L.A. Sitch (eds.). *Rev. Advances in Plant Biotechnol.* CIMMY and IRRI. pp. 275-286.

Dogo, M., Toyota, H., Matsuda, K., Bingo, M., Naoki, Y., Kato, Y., Matsuda, Y., Tampo, H. and Ouchi, S. (1997). control of bacterial wilt of tomato in hydroponic culture by 3-indolpropionic acid and its detoxification in tomato plants. *Ann. Phytopathol. Soc. Japan,* **63** (5): 406-408.

Dogo, M., Zeng, X.M. and Liu, Q.G. (1999). Biological control of tomato bacterial wilt with avirulent bacteriogenic strain of *Ralstonia solanacearum. J. South China Agri. Univ.* **20**(4): 1-4.

Doijode, S. D. (2001). *Seed storage of horticultural crops* (pp. 157). Haworth Press: ISBN 1-56022-901-2

Drenth, A. (2004). Fungal epidemics - does spatial structure matter? *New Phytologist* (Blackwells) **163**(1): 4–7.

Drigues, P., Demery-Lafforgue, D., Trigalet, A., Dupin, P., Samain. D. and Asselineau, J. (1985). Comparative studies of lipopolysaccharide and exopolysaccharide from a virulent strain of *Pseudomonas solanacearum* and from 3 avirulent mutants. *J. Bacteriol.* **162**: 504-509.

Duffy, B.K., Defago, G. (1999). Environmental factors modulating antibiotic and siderophore biosynthesis by *Pseudomonas fluorescens* biocontrol strains. *Appl. Environ. Microbiol.* **65**: 2429– 2438.

Duijff, B.J., Gianinazzi-Pearson, V. and Lemanceau, P. (1997). Involvement of the outer membrane lipopolysaccharides in the endophytic colonization of tomato roots by biocontrol *Pseudomonas fluorescens* strain WCS417r. *New Phytopathol.* **135**: 325-334.

Dukes, P.D. and Others (1965).The identification and persistence of an indigenous race of *Pseudomonas solanacearum* in soil of Georgia. *Plant Report.* **49**: 589-590.

Dunne, C., Monne-Loccoz, Y., de Bruijn, F.J. and O'Gara, F. (2000). Over production of an inducible extracellular serine protease improves biological control of *Pythium ultimum* by *Stenotrophomonas maltophilia* strain W81. *Microbiology,* **146**: 2069-2078.

Düring, K. (1996). Genetic engineering for resistance to bacteria in transgenic plants by introduction of foreign genes. *Mole. Breeding* **2:** 297-305.

Dutta, A.K.and Verma, S.P. (1969). Control of bacterial wilt of eggplant with streptocycline. *Hind. Antibiot. Bull.* **11**: 260-261.

Dutta, S. Chakraborty, N., Ray, S. and Khatua, D.C. (2009). Integrated approach for management of bacterial wilt of tomato (S-02) *5th Int. Conf. Plant Pathol. Globalized Era*, Nov. 10-13, New Delhi, India 174 p

Eden-Green, S. J. (1994). Diversity of *Pseudomonas solanacearum* and related bacteria in south east Asia: new directions for moko disease. In: Hayward, A.C. and Hartman, G. L. Bacterial wilt: the disease and its causative agent, *Pseudomonas solanacearum*. 25-34.

Edreva, A. (2004). A novel strategy for plant protection: Induced resistance. *J. Cell and Mol. Biol.* **3**: 61-69.

Edward, F., Domino, E, H, and Tsenge, D.(1993). T, The New England Journal of Medicine, **329**: 437 August 5, 1993 Number 6.

El-Fallal, A.A. and Moussa, Z. (2008). Prospects for biocontrol of brown rot disease of potato *in vitro* and under greenhouse conditions. *Plant Pathol. J. Faisalabad,* **7**: 54-64.

Elphinstone, J. G. (2006). The current bacterial wilt situation: A global overview. In: Bacterial wilt disease and the *Ralstonia solanacearum* species complex, APS Press pp. 9–28.

Elphinstone, J. G., Hennessy, J., Wilson, J. K. and Stead, D. E. (1996). Sensitivity of different methods for the detection of *Ralstonia solanacearum* in potato tuber extracts. *EPPO Bull.* **26:** 663–678.

Elphinstone, J.G. (1996). Survival and possibilities for extinction of *Pseudomonas solanacearum* (Smith) Smith in cool climates. *Potato Res.* **39:** 403-410.

Elphinstone, J.G. (2005). The current bacterial wilt situation: a global overview. In: Bacterial wilt: The disease and the Ralstonia solanacearum species complex. C. Allen, P. Prior and C.A. Hayward eds. APS Press, St. Paul, Minnesota, USA.

Elphinstone, J.G. (2005). The current bacterial wilt situation: A global overview. In: Bacterial wilt disease and the *Ralstonia solanacearum* species complex, pp. 9-28. Edited by C. Allen, P. Prior and A.C. Hayward. St Paul, MN: APS Press.

Elphinstone, J.G. and Aley, P. (1993). Integrated control of bacterial wilt of potato in the tropics of Peru. In: Hartman, G.L. and Hayward, A.C. (eds): Bacteril Wilt. Proceeding of an International Conference held at Kaohsiun, Taiwan, 28-31 Oct.1992. Canberre, Australia, *ACIAP Proceeding* 45.

Elphinstone, J.G., Stanford, H.M. and Stead, D.E. (1998). Detection of *Ralstonia solanacearum* in potato tubers, *Solanum dulcamara*, and associated irrigation water In: Prior P, Allen C, Elphinstone, JG (eds.). *Bacterial Wilt Disease*: *Molecular and Ecological Aspects*. Springer, Berlin, Germany, pp. 133-139.

El-Raheem R. Abd. El-Shanshoury, Soad M. Abu El-Sououd, Omima A. Awadalla, Nabila B. and El-Bandy, N.B. (1996). Effects of *Streptomyces corchorusii*, *Streptomyces mutabilis*, Pendimethalin, and Metribuzin on the control of bacterial and *Fusarium* wilt of tomato. *Can. J. Bot.* **74**(7): 1016-1022.

EL-Raheem, A., El-Shanshoury, R., EL-Sououd, S.M.A., Awadalla, O.A. and EL-Brandy, N.B. (1995). Formation of tomatine in tomato plant infected with *Streptomyces* spp. and treated with herbicide, correlated with reduction of *Pseudomonas solanacearum* and *Fusarium oxysporum* f. sp. *lycopersici*. *Acta Microbiol. Polonica,* **44**(3/4): 255-266.

Empig, L.T. and others (1962). Screening tomato, eggplants and pepper varieties and strains for bacterial wilt (*Pseudomonas solanacearum* E.F.S.) resistance. *Philipp. Agric.* **46**: 303-314.

Enfinger, J.M. and Mc Carter, S.M. (1976). Efficacy of selected chemicals for controlling bacterial wilt of tomato. *Proc. Am. Phytopathol. Soc.* **3**: 335.

Enfinger, J.M., Mc Carter, S.M. and Jaworski, C. (1979). Evaluation of chemicals and application methods for control of bacterial wilt of tomato transplant. *Phytopathology,* **69**(6): 637-40.

Englyst, H.N., Kingman, S.M. and Cummings, J.H. (1992). Classification and measurement of nutritionally important starch fractions. *Eur. J. Clin. Nutr.* **46**: S33-S50.

EPPO (2006). EPPO Standards PM 10/1(1). Disinfection procedure in potato production. *Bull. OEPP/EPPO Bull.* **36:** 463-466.

EPPO. (2004). EPPO Standards PM 7/21. Diagnostic Protocols for Regulated Pests: *Ralstonia solanacearum. EPPO Bull.* **34**: 173-178.

EPPO/CABI. (2006). Distribution maps of plant diseases: *Ralstonia solanacearum* (2003-2006). *http://www cabi org/DMPD* 2006

Ercolini, D., Villani, F., Aponte, M. and Mauriello, G. (2006). Fluorescence in situ hybridisation detection of *Lactobacillus plantarum* group on olives to be used in natural fermentations. *Inte. J. Food Microbiol.* **112:** 291–296.

Erinle, I.D. (1976). A preliminary report on bacterial wilt in Nigeria. *Samaru Miscellaneous Paper No.* 19 pp.

Erocolani, G.L. (1984). Infectivity titration with bacterial plant pathogen. *Annu. Rev. Phytopathol.* **22**: 35-52.

Erocolani, G.L. Casarini, B. and Silivestri, G.C. (1961).Notes on tomato bacterial diseases on the plants of padua. *Indian Phytopath.* **11**: 105-108.

Evangelia Mourvaki, Stefania Gizzi, Ruggero Rossi and Stefano, R. (2005). Passion flower fruit—a "new" source of lycopene? *J. Medicinal Food.* **2005**: 104-106.

Fahy, P.C. and Persley, G.J. (1983). Plant bacterial diseases. A diagnostic guide. Academic Press, London, New York, Sydney. pp. 393.

Fall, P. A., Fredrikson M., Axelson, O. and Granérus, A. K. (1999). Nutritional and occupational factors influencing the risk of parkinson's disease: A case-control study in Southeastern Sweden. *Movement Disorders* **14** (1): 28-37.

Falla, T.J., Karunaratne, D.N. and Hancock, R.E.W. (1996). Mode of action of the antimicrobial peptide indolicidin. *J. Biol. Chem.* **271:** 19298-19303.

Farag, N.S., Eweda, W.E., Mostafa, M.I. and Balabel, N.M. (2004). Preliminary observations on the bacteriology and pathology of *Ralstonia solanacearum*. *Egyptian J. Agri. Res.* **82**: 1519-1523.

Farag, N.S., Fawzi, F.G., El-Said, S.I.A. and Mikhail, M.S. (1986).Streptomycin in relation to potato brown rot control. *ActaPhytopathol. Entomol. Hung.* **21**: 115-122.

Farag, N.S., Lashin, S.M., All-Abdel, R.S., Shatta, H.M., Seif-Elyazal, A.M. (1982). Antibiotics and control of potato black leg and brown rot diseases. *Agric. Res. Rev.* **60**: 149-166.

Fegan, M. and Prior, P. (2005). How complex is The *Ralstonia solanacearum* species complex, In: Allen C, Prior P and Hayward C (Eds.), Bacterial wilt: The disease and the *Ralstonia solanacearum* species complex, APS Press, St. Paul, Minnesota, USA. pp. 449-462.

Fegan, M. and Prior, P. (2004). Recent developments in the phylogeny and classification of *Ralstonia solanacearum*. Presentation at the 1st*International Tomato Symposium.* Orlando, FL.

Fegan, M. andPrior, P. (2006). Diverse members of the *Ralstonia solanacearum* species complex cause bacterial wilts of banana. *Austr. Plant Pathol.* **35**: 93-101.

Fegan, M., and Prior, P. (2005). Bacterial wilt disease and the *Ralstonia solanacearum* species complex. C. Allen, P. Prior, and C. Hayward, eds. American Phytopathological Society, St. Paul, MN.

Felix, S. (1973). A method of controlling bacterial wilt and eelworm on tomatoes. *Revue Agricole Sucriere de I'lle Maurice,* **57**(2): 50-53.

Feng, L.L., Qu, D.Y., Jin, L.P. and Lian ,Y. (2003). Genetic analysis of resistance to bacterial wilt (*Ralastonia solanacearum*) an eggplant (*Solanum melongena* L.). *Acta Horti. Sin.* **30**(2): 163-166.

Fessehaie, A., De Boer, S.H. and Levesque, C.A. (2003). An oligonucleotide array for the identification and differentiation of bacteria pathogenic on potato.*Phytopathology*, **93**: 262–269.

Fitzgerald, H. A., Chern, M. S., Navarre, R., and Ronald, P. C. (2004). Overexpression of (At) NPR1 in rice leads to A BTH- and environment induced lesion-mimic/cell death phenotype. *Mol. Plant-Microbe Intera.* **17**: 140-151.

Flavier, A.B., Ganova-Raeva, L.M., Schell, M.A., and Denny, T.P. (1997a). Hierarchical autoinduction in *Ralstonia solanacearum*: control of N-acylhomoserinelactone production by a novel autoregulatory systemresonsive to 3-hydroxypalmitic acid methyl ester. *J. Bacteriol.* **179**: 7089-7097.

Flavier, A.B., Ganova-Raeva, L.M., Schell, M.A., and Denny, T.P. (1997a). Hierarchical autoinduction in *Ralstonia solanacearum*: control of N-acylhomoserinelactone production by a novel autoregulatory systemresonsive to 3-hydroxypalmitic acid methyl ester. *J. Bacteriol.* **179**: 7089-7097.

Flavier, A.B., Clough, S.J., Schell, M.A. and Denny, T.P. (1997a). Identification of 3-hydroxypalmitic acid methyl ester as a novel autoregulator controlling virulence in *Ralstonia solanacearum*. *Mol. Microbiol.* **26**: 251-259.

Flor, H.H. (1971). Current status of the gene-for-gene concept. *Annu. Rev. Phytopathol.* **9**: 275-296.

Fock, I., Collonnier, C., Luisetti, J., Purwito, A., Souvannavong, V., Vedel, F., Servaes, A., Ambroise, A., Kodja, H., Ducreux, G. and Sihachakr, D. (2001). Use of *Solanum stenotomum* for introduction of resistance to bacterial wilt in somatic hybrids of potato. *Plant Physiol. and Biochem.* **39**: 899-908.

Fogoonee, I. (1978). Nematodes, *Meloidogyne incognita* and the growing of tomatoes (*Lycopersicon*). *Revue Agricole Sucriere de I'lle Maurice* **57**(2): 50-53.

François, P., Bento, M., Vaudaux, P., and Schrenzel, J. (2003). Comparison of fluorescence and resonance light scattering for highly sensitive microarray detection of bacterial pathogens. *J. Microbiol. Methods* **55**: 755-762.

Franke-Whittle, I.H., Klammer, S.H., and Insam, H. (2005). Design and application of an oligonucleotide microarray for the investigation of compost microbial communities. *J. Microbiol. Methods* **62**: 37-56.

Freedman, N.D., Park, Y., Subar, A.F. et al. (2008). Fruit and vegetable intake and head and neck cancer risk in a large united states prospective cohort study. *Int. J. Cancer,* **122** (10): 2330–2336.

French E.R. (1998). Recent Spread of bacterial wilt of potato: regional perspectives and experiences, Latin America, Pages 426-427. In: Prior, P., C. Allen and J. Elphinstone (eds.). Bacterial Wilt Disease: Molecular and Ecological Aspects. Reports of the Second International Bacterial Wiltsymposium held in Gosier, Guadeloupe, France, June 22-27, 1997. Springer Verlag. Berlin Heidelberg, INRA, Paris. 447 pp.

French, E. B., Gutarra, L., Aley, P. and Elphinstone. J. (1995). Culture media for *Ralstonia solanacearum* isolation, identification and maintenance. *Fitopatologia,* **30**: 126-130.

French, E. R., and Sequeira, L. (1970). Strains of *Pseudomonas solanacearum* from Central and South America: A comparative study. *Phytopathology,* **60**: 506-512.

French, E. B., Gutarra, L., Aley, P. and Elphinstone. J. (1995). Culture media for *Ralstonia solanacearum* isolation, identification and maintenance. *Fitopatologia,* **30**: 126-130.

Frey, P., Prior, P., Marie, C., Kotouiansky, A., Trigalet-Demery, D. and Trigalet, A. (1994). Hrp mutants of *Pseudomonas solanacearum* as potential biocontrol agents of tomato bacterial wilt. *Appl. Environ. Microbiol.* **60**: 3175-3181.

Friedman, M. (1997). Chemistry, biochemistry, and dietary role of potato polyphenols. A review. J. *Agric. Food Chem.*, **45** (5): 1523–1540.

Fuchsia, D. (2006). *Revolutionary Chinese Cookbook: Recipes from Hunan Province*, Ebury Press, pp. 202

Fucikovsky, L. (1980). Methods for diagnosing *Pseudomonas solanacearumi.* In: Developments in control of potato bacterial diseases: Report of a Planning Conf. Int. Potato Center (CIP), Lima, Peru. p. 40-48.

Fukuta, S., Lida, T., Mizukami, Y., Ishida, A., Ueda, J., Kanbe, M., and Ishimoto, Y. (2003). Detection of *Japanese yam mosaic virus* by RT-LAMP. *Arch. Virol.* **148**: 1713-1720.

Fuqua, W.C., Winans, S.C. and Greenberg, E.P. (1994). Quorum-sensing in bacteria: The Luxr-Lux I family of cell density-responsive transcriptional regulators. *J. Bacteriol.* **176:** 269–275.

Furuya, N. and Matsuyama, N. (1991). Biological control of the bacterial wilt of tomato with antibiotic producing strain of *Pseudomonas glumae. J. Faculty of Agri. Kyushu Univ.* **37**(2): 159-171.

Furuya, N., Kushima, Y., Tsuchiya, K., Matsuyama, N. and Wakimoto, S. (1991). Protection of tomato seedlings by pre-treatment with *Pseudomonas glumae* from infection with *Pseudomonas solanacearum* and its mechanism. *Ann. Phytopathol. Soc. Japan,* **57**(3): 363-370.

Furuya, N., Yamasaki, S. Mishioka, M., Shiraishi, I., IIyama, K. I. and Matsuyama, N. (1997). Antimicrobial activities of *Pseudomonas* against plant parasitic organisms and efficacy of *Pseudomonas aeruginosa* ATCC 7700 against bacterial wilt of tomato. *Ann. Phytopathol. Soc. Japan,* **63**(6): 417-424.

Gabriel, D.W. and Rolfe, B.G. (1990). Working models of specific recognition in plant-microbe interaction. *Annu. Rev. Plant Physiol.* **28**: 365-391.

Gabriel, D.W., Allen, C.and Schell, M. *et al.* (2006). Identification of open reading frames unique to a select agent: *Ralstonia solanacearum* race 3 biovar 2. *Mol. Plant-Microbe Interact.* **19**: 69-79.

Gadewar, A.V., Trivedi, T. P. and Shekhawat, G.S. (1991). Potato in Karnataka. *Tech. Bull.* No. 17, Central Potato Research Institute (CPRI), Shimla, India. 33 p.

Gallardo, P.B.; Panno, L.C. (1989) Biological control of bacterial wilt of potato induced by Pseudomonas solanacearum. *Revista de Microbiol.* **20**: 18-26.

Gallegly, M.E. (1949). Host nutrition and predisposition in relation to development of bacterial wilt of tomato. *Phytopathology,* **39**: 7.

Gallegly, M.E. Jr. and Walker, J.C. (1949a). Plant nutrition in relation to disease development V bacterial wilt of tomato. *Amer. J. Bot.* **36**: 613-623.

Gallegly, M.E. Jr. and Walker, J.C. (1949b). Relation of environmental factors of bacterial wilt of tomato. *Phytopathology,* **39:** 936-946.

Gao, G., Qu, D. Y., Lian, Y., Jin, L.P., and Feng, L.X. (2000). Identification of molecular markers linked with resistance to bacterial wilt (*Ralstonia solanacearum*) in diploid potato. *Acta Hort. Sin.* **27**: 37-41.

Garbeva, P., van Overbeek, L.S., van Vuurde, J.W.L. and Van Elsas, J.D. (2001). Analysis of endophytic bacterial communities of potato by plating and denaturing gradient gel electrophoresis (DGGE) of 16S rDNAbased PCR fragments. *Microbial Ecol.* **41**: 369-383.

Gargi Chakravarty and M.C. Kalita (2011). Management of bacterial wilt of brinjal by *P. fluorescens* based bioformulation. *ARPN J. Agri. and Biol. Sci.* **6**(3): 1-11.

Gassmann, W., Hinsch, M.E. and Staskawicz, B.J. (1999). The *Arabidopsis* RPS4 bacterial-resistance gene is a member of the TIR-NBS-LRR family of disease resistance genes. *Plant J.* **20**: 265–277.

Gautam, G. and Parkash, S. (2006). Evaluation of bacterial wilt resistant genotypes of tomato for yield and other traits under different growing seasons and transplanting dates in mid hills of Himachal Pradesh. *Himachal J. Agric. Res.* **32** (1): 10 (abstract)

Geetha, P. T. and Peter, K. V. (1993). Bacterial wilt resistance in a few selected lines and hybrids of brinjal *Solanummelongena* L. *J. Trop. Agri.* **31**(2): 274-276.

Genin, S., and Boucher, C. (2004). Lessons learned from the genome analysis of *Ralstonia solanacearum*. *Annu. Rev. Phytopathol.* **42**: 107-134.

Genin. S. and Denny, T.P. (2012). Pathogenomics of the *Ralstonia solanacearum* species complex. *Annu Rev Phytopathol.* **50:** 67-89.

Ger, M.J., Chen, C.H., Hwang, S.Y., Huang, H.E., Podile, A.R., Dayakar, B.V., and Feng, T.Y. (2002). Constitutive expression of *Hrp* genen transgenic tobacco plant enhances resistance against virulent bacterial pathogens by induction of a hypersensitive response. *Mol. Plant-Microbe Interact.* **15**: 764-773.

Germani G. and Plenchette, (2004). Potential of *Crotolaria* species as green manure crops for management of pathogenic nematodes and beneficial mycorrhizal fungi. *Plant and Soil,* **266**: 333-342.

Gerth van Wijk, R., Toorenenbergen, A.W. and Dieges, P.H. (1989). Occupational pollinosis in commercial gardeners. *Ned. Tijdschr Geneeskd*: **133**(42): 2081-2083.

Ghufran, S.M. and Chakravarti, B.P. (1960). Potato wilt in Ranchi. *Proc. BiharAcad. Agr. Sci.* VIII and IX.p. 126-127.

Gilbert, J.C., Tanaka, J.S. and Takeda, K.Y. (1974). Kewalo tomato. *Hort. Sci.* **9**: 481-482.

Gillings, M. and Fahy, P. (1993). Genetic diversity of *Pseudomonas solanacearum* biovar 2 and biovar N2 assessed using restriction endonuclease analysis of total genomic DNA. *Plant Pathol.* **42**: 744-753.

Girard, J.C., Marchand, J.L. and Mitchell, N.O.R. (1989). Search for bacterial wilt resistance tomato varieties in reunion in tomato and pepper production in tropics. *Proc. Int. Symp. Integrated Manage. Practices.* Taiwan, 21-26 March 1988 (Edited by Green, S.K.; Griggs, T.D.; Mc Cleen, B.T.)

Glick, B.R. (1995). The enhancement of plant growth by free– living bacteria. *Can. J. Microbiol.* **4**: 109-117.

Gnanamanickam, S.S., Lokeswari, T.S. and Nandini. K.R. (1979). Bacterial wilt of banana in South India. *Plant Dis. Rep.* **63**: 525-528.

Godwa, T.K.S., Shetty, K.S. and Patil, R.B. (1975). Antimicrobial activity of chebula fruit (*Terminalia Chebula* Retz.). *Curr. Res.* **4 (7)**: 117.

Godwa, T.K.S., Shetty, K.S., Balasubramanya, R.H., Shetty, K.P.V. and Patil, R.B. (1974). Studies on the bacterial wilt of solanaceous crops caused by *Pseudomonas solanacearum* E.F. Smith in the wilt sick soil. *Mysore J. Agri. Sci.* **8**(4): 560-566

Goldfrey, S.A. (1973). Tomato production in Sierra Leone: Problem and future research. *Acta Hort.* No. **33**: 97-104.

Gonzalez, E.T.and Allen, C. (2003). Characterization of a *Ralstonia solanacearum* operon required for polygalacturonate degradation and uptake of galacturonic acid. *Mol. Plant-Microbe Intera.* **16**: 536-544.

Goodman, R.N. (1959). The influence of antibiotics on plants and plant disease control. In: Antibiotics: Their chemistry and non-medical uses. H.S. Goldberg, ed. D. van Nostrand and Company, Inc. Princeton, NJ. pp. 322-448.

Gopal Krishnan, C. and Krishna, M. (2010-11). In Annual Report. *Phytophthora, Fusarium* and *Ralstonia* Diseases of Horticultural Crops. Indian Institute of Spices Research, pp. 91-92.

Gopalakrishnan, T.R., Singh, P.K., Sheela, K.B., Shankar, M.A., Kutty, P.C.J. and Peter, K.V. (2005). Development of bacterial wilt resistant varieties and basis of resistance in eggplant (*Solanum melongena* L.), In: Allen C., Prior P. and Hayward A. (eds.) Bacterial wilt disease and the *Ralstonia solanacearum* species complex, APS Press, St Paul, pp. 293-300.

Gopalakrishnan, T. R.and Peter K. V. (1991). Screening and selection for bacterial wilt resistance in chilli. *The Indian J. Genet. and Plant Breeding,* 332-334

Gopalkrishnan, T.R. (1996). Ujwala, a bacterial wilt resistant chilli variety. ICAR News (Oct-Dec, 1996) pp. 23.

Gopinath G. and Madalageri B.B. (1986). Bacterial wilt (*Pseudomonas solanacearum* Smith, E.F.) resistance in eggplant. *Veg. Sci.* **13**: 189-195.

Gopinath, G. (1983). Studies on the inheritance of bacterial wilt *(Pseudomonas solanacearum* E.F. Smith) resistance and characters of agronomic importance in brinjal *(Solanum melongena* L.). M.Sc. Thesis. University of Agricultural Sciences, Bangalore, India.

Gorissen, A., Overbeek, L. S. and Elsas, J. (2004). Pig slurry reduces the survival of *Ralstonia solanacearum* biovar in soil. *Can. J. Microbiol.* **50**: 587-593.

Gorris, M. T., Alarcón, B., López, M. M. and Cambra, M. (1994). Characterization of monoclonal antibodies specific for *Erwiniacarotovora* subsp. *atroseptica* and comparison of serological methods for its selective detection on potato tubers. *Appl. Environ. Microbiol.* **60:** 2076-2085.

Goth, R.W., Haynes, K.G.and Barksdale, T.H. (1991). Improvement of levels of bacterial wilt resistance in eggplant through breeding. *Plant Dis.* **75**: 398-401.

Gousset, C., Collonnie, C., Mulya, K., Mariska, I, Rotino, G. L., Besse, P., Servaes, A., Sihachakr, D. (2005). *Solanum torvum*, as a useful source of resistance against bacterial and fungal diseases for improvement of eggplant (*Solanum melongena* L.). *Plant Sci.* **168**: 319-327.

Gousset, C., Lian, Y., Mariska, I., Rotino, G., Servaes, A. and Sihachakr, D. (2004). Determination of the inheritance of resistance to bacterial wilt (*Ralstonia solanacearum*) in eggplant (*Solanum melongena*). Proc. of the 12th Eucarpia Meeting on Genetics and Breeding of *Capsicum and Eggplant*, p. 183.

Gowha, P.H.R. and Shivasbankara, K.T. (1990). Intervarietal hybridization in brinjal for bacterial wilt resistance. *Curr. Res. United Agric. Sci.* **19**(4): 70-71.

Graham ,J. and Lloyd, A.B.(1978). *Solanumcinereum* R. Br., a wild host of *Pseudomonas solanacearum* biotype II.*J. Aust.Insti. Agric. Sci.* **44**: 124-126.

Graham, J. and Lloyd, A.B. (1979). Survival of potato strain (race 3) of *Pseudomonas solanacearum* in the deeper soil layers. *Aust. J. Agric. Res.* **30**: 489-496.

Graham, J., Jones, D.A. and Lloyd, A.B. (1979). Survival of *Pseudomonas solanacearum* race 3 in plant debris and in latently infected potato tubers. *Phytopathology,* **69**: 1100-1103.

Graham, K.M. and Yap, T.C. (1976). Studies on bacterial wilt inheritance of resistance to *Pseudomonas solanacearum* in tomato. *Malays Agric. Res.* **5**: 1-8.

Granada, G.A. (1988). Latent infection of *Pseudomonas solanacearum* in potato and symtomless plant in bacterial diseases of the potato. pp. 93-108, International Potato Centre, Lima, Peru.

Granada, G.A. and Sequeira, L. (1983a). A new selective medium for *Pseudomonas solanacearum. Plant Dis.* **67**: 1084-1088.

Granada, G.A. and Sequeira, L. (1983b). Survival of *Pseudomonas solanacearum* in soil, rhizosphere, and plant roots. *Can. J. Microbiol.* **29**: 443-440.

Grant, M.R., Godiard, L., Straube, E., Ashfield, T., Lewald, J., Sattler, A., Innes, R.W. and Dangl, J.L. (1995). Structure of the *Arabidopsis* RPM1 gene enabling dual specificity of disease resistance. *Science* **269**: 843-846.

Green, L.S., Yee, B.C., Buchanan, B.B., Kamide, K., Sanada, Y. and Wada, K. (1991). Ferredoxin and ferredoxin-NADP reductase from photosynthetic and non photosynthetic tissues of tomato. *Plant Physiol.* **96**: 1207-1213.

Grey, B.E. and Steck, T.R. (2001). The viable but nonculturable state of *Ralstonia solanacearum* may be involved in long-term survival and plant infection. *Appl. Environ. Microbiol.* **67:** 3866–3872.

Griep, R.A., van Twisk, C., van Beckhoven, J.R.C.M., van der Wolf, J.M. and Schots, A. (1998). Development of specific recombinant monoclonal antibodies against the lipopolysaccharide of *Ralstonia solanacearum* race 3. *Phytopathology,* **88:** 795–803.

Grimault, V. and Prior, P. (1994a). Grafting tomato cultivars resistant or susceptible to bacterial wilt: analysis of resistance mechanisms. *J. Phytopathol.* **141**(3): 330–334.

Grimault, V., Anaïs, G.and Prior, P. (1994d). Distribution of *Pseudomonas solanacearum* in the stem tissues of tomato plants with different levels of resistance to bacterial wilt. *Plant Pathol.* **43**: 669-674.

Grimault, V., and Prior, P. (1994d). Invasiveness of *Pseudomonas solanacearum* in tomato, eggplant and pepper: A comparative study. *Eur. J. Plant Pathol.* **100**: 259-267.

Grimault, V., Prior, P. and Anals, G. (1995). A monogenic dominant resistance of tomato to bacterial wilt in hawaii 7996 is associated with plant colonization by *Psedomonas solanacearum*. *J. Phytopathol.* **143**: 349-352.

Grimault, V. and Prior, P. (1993). Bacterial wilt resistance in tomato associated with tolerance of vascular tissues to *Pseudomonas solanacearum. Plant Pathol.* **42**(4): 589-594.

Grimault, V., Anais, G. and Prior P. (2007). Distribution of *Pseudomonas solanacearum* in the stem tissues of tomato plants with different levels of resistance to bacterial wilt. *Plant Pathol.* **43**(4): 663–668.

Guan, W. Zhao, X. Hassell, R. and Thies, J. (2012). Defense mechanisms involved in disease resistance of grafted vegetables. *Hort. Science,* **47**(2): 164-170

Gueneron, M., Timmers, A.C.J., Boucher, C. and Arlat, M. (2000). Two novel proteins, Popb, which has functional nuclear localization signals, and Popc, which has a large leucine-rich repeat domain, are secreted through the hrp-secretion apparatus of *Ralstonia solanacearum. Mol. Microbiol.* **36**: 261-277.

Guidot, A., Elbaz. M., Carrère, S., Siri, M.I., Pianzzola, M.J., Prior, P. and Boucher, C. (2009). Specific genes from the potato brown rot strains of *Ralstonia solanacearum* and their potential use for strain detection. *Phytopathology*, **99**(9): 1105-1112.

Guidot, A., Prior, P. Schoenfeld J., Carrere, S. Genin, S. and Boucher, C. (2007). Genomic structure and phylogeny of the plant pathogen *Ralstonia solancearum* inferred from gene distribution analysis. *J. Bacteriol.* **189**: 377-387.

Gunawan, O.S. (1988a). Control of bacterial wilt (*Pseudomonas solanacearum* E.F. Smith) with Agrimycin 15/1.5WP on tomato Dago Bandung. *Bull. Penetition Horti.* **17**(3): 41-16.

Gunawan, O.S. (1988b). Resistance of tomato cultivars to bacterial wilt (*Pseudomonas solanacearum* E.F. Smith) at Lembang. *Bull. Penetition Hort.* **17**(1): 13-16.

Gunawan, O.S. (1988c). The yield of tomato cultivars caused by bacterial wilt (*Pseudomonas solanacearum* E.F.Smith) at Lembang. *Bull. Penetition Hort.* **16**(4): 25-26.

Gunawan, O.S. (1988d). The Effect of spraying intervals of streptomycin sulfate/oxytetracycline 15/1.5WP on bacterial wilt of tomato, *Pseudomonas solanacearum. Bull. Penetition Horti.* **17**(1): 13-16.

Gunawardena, U., and Seneviratne, S. N. de S. (1980). New hosts of *Pseudomonas solanacearum* . In: Proc. of the Thirty-Sixth Ann. Sessions of the Sri Lanka Assoc. for the Advancement of Sci. 34.

Guo J.H., Qi H.Y., Guo Y.H., Ge, H.L., Gong L.Y., Zhang, L.X. and Sun P.H. (2004). Biocontrol of tomato wilt by growth promoting rhizobacteria. *Biol. Cont.* **29**: 66-72.

Gupta, S.K., Dohroo, N.P. and Shyam, K.R. (1998). Occurrence of bacterial wilt of tomato in Himachal Pradesh. *Plant Dis. Res.* **13** (2): 174.

Habashy, W.H.S., Fawzi, F.G., El-Huseiny, T.M. and Neweigy, N.A. (1993). Bacterial wilt of potatoes. II. Sensitivity of the pathogen to antibiotics and pathogenesis by streptomycin-resistant mutants. *Egyptian J. Agri. Res.***71**: 401-412.

Hacisalihoglu, G., Ji, P., Longo, L.M. Olson, S. and Momol, T.M. (2007). Bacterial wilt induced changes in nutrient distribution and biomass and the effect of acibenzolar-*s*-methyl on bacterial wilt in tomato. *Crop Protec.* **26**(7): 978–982.

Hacisalihoglu, G., Momol. M.T., Wen, A., and Olson, S. (2010). Physiological responses of Ca-efficient and Ca-inefficient tomato genotypes to differential BW conditions. *Comm. Soil & Plant Analysis* **41**: 1426-1432

Hadidi, A., Czosnek, H., and Barba, M. (2004). DNA microarrays and their potential applications for the detection of plant viruses, viroids, and phytoplasmas. *J. Plant Pathol.* **86**: 97-104.

Hallmann, J., Quadt-Hallmann, W.F., Mahaffee, A. and Kloepper, J.W. (1997). Bacterial endophytes in the agricultural crops. *Can. J. Microbiol.* **43**: 895–914.

Hamilton, J.L. and Lowe, R.H. (198)1. Organic matter and N effects on soil nitrite accumulation and resultant nitrite toxicity to tobacco transplants. *Agron. J.* 73: 787-790

Hanson, P.M., Licardo, O, Hanudin, Wang J.W. and Chen, J.T. (1988a). Diallel analysis of bacterial wilt resistance in tomato derived from different sources. *Plant Dis.* **2**(1): 74-78.

Hanson, P.M., Wang, J.F., Licardo, O., Hanudin S.Y.M., Hartman G.I., Lin, Y.C. and Chen, J.T. (1988b). Variable reaction of tomato lines to bacterial wilt evaluated at several location in Southeast Asia. *Hort Science,* **31**(1): 143-146.

Hanudin (1987). Controlling the incidence of bacterial wilt (*Pseudomonas Solanacearum* E.F. Smith) on tomato plants by some plant extracts. *Bull. Penetition Horti.* **15**(4): 60-66.

Hanudin and Gaos, M.A.H. (1993). Screening of eggplant accessions forresistance to bacterial wilt. pp. 191-192.

Hanudin and Machand (1994). Effects of bactericide "Terlai" and *Pseudomonas fluorescens* on bacterial wilt of tomato. *Bacterial Wilt Newslett. No.* **10**: 10-12.

Hari Kishore and Pushkarnath. (1963). Brown rot of potatoes in India and its control. *Indian Potato J.* **6:** 3-10.

Harish Babu, B.N., Mahesh, P.A. and Venkatesh, Y.P. (2008). A cross-sectional study on the prevalence of food allergy to eggplant (*Solanum melongena* L.) reveals female predominance. *Clini. and Experi. Allergy,* **38**(11): 1795-1802,

Harris, D.C. (1976). Media for estimating a strain of *Pseudomonas solanacearum* in Kenyan soils by the dilution plate technique. *In: Proceeding of the First International Planning and Conference and Workshop on Ecology and Control of Bacterial wilt caused by Pseudomonas solanacearum* (ed. Sequira, L. and Kelman, A.) North Carolina State University, pp. 48-149.

Hartman, G.L. Hong W.F. and Wang, T.C. (1991). Survey of bacterial wilt on fresh market hybrid tomato in Taiwan. *Plant Protection Bull. Taiwan,* **33**(2): 197-203.

Hasabi, V. (2014). Study on the Response of some Solanaceous Plants to *Ralstonia solanacearum* Biovars 2A and 2T. *Turkish J. Agri. - Food Sci. and Technol.* **3**(2): 80-83

Hashimoto, N., Matsumoto, S., Yoshikawa, M., Horita, M. and Tsuchiya, K. (2001). Varietal resistance among red pepper and sweet pepper cultivars to *Ralstonia solanacearum* isolated Inkyoto Prefecture. *Japan J. Phytopathol.* **67**: 201–202.

Hauck R.D. (1984). Significance of nitrogen fertilizer microsite reactions in soil, p. 507-519. In: Hauck R.D., Beaton J.D, Goring C.A.I., Hoeft R.G., Randall G.W. and Russel D.A. (eds.), Nitrogen in crop production. ASA-CSSA-SSSA, Madison, WI, USA

Hauck, R. D. and Stephenson, H.F. (1965). Nitrogen sources, nitrification of nitrogen fertilizers. effect of nitrogen source, size and pH of granule, and concentration. *J. Agri. and Food Chem.* **13**: 486.

Hayward A.C. (1964). Characteristics of *Pseudomonas solanacearum. J. Appl. Bact.* **27**(32): 265-277.

Hayward A.C. (1985). Bacterial wilt caused by *Pseudomonas solanacearum* in Asia and Australia: an overview pp 15-24. *In Bacterial wilt disease in Asia and the South Pacific. ACIAR Proceeding* No.13.

Hayward, A.C. (1994). The hosts of *Pseudomonas solanacearum*.In: Bacterial Wilt: the Disease and its Causative Agent, *Pseudomonas solanacearum*, pp. 9-24. Edited by A.C. Hayward & G.L. Hartman. Wallingford: AB International.

Hayward, A.C., H.M. El-Nashaar, U. Nydegger, and L. De Lindo. (1990). Variation in nitrate metabolism in biovars of *Pseudomonas solanacearum. J. Appl. Bacteriol.* **69**: 269-280.

Hayward, A.C. (1987). Prospects for the integrated control of bacterial wilt (*Pseudomonas solanacearum). In Plant Pathology Bacteria* (ed. Civerodo, E.L.; Colmer, A; Davis, R.E. and Gillaspic, A.G.) Martinus Nijhoff Publisher. Dordrecht/Boston/Lancester.

Hayward, A.C. (1991). Biology and epidemiology of bacterial wilt caused by *Pseudomonas solanacearum. Annu. Rev. Phytopathol.* **29**: 65-87.

Hayward, A.C. (1995). *Pseudomonas solanacearum*. In: U.S. Singh, R.P. Singh and K. Kohmoto (eds.). *Pathogenesis and host specificity in plant diseases: histopathological, biochemical, genetic and molecular bases*. Vol.1 Prokaryotes (pp. 139-151). Tarrytown, N.Y.; Elsevier Science. Inc.

Hayward, A.C. and Hartman, G.L., eds. (1994). Bacterial wilt: The disease and its causative agent, *Pseudomonas solanacearum*. CABI, UK.

Hazarika, K., Rahman, F. and Bora, L.C. (2006). Avoidable yield loss of jute by combined infestations of *Meloidogyne incognita* and *Ralstonia solanacearum. Ann. Plant Protec. Sci.* **14** (2): 435-438.

He, L. Y. S., and L., Kelman, A. (1983). Characteristics of strains of *Pseudomonas solanacearum* from China. *Plant Dis.* **67**: 1357-1361.

He, L.Y. (1985). Bacterial wilt in the people's Republic of China pp. 40-48. In: *Bacterial wilt disease in Asia and the South Pacific. ACIAR Proceeding* No.13.

He, S.Y., Huang, H.C. and Collmer, A. (1993). *Pseudomonas syringae* pv. *syringae* harpin Pss: A protein that is secreted via the hrppathway and elicits the hypersensitive reponse in plants. *Cell,* **73:** 1255-1266.

He, S.Y., Nomura, K. and Whittam, T.S. (2004). Type III protein secretion mechanism in mammalian and plant pathogens. *Mol. Cell Res.* **1694**: 181-206.

Heim, S., Lleo, M.D.M., Bonato, B., Guzman, C.A. and Canepari, P. (2002). The viable but nonculturable state and starvation are different stress responses of *Enterococcus faecalis*, as determined by proteome analysis. *J. Bacteriol.* **184**: 6739-6745.

Henderson, W.R. and Jenkins, S.F.Jr. (1971). New tomatoes resistance to bacterial wilt. *Res. and Farm.* **29**(3/4): 10.

Henderson, W.R. and Jenkins, S.F.Jr. (1972). Venus and Saturn. *N.C. Agric. Exp. Stn. Bull. 444.*

Hendrick, C.A.and Sequeira, L. (1984). Lipopolysaccharide-defective mutants of the wilt pathogen *Pseudomonas solanacearum. Appl. Environ. Microbiol.* **48**: 94-101.

Hermsen, J.G. Th. (1987). Efficient utilization of wild and primitive species in potato breeding. In: G.J. Jellis and D.E. Richardson (Eds.). the production of new potato varieties. Cambridge University Press. pp. 172-185.

Hiddink, G.A., Termorshuizen, A.J., Raaijmakers, J.M. and van Bruggen, A.H.C. (2005). Effect of mixed and single crops on disease suppressiveness of soils. *Phytopathology,* **95**: 1325- 1332.

Hikichi, Y., Y. Nakazawa-Nasu, S. Kitanosono, K. Suzuki, and T. Okuno. 1999. The 16 behavior of genetically lux-marked Ralstonia solanacearum in grafted tomato cultivars 17 resistant or susceptible to bacterial wilt. *Ann. Phytopathol. Soc. Jpn.* **65**: 597-603.

Hikichi, Y., Yoshimochi, T., Tsujimoto, S. Shinohara, R., Nakaho, K., Kanda, A., Kiba, A., Ohnishi, *K.* (2007). Global regulation of pathogenicity mechanism of *Ralstonia solanacearum. Plant Biotechnol.* **24**:149-154.

Himathongkham, S. and Riemann, H. (1999). Destruction of *Salmonella typhimurium*, *Escherichia coli* O157:H7 and *Listeria monocytogenes* in chicken manure by drying and/or gassing with ammonia. *FEMS Microbiol. Lett.* **171**: 179-182.

Hinata, K. (1986). Eggplant (*Solanum melongena* L.). In: Y.P.S. Bajaj (Ed). Biotechnology in Agricultural and Forestry, Springer Verlag, Berlin, Heidelberg. 2(1): 363-370.

Hirai, T., Heymann, J. A. W., Shi, D., Sarker, R., Maloney, P. C. and Subramaniam, S. (2002). Three-dimensional structure of a bacterial oxalate transporter. *Nat. Struct. Biol.* **9:** 597–600.

Hoitink, H.A.J. and Boehm, M.J. (1999). Biocontrol within the context of soil microbial communities: A substrate-dependent phenomenon. *Annu. Rev. Phytopathol.* **37**: 427-446.

Hoque, M.O., Hoq, M.L. and Chowdhary (1981). Screening tomato varieties for resistance to bacterial wilt. *Bangladesh J. Agric. Res.* **6**(1): 55.

Horita, M. and Tsuchiya, K. (2001). Genetic diversity of Japanese strains of *Ralstonia solanacearum*. *Phytopathology,* **91:** 399-407.

Horita, M., Kazutaka, Y. and Kenichi, T. (2004). PCR based specific detection of *Ralstonia solanacearum* race 4 strains. *J. Plant Pathol.* **70:** 278-283.

Hossain, M. M., Islam, J. and Rahman, S. (1991). A note on the incidence on bacterial wilt of tomato. *Bangladesh J. Agri. Sci.* **18**(2): 291-292.

Hsu S.T., 1991. Ecology and control of *Pseudomonas solanacearum* in Taiwan.*Plant Protection Bulletin Taiwan* **33**: 72-79.

Huang, J.W.(1991) Control of soil borne crop diseases by soil amendments. *Plant Protection Bulletin* (Taipei) 33(1):113-123 *Fide Rev.Plant Path.*(1992) 71(8):4568

Huang, H.E., Liu, C.A., Lee, M.J., Kuo, C.G., Chen, H.M., Ger, M.J., Tsai, Y.C., Chen, Y.R., Lin, M.K., and Feng, T.Y. (2007). Resistance enhancement of transgenic tomato to bacterial pathogens bythe heterologous expression of sweet pepper ferredoxin-I protein. *Phytopathology,* **97**: 900-906.

Huang, J. and Schell, M.A. (1995). Molecular characterization of the eps locus of *Pseudomonas solanacearum* and its transcriptional regulation at a single promoter. *Mol. Microbiol.* **16**: 977-989.

Huang, J., Carney, B.F., Denny, T.P., Weissinger, A.K. and Schell, M.A. (1995). A complex network regulates expression of eps and other virulence genes of *Pseudomonas solanacearum. J. Bacteriol.* **177**: 1259-1267.

Huang, Q. and Allen, C. (1997). An exo-poly-alpha-D-galacturonosidase, PehB, is required for wild-type virulence of *Ralstonia solanacearum. J Bacteriol.* **179**: 7369-7378.

Huang, Q. and Lakshaman, D.K. (2010). Effect of clove oil on plant pathogenic bacteria and bacterial wilt of tomato and *Geranium. J. Plant Pathol.* **92**(3): 701-707.

Huang, T.C. and Burr, T.J. (1999). Characterizationof plasmids that encode streptomycin-resistance in bacterial epiphytotics of apple. *J. Appl. Microbiol.* **86**: 741-751.

Huang, Y., Xu, C., Ma, L., Zhang, K., Duan, C. and Mo, M. (2010). Characterization of volatiles produced from *Bacillus megaterium* YFM 3.25 and their nematicidal activity against *Meloidogyne incognita. Eur. J. Plant Pathol.* **126:** 417-422.

Huckelhoven, R. (2007). Cell wall–associated mechanisms of disease resistance and susceptibility. *Annu. Rev. Phytopathol.* **45**: 101-127.

Hudelson B.D., (1999). Southern wilt. University of Wisconsin Garden Facts, May 11, 1999.

Hudelson, B.D. (2005). University of Wisconsin pest alert - *Ralstonia* wilt. From Wikipedia, the free encyclopedia.

Hus, S.T. (1991). Ecology and control of *Pseudomonas solanacearum* in Taiwan. *Plant Prote. Bull. (Taipei),* **33**(1): 72-79.

Husain, A. and Kelman, A. (1958). Relation of slime production to mechanism of wilting and pathogenecity of *Pseudomonas solanacearum. Phytopathology,* **48**: 155-165.

Hussain, M. Zakir, M.A. Rahman, and Bashar, M.A. (2005). Screening of brinjal accessions for bacterial wilt caused by *Ralstonia solanacearum. Bangladesh J. Bot.* **34**(1): 53-58.

Hussain, Z. and Bora, B.C. (2008a). Integrated management of *Melodogyne incognita* and *Ralstonia solanacearum* complex in brinjal. *Indian J. Nematol.* **38**(2):159-164.

Hussain, Z. and Bora, B.C. (2008b). Species and race of root knot nematode prevalence in Nagaon District of Assam. *Indian J. Nemato* **38**(1): 104

Hutagalung, L. (1988). Garlic Bulb as a material for suppressing the incidence of bacterial wilt (*Pseudomonas solanacearum* E.F. Smith) on tomatoes. *Bull. Penelitian Hort. Kultura,* **16**(1): 84-93.

Hutchinson, C. M. (1913). Rangpur Tobacco Wilt: SAL 1&2 » Stacks » 616.015 .I39 V.1

Hutchinson, C.M. (1913b). Agriculture bacteriology. *India Board Sci. Advice Ann. Rep.* 1911/1912:128-131.

Huynh, T.V., Dahlbeck, D. and Staskawicz, B.J. (1989). Bacterial blight of soybean: Regulation of a pathogen gene determining host cultivar specificity. *Science,* **245**: 1374-1377.

Hylla, S., Gostner, A., Dusel, G., *et al.* (1998). Effects of resistant starch on the colon in healthy volunteers: Possible implications for cancer prevention. *Am. J. Clin. Nutr.* **67** (1): 136-142.

Iizuka, H., Tataki, H., Kurihara, K., Suda, T. and Hashino, F. (1988). The integrated control of soil borne disease of forced tomato. *Gumma J. Agri. Res. A (General),* **5**: 55-58.

Imazaki, I. and Nakaho, K. (2010). Pyruvate-amended modified SMSA medium: improved sensitivity for detection *of Ralstonia solanacearum. J. Gen. Plant Pathol.* **76** (1): 52.

In Moo Seong, Choi Eunjoo, ChoijaeEul (1996). Resistance of tomato cultivars to bacterial wilt caused by *Pseudomonas Solanacearum* and effect on soil sterilization. *RDA J. Agri. Sci. Crop Prote.* **38**(1): 473-476.

Ishikawa, R.F. and Matura, K. (1996). Antibacterial activity of validamycin against *Pseudomonas solanacearum* and its efficacy against tomato bacterial wilt. *Ann. Phytopathol. Soc. Japan,* **62**(5): 478-482.

Ishihara T, Mitsuhara I, Takahashi H, Nakaho K (2012). Transcriptome Analysis of Quantitative Resistance-Specific Response upon *Ralstonia solanacearum* Infection in Tomato. *PLoS ONE* **7**(10): e46763.doi:10.1371/journal.pone.0046763.

Islam, T.M.D. and Toyata, K. (2004). Suppression of bacterial wilt of tomato by *Ralstonia solanacearum* by incorporation of composts in soil and possible mechanisms. *Microbes and Environ.* **19**(1):153-160.

Isman, M.B. (2000). Plant essential oils for pest and disease management. *Crop Prot.,* **19**: 603–608.

Ito, S., Ushijima, Y., Fujii, T., Tanaka, S., Kameya-Iwaki, M., Yoshiwara, S. and Kish, F. (1998). Detection of visible cells of *Ralstonia solanacearum* in soil using a semi selective medium and a PCR technique. *J. Phytopath.* **146**: 379-384.

Jackson, M.T and Gonzdlez, L.C. (1981). Persistence of *Pseudomonas solanacearum* (Race 1) in a naturally infested soil in Costa Rica. *Phytopathology,* **71**: 690-693.

Jacobs, M.J., Bugbee, W.M. and Gabrielson D.A. (1985). Enumeration, location and characterization of endophytic bacteria within sugar beet roots. *Can. J. Bot.* **63**: 1262–1265.

Jakobi, M., Winkelmann, G., Kaiser, D., Kempler, C., Jung, G., Berg, G. and Bahl, H. (1996). Maltophilin: A new antifungal compound produced by *Stenotrophomonas maltophilia* R3089. *J. Antibiot.* (Tokyo), **49**: 1101-1104.

James D, Girija D, Mathew S, Nazeem PA, Babu TD, Varma A (2003). Detection of *Ralstonia solanacearum* race 3 causing bacterial wilt of solanaceous vegetables in Kerala, using random amplified polymorphic DNA (RAPD) analysis. *J. Trop. Agri.* **41**: 33-37.

James, T.Y., Kauff, F., Schoch, C.L.*et al.* (2006). Reconstructing the early evolution of the fungi using a six gene phylogeny. *Nature,* **443**: 818–822.

Janse, J.D (1988). A detection method for *Pseudomonas solanacearum* in symptomless potato tubers and some data on its sensitivity and specificity. ***EPPO Bulletin*** **18**(3): 343–351.

Janse, J. D. (1991). Infra and intraspecific classification of *Pseudomonas solanacearum* strains, using whole cell fatty acid analysis. *Syst. Appl. Microbiol.* **14**: 335-345.

Janse, J. D. (1996). Potato brown rot in Western Europe – History, present occurrence and some remarks on possible origin, epidemiology and control strategies. *EPPO Bull.* **26**: 679–695.

Janse, J.D. (2006). Phytobacteriology: Principles and Practice. CABI Publishing Wallingford, Oxon, UK, pp. 137.

Janse, J.D., Arulappan, F.A.X., Schans, J., Wenneker, M. and Westerhuis, W. (1998). Experiences with bacterial brown rot *Ralstonia solanacearum* biovar 2, race 3 in The Netherlands. In: Prior P., Allen C., Elphinstone J.G. (eds.) Bacterial Wilt Disease. *Molecular and Ecological Aspects*, Springer-Verlag, Heidelberg, Germany, pp. 146-152.

Jaworski, C.A. and Morton, D.J. (1964). An epiphtotic of *Pseudomonas solanacearum* in tomatoes on newly –cleared klej sand in relation to potassium, calcium and magnesium levels. *Plant Dis. Rep.* **48**: 88-89.

Jaworski, C.A. Pathak, S.C. Gitaitis, R.D. and Ghate, S.R. (1988). Bacterial wilt research for 1966 through 1986 in South Eastern Coastal Plain at Tifton, Georgia, U.S.A. *Bacterial Wilt Newslett. ACIAR No* **4**: 1-2.

Jaworski, C.A. Pathak, S.C., Pathak, Ghate, S.R. Gitaitis, R.D. and Widrleechner, M.P. (1987).A1565-2-4-BWT, GA 219-1-2 BWT, GA 1095-1-4 BWT, and GA 1405- 1-2 BWT Bacterial –tolerant tomato. *Hort. Science,* **22**: 324-325.

Jaxon, T.C.D. (2007). Host response to antagonistic bacteria and *Ralstonia solanacearum* (Smith) in brinjal: Identification of *Pseudomonas fluorescens* and detection of Phl D gene. *M.Sc. Dissertation* Submitted to Allahabad Agricultural Institute, Deemed University, p. 86.

Jay Shree, J. Raman and. Verma, J.P. (2002). Current scenario of bacterial diseases in India and their management. *Ann. Rev. Plant Pathol.* **1**: 27-50.

Jaya Prakash, M.G. and Rajan, K.M. (1975). Control of bacterial wilt of tomato (Abstract). *Indian J. Mycol. Plant Pathol.* **5**(1): 30.

Jennings P.H., Brannaman B.L. and Zscheile F.P.Jr. (1969): *Phytopathalogy* **59:** 963-67

Jetiyanon, K. and Kloepper, J.W. (2002). Mixtures of plant growth-promoting rhizobacteria for induction of systemic resistance against multiple plant diseases. *Biol. Contr.* **24**: 285–291.

Jeun, Y. C., Park, K. S. and Kim, C. H. (2001). Different mechanisms of induced systemic resistance and systemic acquired resistance against *Colletotrichum orbiculare* on the leaves of cucumber plant. *Mycobiology,* **29**: 19-26.

Ji, P., Momol, T., Olson, S.M., Hong, J., Pradhanang, P., Narayanan A. and Jones, J.B. (2004). New tactics for bacterial wilt management on tomatoes in the Southern US. *Ralstonia solanacearum* - North Carolina state university www.cals.ncsu.edu/course/pp. 728/Ralstonia/Ralstonia_solanacearum.htm

Ji, P., Momol, M.T., Olson, S.M. and Pradhanang, P.M. (2005). Evaluation of thymol as biofunmigant for control of bacterial wilt of tomato under field conditions. *Plant Dis.* **89**: 497–500.

Ji, P., C. Allen, A. Sanchez-Perez, J. Yao, J. Elphinstone, J. B. Jones, and M. T. Momol. (2007a). New diversity of *Ralstonia solanacearum* strains associated with vegetable and ornamental crops in Florida. *Plant Dis.* **91**: 195-203.

Ji, P., Momol, M. T., Olson, S.M. and Pradhanang, P.M. (2005). Evaluation of thymol as biofumigant for control of bacterial wilt of tomato under field conditions. *Plant Dis.* **89**: 497-500.

Ji, P., Momol, M.T., Rich, J.R., Olson, S.M., and Jones, J.B. (2007b). Development of an integrated approach for managing bacterial wilt and root-knot on tomato under field conditions. *Plant Dis.* **91**: 1321-1326.

Ji, P., Momol, T., Olson, S.M., Hong, J., Pradhanang, P., Narayanan A. and Jones, J.B. (2004). New tactics for bacterial wilt management on tomatoes in the Southern US. *Ralstonia solanacearum* - North Carolina state university www.cals.ncsu.edu/course/pp. 728/Ralstonia/Ralstonia_solanacearum.htm

Ji, P., Momol, M.T., Olson, S.M. and Pradhanang, P.M. (2005). Evaluation of thymol as biofunmigant for control of bacterial wilt of tomato under field conditions. *Plant Dis.* **89**: 497–500.

Johns, T. (1990). With bitter herbs they shall eat it: Chemical ecology and the origins of human diet and medicine, The University of Arizona Press, Tucson 1990, ISBN 0-8165-1023-7, p. 82-84

Jones, J.B., Jones, J.P., Stall, R.E., and Zitter, T.A. (1991). Compendium of tomato diseases. *Amer. Phytopathol. Soc.*, St. Paul, MN.

Julien, M. (1973).Culture *in vitro* of cells from leaf tissue of *Asparagus officinalis* the parenchymatous cells of differentiating culture. *Comptes Rendus,* **276**(5): 733-736.

Jyothi, A.R., Abraham, K. and Mathew, J. (1993). Bacterial wilt of chillies in Kerala and reaction of certain chilli accessions to *Pseudomonas solanacearum* (Smith) Smith. *ACIAR Bacterial Wilt Newslet.* **9**: 6.

Kaan, F., Messiaen, C.M. and Beramis, M. (1971). The effects of clinmate, soil and cultural condition on the devlopment of tomato bacterial wilt in guadeloupe. *Nouvelles Maraiche res et. Vivrieres de IIHRA aux. Antilles No.* **1**: 28-29.

Kabashima, K. and Miyachi, Y. (2004). Contact dermatitis due to eggplant contact dermatitis. **50**(2): 101-102.

Kalloo, G., Kapoor, K. S., Banarjee, M.K., Singh, B., Raj Kumar, Singh, M., Ram, D., Prakash, C., Kumar, S., Singh, N., De, M. and Verma, A. (1998). *Annual Report* (1996-97) Part-1 ICAR Directorate of Vegetable Research, Varanasi.

Kanda, A., Yasukochi, M., Ohnishi, K., Kiba, A., Okuno, T. and Hikichi, Y. (2003). Ectopic expression of *Ralstonia solanacearum* effector protein popa early in invasion results in loss of virulence. *Mol. Plant-Microbe Intera.* **16**: 447-455.

Kang, J., Man., Lee, M.H., Shim, J.K., Sang, T.S., Rosemary, S., Cho, M.S., Jang, H.H. and Dong, S.P. (2007). PCR-based specific detection of *Ralstonia solanacearum* by amplification of cytochrome c1 signal peptide sequences. *J. Microbiol. Biotechnol.* **17**(11): 1765–1771.

Kang, Y., Mao, G., Lu, C. and He, L. (1995). Biological control of bacterial wilt of tomato by extracellular protein defective mutant of *Pseudomonas solanacearum. Acta Phytophyl. Sin.* 22: 287–288.

Kang, Y.W., Liu, H.L., Genin, S., Schell, M.A. and Denny, T.P. (2002). *Ralstonia solanacearum* requires Type 4 pili to adhere to multiple surfaces and for natural transformation and virulence. *Mol. Microbiol.* **46**: 427-437.

Kanthaswamy, V., Natarajan, S., Srinivasan, K. and Ananthalakshmi, A. (2003). Genetic studies in brinjal (*Solanum melongena* L.). *S. Indian Hort.* **51**: 144-148.

Kao, C.C. and Sequeira, L. (1991). A gene cluster required for coordinated biosynthesis of lipopolysaccharide and extracellular polysaccharide also affects virulence of *Pseudomonas solanacearum. J. Bacteriol.* **173**: 7841-7847.

Kao, C.C. Barlow, E. and Sequeira, L. (1992). Extracellular polysaccharide is required for wilt type virulence of *Pseudomonas solanacearum. J. Bacterial.* **174**(3): 1068-1071.

Kao, C.C., Gosti, F., Huang, Y. and Sequeira, L. (1994). Characterization of a negative regulator of exopolysaccharide production by the plant-pathogenic bacterium *Pseudomonas solanacearum. Mol. Plant-Microbe Intera.***7**: 121-130.

Kapoor, A.S., Sugha, S.K. and Singh, D. (1991). Varietal resistance to bacterial wilt in tomato. *Indian Phytopath.* **44**(2): 224-225.

Kartan, J., Greenberger, A., Alon, H. and Grimtein, A. (1976). Solar heating by polyethylene mulching for the centre of diseases caused by soil borne pathogen. *Phytopathology,* **76**: 683.

Kaushik, V.K., Sharma, S.K. and Kumar, S. (2009). Environment-friendly system of cropping. *J. Env. Res. Develop.* **4**: 222-232.

Kayamura, T., Komaki, K. and Kiyota, H. (1995). Race of *Pseudomonas solanacearum* isolated from tomato and eggplant in Kumalmoto Prefecture. *Proc. Asso. Plant Protec. of Kyushu,* **41**: 21-23.

Keen, N.T. (1990). Gene-for-gene complementarity in plant-pathogen interactions. *Annu. Rev. Genet.* **24**: 447-463.

Kelaniyangoda, D.B. (1992). Study of micriflora in green manure and control of selectected antagonist against bacterial wilt caused by *Pseudomonas solanacearum.*1'64-73. SAVERNET Bacterial wilt Training Centre, AVROC Taiwan.

Kelaniyangoda, D.B., Tharmarajahi, S.K. and Heart, L.G. (1994 and 1995). Effect of greenmanure, organic manure and chemicals on the incidence of bacterial wilt (*Ralstonia (Pseudomonas) solanacearum*) in potato. *J. of Trop. Agri.* **150:** 35-47.

Kelman A., Hartman G.L. and Hayward, A.C. (1994). Introduction. In: Hayward A.C., Hartman G.L. (eds) Bacterial Wilt: The Disease and Its Causative Agents, *Pseudomonas solanacearum*. CAB International, UK, pp 1–7

Kelman, A. (1953). The bacterial wilt caused by *Pseudomonas solanacearum*. A literature review and bibliography. *Tech. Bull. N.C. Agric. Exp. Stat.* 99, pp. 194.

Kelman, A. (1954). The relationship of pathogenicity in *Pseudomonas solanacearum* to colony appearance on a tetrazolium medium. *Phytopathology,* **44**: 693-695.

Kelman, A. (1956). Factors influencing viability and variation in cultures of *Pseudomonas solanacearum. Phytopathology,* **46**: 16–17.

Kelman, A. (1976). Mission of the conference, In: *Proc. First Int. Planning and Conf. and Workshop on Ecology and Control of Bacterial wilt caused by Pseudomonas solanacearum* (ed. Sequira, L. and Kelman, A.) North Carolina State University, pp. 6-21.

Kelman, A. and Hruschka, J. (1973). The role of mortality and aerotaxis in the selective increase of avirulent bacteria in still broth cultures of *Pseudomonas solanacearum. J. Gen. Microbiol.* **76**: 177-188.

Kelman, A. and Jensen, J.H. (1951). Maintaining virulence in isolates of *Pseudomonas solanacearum. Phytopathology,* **41**: 185-187.

Kelman, A. and Person, L.H. (1961). Strains of *Pseudomonas solanacearum* differing in pathogenicity to tobacco and peanut. *Phytopathology*, **51**: 158-161.

Kelman, A. and Sequeira, L. (1961). Root infection in relation to spread of *Pseudomonas solanacearum. Phytopathology,* **52**: 16.

Kelman, A. and Sequeira, L. (1965). Root-to-root spread of *Pseudomonas solanacearum. Phytopathology,* **55**: 304-309.

Keshwal R.L. Joshi, I.K. and Kulkarni, S.N. (1975). Control of bacterial wilt of tomato by cultural practices. *Indian J. Mycol. Plant Pathol.* **5**(1): 32.

Keshwal, R.L., Khare, U.K. and Singh, R.P. (2000). Effect of physical properties of soil on wilt incidence and population of *Ralstonia solanacearum. Ann. Plant. Protec. Sci.* **8**(1): 40-43.

Keshwal, R.L. (1976). Survival of the soil and virulence of *Pseudomonas solanacearum*. In: Sequeira, L. and A. Kelman, *Proc. First Int. Conf. and Workshop on the Ecology and Control of Bacterial Wilt* caused by *Pseudomonas solanacearum*. Releigh, N.C. July 18-24, 1976.

Keshwal, R.L. and Joshi, L.K. (1979). Reaction of tomato, brinjal and chilli varieties to *Pseudomonas solanacearum* E.F.Smith. [India]. *Plant Dis.* **10**(2): 190-191.

Kessman, H., Staub, T., Hofman, C., Maetzke, T., Herzog, J.(1994). Induction of systemic acquired disease resistance in plants by chemicals. *Annu. Rev. Phytopathol.* **32**: 439-459.

Khakvar, R., Sijan, K., Wong, M.Y., Radu, S. and Thong, K.L. (2008). Improving a PCR-based method for identification of *Ralstonia solanacearum* in natural sources of West Malaysia. *Am. J. Agric. Bio. Sci.* **4**: 490-493.

Khan, A. A., Badshah, K. and Khan, M.W. (1998). Resistance of aubergine cultivars to root-knot nematodes. *Tests-of- Agrochemicals-and-Culti.* **19**: 40-41.

Khan, A.N.A. (1974). Studies on *Pseudomonas solanacearum* E.F. Smith causing wilt of brinjal, potato and tomato in Mysore state. *J. Agric. Sci.* **8**: 477-478.

Khan, A.N.A., Shetty, K.S. and Patil, R.B. (1980). Occurrence of bacterial wilt of chilli in Karnataka and other its relationship to the wilt of other solanaceous vegetables. *Indian Phytopath.* **32**: 507-512.

Khoury, W.E. and Makkouk, K. (2010). Integrated plant disease management in developing countries. *J. Plant Pathol.*92:(4) Supplement), S4.35-S4.42 *Edizioni ETS Pisa, 2010* S4.35

Kim, W.S. and Geider, K. (2000). Characterization of a viral eps-depolymerase, a potential tool for control of fire blight. *Phytopathology,* **90**: 1263-1268.

King, Y., Mao, G., Lu.C. and He, L.Y. (1995). Biological control of bacterial wilt of tomato by using exracellular protein defective mutant of *Pseudomonas solanacearum*. *Acta Phytophyl. Sin.* **22**(3): 287-288.

Király, Z., Elzahaby, H.M. and Klement, Z. (1997). Role of extracellular polysaccharide (eps) slime of plant pathogenic bacteria in protecting cells to reactive oxygen species. *J. Phytopathol.* **145**: 59-68.

Kishan–Swaroop, Suryanarayan M.A., Sharma, T.V.R.S., Anil Kumar, Swaroop, K. and Kumar, A. (2000 a). Screening brinjal (*Solananum melongena* L) varieties for bacterial wilt resistance and higher yield in Andaman and Nicobar Island. *J. Trop. Agri.* **38**(1-2): 90-91.

Kishan–Swaroop, Suryanarayan, M.A., Sharma, T.V.R.S., Anil Kumar, Swaroop, K. and Kumar, A. (2000 b). Performance of bacterial wilt resistant brinjal varieties/lines under two locations during rainy season in Andaman & Nicobar. *Madras Agri. J.* **87**(4-6): 356-358.

Kishore, H and Pushkarnath. (1963). Brown rot of potatoes in India and its control. *Indian Potato J.* **6**: 3-10.

Kishore, V, Sunaina, V and Shekhawat., G.S. (1991). Occurrence of *Pseudomonas solanacearum* race 1 in High Hills, a New Report. *J. Indian Potato Assoc.* **18**: 106-107.

Kishore, V., Sunaina, V. and Shekawat, G. S. (1992). Relative effectiveness of various crop rotations on the incidence of bacterial wilt. *Indian J. Agri. Sci.* **51:** 49.

Kishore, V., Sunaina, V. and Shekhawat, G.S. (1994). Estimation of infection by *Pseudomonas solanacearum* in weed and non-host plants. In: Potato: Present And Future. *Proc. Nat. Symp.* held at Modipuram during 1-3 March 1993 (Edited by Shekhawat, G.S; Khurana, S.M.P; Pandey, S.K. and Chandla V.K.).

Kishun, R. and Sohi, H.S. (1982).Vertical distribution of *Pseudomonas solanacearum* in cultivated and fallow land. *Zbl. Mikrobiol.* **137**: 643-645.

Kishun, R. (1987a). Bacterial diseases of vegetable crops. In: Advances in Horticulture, Vol. 6, Vegetable Crops Ed. K.L. Chadha and G. Kalloo, Malhotra Publishing House, New Delhi, p.69.

Kishun, R. (1987b). Loss in yield of tomato due to bacterial wilt caused by *Pseudomonas solanacearum. Indian Phytopath.* **40**(2): 152-155.

Kishun, R. and Chand, R. (1990). Efficiency of different methods of inoculation and inoculum concentration for inducing bacterial wilt in tomato. *Plant Dis. Res.* **5**(2): 126-131.

Kishun, R., Sohi, H.S. and Rao, M.V.B. (1980).Two collateral hosts of *Pseudomonas solanacearum*. *Curr. Sci.* **49**: 639.

Kissel, D. E., Sander, D. H. and Ellis, R. (1985). Fertilizer-plant interaction in alkaline soils. In: Engelstad, O.P. (Ed.), Fertilizer Technology and Use. *Soil Science Society of America*, Madison, WI, pp. 153-196

Kline, W.L., Roberts, W.J., Konia, S.T. and Johnson, S.A. (1999). Soil solarization to eliminate diseases from greenhouse. *Plasticulture*. **118**: 35-41.

Kloepper, J. W. (1993). Plant growth-promoting rhizobacteria as biological control agents. In: Metting F.B. (ed) *Soil microbial ecology-applications in agricultural and environmental management. Marcel Dekker Inc.*, New York, pp. 255–274.

Kloos, J.P., Fernandez, B.B., Tumapon, A.S. and Villanueva, L. (1991). Effects of short crop rotation on the incidence of potato bacterial wilt caused by *Pseudomonas solanacearum* E.F. Smith. *Asian Potato J.* **2**(1): 1-3.

Kobayashi, D.Y. and Palumbo, J.D. (2000). Bacterial endophytes and their effects on plants and uses in agriculture. In: Bacon C.W., White J.F. (eds) *Microbial endophytes. Marcel Dekker Inc.*, New York, pp 199-233.

Kojima, E.S. and Buddenhagon, I.W. (1969). Antigenic relationship of strains of *Pseudomonas solanacearum. Phytopathology,* **59**: 1035-1036. (Abstr.).

Komarova, M.S. and Korunets, I.V. (1997). Biological means for control of bacterial disease of tomatoes. *Zashchita Karantin Rasteni No.* **4**: 27.

Kostlanova, N., Mitchell, E.P., Lortat-Jacob, H. Oscarson S., Lahmann M., Gilboa-Garber N., Chambat G., Wimmerova M. and Imberty A. (2005). The fucose-binding lectin from *Ralstonia solanacearum*: a new type of betapropeller architecture formed by oligomerization and interacting with fucoside, fucosyllactose, and plant xyloglucan. *J. Biol. Chem.* **280**: 27839-27849.

Koul, O., Walia, S., and Dhaliwal, G.S. (2008). Essential oils as green pesticides: Potential and constraints. *Biopestic. Int.* **4**(1): 63-84.

Kovacs, N. (1956). Identification of *Pseudomonas pyocyanea*by the oxidase reaction. *Nature* London, **178**: 703.

Kratky, B.A. and Ko, W.H. (1974). A Virgin –soil technique for controlling bacterial wilt of greehouse-grown tomatoes. *Plant Dis. Repor.* **58**(1): 86-87.

Krausz, J.P. and Thurston, H.D. (1975). Break down of resistance to *Pseudomonas solanacearum*in tomato. *Phytopathology,* **65**(11): 1272-1274.

Kubota, R., Vine, B. G., Alvarez, A. M., and Jenkins, D.M. (2008). Detection of *Ralstonia solanacearum* by loop-mediated isothermal amplification. *Phytopathology,* **98**: 1045-1051.

Kuc J. (2001). Concepts and direction of induced systemic resistancein plants and its application. *Eur. J. Plant Pathol.* **107**: 7-12.

Kucharek, T. (1998). Bacterial wilt of row crops in Florida. 9 P. University of Florida, Cooperative Extension Service, Institute of Food and Agricultural Sciences, Florida, U. S. Cir. 1207.

Kulha, C.S., Prashar, A. and Sood, A.K. (1994). Survey for the occurrence of bacterial wilt of tomato in soil-hill sub-humid zone of Himachal Pradesh. *Plant Dis. Res.* **9**(2): 209-211.

Kumar, R. and Singh, M. (2006). Genetic resources, chromosome engineering, and crop improvement. Edited by Ram J. Singh Vegetable Crops, Vol. 3, CRC *Press* Pages 473–496.

Kumar, S. and Sharma, J.P. (1979). *In: Annual Report*, ICAR Research Complex for NEH Region, Shillong. p142

Kumar, V. Singh, B.M. and Sugha, S.K. (1997). Reaction of tomato genotype to bacterial wilt and *Phytophthora*fruit rot. *Plant Dis. Res.* **12**(1): 90-94.

Kyungseok, Diby Paul, Yong, K.K., Ki Woong N., Young Kee L., Hyo Won Choi and Sang, Y. L. (2007). Induced systemic resistance by *Bacillus vallismortis* EXTN-1 suppressed bacterial wilt in tomato caused by *Ralstonia solanacearum. Plant Pathol. J.* **23**(1): 22-25.

Lader, E.S. and Botstein, D. (1989). Mapping mendelian factors under lying quantitatively traits using rflps linkage maps. *Genetics,* **121**: 1855-199.

Laferriere, L.T., Hegeson, J.P. and Allen, C. (1999). Fertile *Solanum tuberosum* + *S. commersomii* somatic hybrids as souece of resistance to *Ralstonia solanacearum. Tag Theor. Appl. Genet.* **98**(8): 1272-1278.

Lafortune, D., Béramis, M., Daubèze, A.-M., Boissot, N., and Palloix, A. (2005). Partial resistance of pepper to bacterial wilt is oligogenic and stable under tropical conditions. *Plant Dis.* **89**: 501-506.

Lasso, R. (1974). Development of tomato varieties resistance to *Pseudomonas solanacearum* by using interspecific crosses. *Fitotecnia Latinoamericana,* **19**(1): 45-51.

Lavie, M., Shillington, E., Eguiluz, C., Grimsley, N. and Boucher, C. (2002). Popp1, a new member of the Yopj/Avrrxv family of type III effector proteins, acts as a host-specificity factor and modulates aggressiveness of *Ralstonia solanacearum. Mol.Plant- Microbe Interact.* **15**: 1058-1068.

Layne, R.E. and Mckeen, C.D. (1967). Southern bacterial wilt of field tomatoes in South Western Ontario, Canad. *Plant. Dis. Surv.* **47**: 97-8.

Lazarovits, G., Tenuta, M.L. and Conn, K.L. (2001). Organic Amendments as a disease control strategy for soilborne disease of high-value agricultural crops. *Australasian Plant Pathol.* **30**: 111-117.

Lebeau, A., Daunay, M.C., Frary, A., Palloix, A., Wang, J.F., et al. (2011). Bacterial wilt resistance in tomato, pepper, and eggplant: genetic resources respond to diverse strains in the *Ralstonia solanacearum* species complex. *Phytopathology*, **101:** 154-165.

Lee, T.J., Coyne, D.P., Clemente, T.E. and Mitra, A. (2002a). Partial resistance to bacterial wilt in transgenic tomato plants expressing antibacterial lactoferrin gene. *J. Amer. Soc. Hort. Sci.* **127**: 158-164.

Lee, Y.A., Fan, S.C., Chiu, L.Y.and Hsia, K.C.(2001). Isolation of an insertion sequence from *Ralstonia solanacearum* race 1 and its potential use for strain characterization and detection. *Appl. Environ. Microbiol.* **67**: 3943-3950.

Leksomboon, C. (2009). Evaluation of *Azadirachta indica* as a soil amendment for controlling bacterial wilt of tomato. *Kasetsart J. (Natural Sciences),* **43**(2): 239-244.

Lemaga, B. (1999-2000). Integrated control of potato bacterial wilt in Kabale District, Southwestern Uganda. *CIP Prog. Repor.* p. 141.

Lemaga, B., Kakuhenzire, R., Kassa, B., Ewell, P.T. and Priou, S. (2005). Integrated control of potato bacterial wilt in Eastern Africa: The experience of African highlands initiative. In: Allen C., Prior P. and Hayward A.C. (eds.) Fate of *Ralstonia solanacearum* biovar 2 as affected by conditions and soil treatments in temperate climate zones. *The Amer. Phytopathol. Soc.*, Minnesota, USA, pp. 145-157.

Lemaga, B., Kanzikwera, R., Kakuhenzire, R., Hakiza, J.J. and Maniz, G. (2001). The effect of crop rotation on bacterial wilt incidence and potato tuber yield. *Afri. Crop Sci. J.* **9**(1): 257-266.

Lemaga, B., Siriri, D. and Ebanyat, P. (2001). Effect of soil amendments on bacterial wilt incidence and yield of potatoes in Southwestern Uganda. *Afri. Crop Sci. J.* **9**(1): 267-278.

Lemessa, F. and Zeller, W. (2007). Screening rhizobacteria for biological control of *Ralstonia solanacearum* in Ethiopia. *Biol. Contr.* **42**: 336-344.

Li, H.T., Zou, Q.D., Lu S.W., Mu, X., Xu, W.K.and Xu, K.H. (2002). Resistance heredity of eggplant to *Pseudomona solanaceanum* Smith E.F. _ Genetic analysis of eggplant resistance material LS1934. *Liaoning Agric. Sci.* **3**: 1-3.

Li, M., Wang, Y.Q., Tian, S.B., Luo, Z.Y. and Wang, X.J. (2006). Genetic analysis of resistance to bacterial wilt and identification of an associated AFLP marker in eggplant (*Solanum melongena*), *Acta Hort. Sin.* **33**: 869-872.

Li, M., Wang, Y.Q., Tian, S.B., Luo, Z.Y. and Wang, X.J. (2006). Genetic analysis of resistance to bacterial wilt and identification of an associated AFLP marker in eggplant (*Solanum melongena*), *Acta Hort. Sin.* **33**: 869-872.

Li, P., Zhou, H., Wu, X., Wang, Z., Ho, H, H.,Wu,Y. Mao, Z.and He, Y. (2012). First record and description of *Ralstonia solanacearum*wilt in Patchouli from Yunnan Province, China. *Indian Phytopathol.* **65**(2): 208-210.

Libman, G. and Leach, J.G. (1962). A study of the role of some nematodes in the incidence and severity of southern bacterial wilt of tomato. *Phytopathology,* **52**:1219.

Libman, G., Leach, J.G. and Adams, R.E. (1964). Role of certain plant parasitic nematodes in infection of tomatoes by *Pseudomonas solanacearum*. *Phytopathology,* **54**: 151-153.

Lin, C. H., Hsu, S. T., Tzeng, K. C. and Wang, J. F. (2009). Detection of race 1 strains of *Ralstonia solanacearum* in field samples in Taiwan using a BIO-PCR method. *Eur. J. Plant Pathol.* **124**: 75 -85.

Lin, C.H., Hsu, S.T., Tzeng, K.C., Wang, J.F. (2008). Application of a preliminary screen to select locally adapted resistant rootstock and soil amendment for integrated management of tomato bacterial wilt in Taiwan. *Plant Dis.* **92**(6): 909-916.

Lin, T.C. and Horng, S. (1995). Varietial improvement and regional yield trial of FMTT tomato lines. *Taichung District Agricultural Improvement Station* No 48, 11-6.

Lin, W.C., Lu, C.F., Wu, J.W., Cheng, M.L., Lin, Y.M., Yang, N.S., Black, L., Green, S.K., Wang, J.F., and Cheng, C.P. (2004). Transgenic tomato plants expressing the *Arabidopsis NPR*1 gene display enhanced resistance to a spectrum of fungal and bacterial diseases. *Transgenic Res.* **13**: 567-581.

Lin, Y.H., Xu, J.L., Hu, J.Y., Wang, L.H., Ong, S.L., Leadbetter, J.R. and Zhang, L.H. (2003). Acyl-homoserine lactone acylase from *Ralstonia* strain XJ12B represents a novel and potent class of quorum-quenching enzymes. *Mol. Microbiol.* **47**: 849-860.

Lindhout, P. (2002). The perspectives of polygenic resistance in breeding for durable disease Resistance. *Euphytica,* **124**: 217-226.

Ling, M., Chen, H.P., Lu, M.Q. and Yang, H.M. (1993). Biotypes of *Pseudomonas solanacearum* and its distribution on tomato. *Acta Agri. Z.* **5**(1): 62-64.

Linnaeus, C. (1753). *Species Plantarum*, 1st ed. Stockholm: *L. Salvius*.

Liu, H.L., Kang, Y.W., Genin, S., Schell, M.A. and Denny, T.P. (2001). Twitching motility of *Ralstonia solanacearum* requires a type IV pilus system. *Microbiology,* **147**: 3215-3229.

Liu, H.L., Zhang, S.P., Schell, M.A. and Denny, T.P. (2005). Pyramiding, unmarked deletions in *Ralstonia solanacearum* shows that secreted proteins in addition to plant cell-wall-degrading enzymes contribute to virulence. *Mol. Plant-Microbe Intera.* **18**: 1296-1305.

Liu, L., Kloepper, J.W., Tuzun, S., (1995). Induction of systemic resistance in cucumber by plant growth-promoting rhizobacteria: duration of protection and effect of host resistance on protection and root colonization. *Phytopathology,* **85**: 1064-1068.

Liu, Y., Cady, N.C., and Batt, C.A. (2007). A plastic microchip for nucleic acid purification. *Biomed Microdevices* **9**: 769-776.

Liyuan, H. (1993). Current status of research on bacterial wilt of potato in China. pp. 11-22. In: Integrated Management of Bacterial Wilt: Proceedings of an International Workshop held in New Delhi, India, October 11-16, 1993/edited by Bill Hardy, Edward R. French.-Lima, Peru: International Potato Center, 1995. 196 p.

Locascio, S.J., Stall, R.E. and Stall, W.M. (1998). Bacterial wilt expression in tomato as influenced by cultivars and lime. *Proc. Florida State Hort. Soc.* **101**: 356-358.

Lockhart, D.J., Dong, H., Byrne, M.C., Follettie, M.T., Gallo, M.V., Chee, M.S., Mittmann, M,, Wang, C., Kobayashi, M., Horton, H., and Brown, E.L. (1996). Expression monitoring by hybridization to high densityolig onucleotide arrays. *Nat. Biotechnol.* **14**: 1675-1680.

Loper, J.E. and Schroth, M.N. (1986). Influence of Bacterial sources of indole-3-acetic acid biosynthetic on root elongation of sugar beet. *Phytopathology,* **76**: 386-389.

Lopes, C.A. and Boiteux, L.S. (2004). Biovar-specific and broad-spectrum sources of resistance to bacterial wilt (*Ralstonia solanacearum*) in *Capsicum*. Crop Breeding and Appl. Biotechnol. **4**: 350-355.

Lopes, C.A., QueZado-Soares, A.M. and Milo, P.E. DE (1994). Differential resistance of tomato cultigens to biovar I and III of *Pseudomonas solanacearum*. *Plant Dis.* **78**: 1091-1094.

Lopes, C.A. and Quezado-Soares, A.M. (1995). Stability of resistance in the potato. Achat to bacterial wilt. *Hortic. Bras.* **13**: 57-58.

Lopez, M.M., Gorris, M.T., Llop, P., Cubero, J., Vicedo, B. and Cambra, M. (1997). Selective enrichment improves the isolation, serological and molecular detection of plant pathogenic bacteria. In: Dehne H.W., Adam G., Diekmann M., Frahn J., Mauler A (eds.) Diagnosis and identification of plant pathogens. Kluwer, Dordrecht.

Lopez, M.M. and Biosca, E.G. (2004). Potato bacterial wilt management: New prospects for an old problem. In: Allen C., Prior P. and Hayward A.C. (eds.) Bacterial Wilt Disease and the *Ralstonia* species Complex. APS. Press, St. Paul, Minnesota, USA, pp. 205-224.

López, M.M., Llop, P., Cubero, J., Penyalver, R., Caruso, P., Bertolini, E., Peñalver, J., Gorris, M.T., Olmos, A. & Cambra, M. (2001). Strategies for improving serological and molecular detection of plant pathogenic bacteria. Proceedings of the 10th International Conference on Plant Pathogenic Bacteria. Charlottetown, Canada, Kluwer, pp. 83–85.

Lottmann, J., Heuer, H., Smalla, K. and Berg, G. (1999). Influence of transgenic T4- lysozyme-producing potato plants on potentially beneficial plant-associated bacteria. *FEMS Microbiol. Ecol.* **29**: 365-377.

Lottmann, J., Heuer, H., de Vries, J., Mahn, A., Düring, K., Wackernagel. W., Smalla, K. and Berg, G. (2000). Establishment of introduced antagonistic bacteria in the rhizosphere of transgenic potatoes and their effect on the bacterial community. *FEMS Microbiol. Ecol.* **33**: 41-49.

Lottmann, J., Heuer, H., Smalla, K. and Berg, G. (1999). Influence of transgenic T4- lysozyme-producing potato plants on potentially beneficial plant-associated bacteria. *FEMS Microbiol. Ecol.* **29**: 365-377.

Louws, F. J., Rademaker, J. L.W. and de Bruijn, F.J. (1999). The three Ds of PCR-based genomic analysis of phytobacteria: diversity, detection, and disease diagnosis. *Annu. Rev. Phytopathol.* **37**: 81–125

Loy, A., Lehner, A., Lee, N., Adamczyk, J., Meier, H., Ernst, J., Schleifer, K.H., and Wagner, M. (2002). Oligonucleotide microarray for 16S rRNA gene based detection of all recognized lineages of sulfate-reducing prokaryotes in the environment. *Appl. Environ. Microbiol.* **68**: 5064-5081.

Loy, A., Schulz, C., Lucker, S., Schopfer-Wendels, A., Stoecker, K., Baranyi, C., Lehner, A., and Wagner, M. (2005).16S rRNA gene-based oligonucleotide microarray for environmental monitoring of the betaproteobacterial order "Rhodocyclales".*Appl. Environ. Microbiol.* **71**: 1373-1386.

Lozano, J.C. and Sequeira, L. (1970). Differentiation of races of *Pseudomonas solanacearum* by a leaf infiltration technique. *Phytopathology,* **60**: 833-838.

Lu, M.Q., Chen, Y.H., Lin, M.S. and Chen, H.P. (1992). An experiment using grafting for control of tomato bacterial wilt. *Plant Protec.* **18**(3): 25.

Luckwill, L. C. (1943). The genus *Lycopersicon*: An historical, biological, and taxonomical survey of the wild and cultivated tomatoes. *Aberdeen Univ. Stud.* **120**: 1-44.

Lum, K.Y. and Kelman, A. (1981). Infectivity titrations of *Pseudomonas solanacearum. Phytopathology*, 81:891.

Lum, K.Y. and Wang, H.K. (1976). Control of bacterial wilt of tomatoes in the low lands through grafting. *ARDI Rearch Bull.* **4**(1): 28-33.

Lyon, G.D., Reglinski, T. and Newton, A.C. (1995). Novel disease controlcompounds: The potential to "immunize" plants against infection. *Plant Pathol.* **44**: 407-427.

Machmud, M. and Suryad, Y. (2008). Detection and identification of *Ralstonia solanacearum* strains using the indirect ELISA technique. *Indonesian J. Agri.* **1**(1): 13-21.

Madalageri, B.B. and Dharmatti, P.R. (1991). A new tomato cv. L-15 for North Karnataka. *Karnataka J. Agric. Sci.* **4**: 150-153.

Madalageri, B.B. and Ramanjini Gowda, P.H. (1988). Inheritance of characters in the progeny of *Solanum macrocarpon* and *S. melongena. Proc. Conf. Cytol. Genet.* **1:** 199-202

Madalgeri, B. B., Sulladmath, V.V. and Belkhindi, G. B. (1983). Wilt resistant highyielding hybrid brinjal. *Curr. Res.* **12**(2): 108-109.

Madden, L.; Gareth, H., Frank Van Den, B. (2007). *Study of Plant Disease Epidemics. Am. Phytopathol. Soc.* ISBN 978-0-89054-354-2.

Mäe, A., Montesano, M., Koiv, V. and Palva, E.T. (2001). Transgenic plants producing the bacterial pheromone N-Acyl-homoserine lactone exhibit enhanced resistance to the bacterial phytopathogen *Erwinia carotovora. Mol. Plant-Microbe Intera.* **14**: 1035-1041.

Mahanta, L.C., Dhal, A. and Mohanty, A.K. (1998). Field performance of tomato cultivars in relation to wilt and leaf curl disease in North Central Plateau Zones of Orrisa. *Orissa J. Hort.* 26(1): 34 -39.

Makoto, F. (2003). Detection of a molecular marker linked to bacterial wilt resistance in eggplant. *Bull. Agri., Food and Environ. Sci. Res.* **39**: 6-9.

Mallikarjun, Y. Kenganaland and Byadgi, A.S. (2008). Management of bacterial wilt in tomato. *The Hindu*, Thursday, Oct 02, 2008.

Malya, K., Watneble, M., Goto, M. Takikawa, Y. and Tsuyumu, S.(1996). Suppression of bacterial wilt disease of tomato by root-dipping with *Pseudomonas fluorescences* Pfg-32-the role of antibiotic substances and rhizosphere production. *Ann. Phytopathol. Soc. Japan,* **62**(2): 134-140.

Mangin. B., Thoquet, P., Olivier, J., Grimsley, N. H. (1999). Temporal and multiple quantitative trait loci analyses of resistance to bacterial wilt in tomato permit the resolution of linked loci. *Genetics,* **151**: 1165-1172.

Mao, G.Z. and He, L.Y. (1998). Relationship of wild type strain motility and interaction with host plants in *Ralstonia solanacearum.* In: Allen C., Prior P., Elphinstone J., eds. *Bacterial wilt disease. Molecular and ecological aspects.* Berlin: Springer-Verlag, 184.

Martin, C. and French, E.R. (1985). Bacterial wilt of potato: *Pseudomonas solanacearum. Technical Infor. Bull.* 13.2nd ed., revised. International Potato Center (CIP), Lima, Peru. 16 p.

Martin, C., French, E.R. and U. Nydegger. (1982). Strains of *Pseudomonas solanacearum* affecting solanaceae in the Americas. *Plant Dis.* **66**: 458-460.

Martin, G.B., Brommonschenkel, S.H., Chunwongse, J., Frary, A., Ganal, M.W., Spivey, R., Wu, T., Earle, E.D. and Tanksley, S.D. (1993). Map-based cloning of a 66 protein kinase gene conferring disease resistance in tomato. *Science,* **262**: 1432- 1436.

Mathew, J., S. Beena, A.K. Cherian, and K. Abraham. (1993a). Bacterial wilt of nutmeg *(Myristicafra Srans* L.) incited by *Pseudomonas solanacearum* (Smith) Smith. *ACIAR Bacterial Wilt Newslett.* **9**: 8.

Mathew, J., S. Beena, and A.K. Cherian. (1993b). Bacterial wilt of mulberry *(Morus alba* L.) incited by *Pseudomonas solanacearum* (Smith) Smith from India. *ACIAR Bacterial Wilt Newslett.* **9**: 11.

Mathew, S.K. and Peter, K.V. (2004). Bacterial wilt in warm humid tropics. Kerala Agricultural University, Vellanikkara, Trisur, Kerala 24.

Mathurot, C., Somporn, C. and Lumyong, S. (2009). Screening of siderophore producing bacteria as potential biological control agent for fungal rice pathogens. *World J. Microbiol. Biotechnol.* **25**: 1919-1928.

Matos, F.S.A., C.A. Lopes, and A. Takatsu. (1990). Identificacxa˜o de fonts de resiste^ncia a *Pseudomonas solanacearum* en *Capsicum* spp. *Hort. Brasil.* **8**: 22–23.

Matsuda, K., Toyoda, H., Nishio, H., Nishida, T., Dohgo, M., Bingo, M., Matsuda , Y., Yashida, S., Harada, S., Tanaka, H., Komai, K. and Ouchi, S. (1998). Control of the bacterial wilt of tomato plant by a derivative of 3-indolpropionic acid based on selective actions on *Ralstonia solanacearum. J. Agri. and Food Chem.* **46**(10): 4416-4419.

Matsunaga, H. and Monma, S. (1995). Varietal differences in resistance to bacterial wilt in related species of *Capsicum annuum. Capsicum Eggplant Newsl.* **14**: 60-61.

Matsunaga, H. and Monma,S. (1999). Sources of resistance to bacterial wilt in *capsicum. J. Jpn. Soc. Hort. Sci.* **68**: 753–761.

Mayak, S., Tirosh, T. and Glick, B.R. (1999). Effect of wild type and mutant plant growth promoting rhizobacteria on the rooting of mung bean cuttings. *J. Plant Growth Regul.* **18**: 49-53.

Mazumder, N. (1998). Managing *Ralstonia solanacearum* wilt in tomato. *J. Mycol. Plant Pathol.* **28**(2): 189-192.

Mazzucchi, V., Manzos, A., Forte, A., Fortea, L. and Mainol, H.P. (2000). Control of *Pseudomonas solanacearum* obligatory in Italy too. *Infor. Agrario,* **56**(10): 97-101.

Mc Carter, S.M and Jaworski, C.A. (1969). Field Studies on spread of *Pseudomonas solanacearum* and tobacco mosaic virus in tomato plant by clipping. *Plant Dis. Repor.* **53**(12): 942-946.

Mc Carter, S.M. (1968). A procedure for infesting field soils with *Pseudomonas solanacearum. Phytopathology,* **66**(7): 799-800.

Mc Carter, S.M. (1976). Persistance of *Pseudomonas solanacearum* in artificially infected soils. *Phytopathology,* **66**(8): 998-1000.

Mc Carter, S.M. and Jaworski, C.A.(1968). Greenhouse studies on the spread of *Pseudomonas solanacerum* in tomato plants by clipping. *Plant Dis. Repor.* **52**: 330-334.

Mc Carter, S.M., Barksdale, T.H. and Jaworski, C.P. (1971). Reduction of bacterial wilt by early harvest of tomato transplants. *Phytopathology,* **61**: 849-851.

Mc Nellis, T.W., Mudgett, M.B., Li, K., Aoyama, T., Horvath, D., Chua, N.H. and Staskawicz, B.J. (1998). Glucocorticoid-inducible expression of a bacterial avirulence gene in transgenic arabidopsis induces hypersensitive cell death. *Plant J.* **14**: 247-257.

McAvoy, T., Freeman, J. H. and Rideout, S.L. (2012). Evaluation of grafting using hybrid rootstocks for management of bacterial wilt in field tomato production. *Hort. Science* **47**(5): 621-625.

McCarter, S.M. (1991). Bacterial wilt. Pages 28-29 in: Compendium of tomato diseases. Jones, J.B., Jones, J.P., Stall, R.E., and Zitter, T.A., eds. APS Press Publisher: St. Paul, M.N.

McGarvey, J.A., Denny, T.P. and Schell, M.A. (1999). Spatial-temporal and quantitative analysis of growth and EPS I production by *Ralstonia solanacearum* in resistant and susceptible tomato cultivars. *Phytopathology*, **89**: 1233-1239.

Meenakumari, K.S., Sivaprasad, P., Sulochama, K.K. and Liza, E. (2003). Suppression of bacterial wilt of chilli and tomato using native isolates of fluorescent pseudomonads. In: Sixth International Workshop on Plant Growth Promoting Rhizobacteria, Indian Institute of Spices Research, India, 5–10 October 2003, pp. 169–175

Mehta, S. and Nautiyal, C.S. (2001). An efficient method for qualitative screening of phosphate solubilising bacteria. *Curr. Microbiol.* **3**: 51-56.

Meiners, U. (2000). Bacterial ring rot: Possibilities and limits of insurance coverage. http://ibws.nexenservices.com/from_july_%2001.htm.

Melton, T.A. and Shew, H. D. (2000). Granville wilt. Tobacco disease information note. TDIN-002. North Carolina State University, College of Agriculture and Life Sciences, Plant Pathology Extension. http:// www.ces.ncsu.edu/depts/pp/notes/Tobacco/tdin002/tdin002.htm.

Meseses, R., Moreir, M., Jimenenez, J.M. and Bustamante, E. (1990). Response of table tomato lines to *Pseudomonas solanacearum* during winter in Costa Rice. *Turrialba,* **40**(2): 222-228.

Messiaen, C.M. (1983). Impressions de voyage en Martinique: la culture de l'aubergine va-t-elle disparaître définitivement de ce pays? *Bulletin Agronomique des Antilles- Guyane* **1**: 17-22.

Messiaen, C.M. (1975). L'aubergine. In: Le Potager tropical, Cultures spéciales. *Presses Universitaires de France, Coll. Techniques vivantes*. pp. 231-249.

Messiha, N.A.S. (2001). Biological control of potato brown rot disease. M.Sc. Thesis, Faculty of Science Zagazig University, Egypt.

Messiha, N.A.S, van Bruggen, A.H.C., Franz, E., Janse, J.D., Schoeman-Weerdesteijn, M.E., Termorshuizen, A.J. and van Diepeningen A.D.(2009). Effects of soil type, management type and soil amendments on the survival of the potato brown rot bacterium *Ralstonia solanacearum. Applied Soil Ecology* **43**(2009) 206–215

Messiha, N.A.S., van Diepeningen, A.D., Farag N.S., Abdallah S.A., Janse, J.D. and van Bruggen, A.H.C. (2007). *Stenotrophomonas maltophilia*: A new potential biocontrol agent of *Ralstonia solanacearum*, causal agent of potato brown rot. *Eur. J. Plant Pathol.* **118:** 211-225.

Messiha, N.A.S., van Diepeningen, A.D., Farag N.S., Abdallah S.A., Janse, J.D. and van Bruggen, A.H.C. (2007). *Stenotrophomonas maltophilia*: A new potential biocontrol agent of *Ralstonia solanacearum*, causal agent of potato brown rot. *Eur. J. Plant Pathol.* **118**: 211-225.

Messilha, N.A.S (2006). Bacterial wilt of potato (*Ralstonia solanacearum*, race 3, biovar 2): disease management. pathogen survival and possible eradication. Ph.D. Thesis. Wageningen University, The Netherlands.

Messilha, N.A.S (2006). Bacterial wilt of potato (*Ralstonia solanacearum*, race 3, biovar 2): disease management .pathogen survival and possible eradication. Ph.D. Thesis. Wageningen University, The Netherlands.

Mew, T.W. and Ho, W.C. (1976). Varietal resistance to bacterial wilt in tomato. *Plant Dis. Repor.* **60**(3): 264-268.

Mew, T.W. and Ho, W.C. (1977). Effect of soil temperature on resistance of tomato cultivars to bacterial wilt. *Phytopathology,* **67**(7): 909-911.

Meyer, D., Cunnac, S.and Gueneron, M. *et al.* (2006). PopF1 and PopF2, two proteins secreted by the type III protein secretion system of *Ralstonia solanacearum*, are translocators belonging to the Hrpf/Nopx family. *J. Bacteriol.* **188**: 4903-4917.

Meyer, Y., Verdoucq, L. and Vignols, F. (1999). Plant thioredoxins and glutaredoxins: identity and putative roles. *Trends Plant Sci.* **4**: 388–394.

Miah, A.J., Rahman, M.Z., Adhikaray, S.K. and Rahman, M. M. (1995). Soil ammendments to control of bacterial wilt to tomato. *Bangladesh J. Plant Pathol.* **11**(1-2): 17-18.

Mian, D., R. Coranson Beaudu, D. Dufeal, L. Grassart, and P. Mention. (2002). Status report on *Ralstonia solanacearum* on *Anthurium* in Martinique [France]. - *Ralstonia solanacearum* sur anthurium a la Martinique [France]. Une bacteriose inquietante. *Phytoma*, 43-45.

Mian, I. H. (1986). Plant parasitic nematodes associated with some crop species in Bangladesh. *Bangladesh J. Plant Path.* **23**: 73.

Miao, L., Shou, S., Cai, J., Jiang, F., Zhu, Z. and Li, H. (2009). Identification of two AFLP markers linked to bacterial wilt resistance in tomato and conversion to SCAR markers. *Mol. Biol. Repor.* **36**(3): 479-486.

Michel, V.V. and Mew, T.W. (1998). Effect of soil amendment on the survival of *Ralstonia solanacearum* in different soils. *Phytopathology,* **88**: 300-305.

Michel, V.V. Hartman, G.L. and Midmore, D.J. (1996). Effect of previous crops on soil population of *Burkholeria solanacearum*, bacteral wilt and yield of tomatoes in Taiwan. *Plant Dis.* **80**(12): 1367-1372.

Michel, V.V., Wang, J.F., Midmore, D.J. and Hartman, G.L. (1997). Effect of intercropping and soil amendments with urea and calcium oxide on the incidence of bacterial wilt of tomato and survival of soil borne *Pseudomonas solanacearum* in Taiwan. *Plant Dis.* **46**(4): 600-610.

Michelmore, R. (1995). Molecular approaches to manipulation of disease resistance genes. *Annu. Rev. Phytopathol.* **15**: 393-427.

Miller, P. (1768). *The gardener's dictionary*, Abridged 8th edn. London.

Milling, A., Babujee, L. and Allen, C. (2011). *Ralstonia solanacearum* extracellular polysaccharide is a specific elicitor of defense responses in wilt-resistant tomato plants. *PLoS ONE*6(1): e15853. doi:10.1371/journal.pone.0015853

Milling, A., Meng, F., Denny, T.P. and Allen, C. (2009). Interactions with hosts at cool temperatures, not cold tolerance, explain the unique epidemiology of *Ralstonia solanacearum*race 3 biovar 2. *Phytopathology,* **99**(10): 1127-1134.

Mimura, Y. and Hirai, M. Y.M. (2009). Pepper accession LS2341 is highly resistant to *Ralstonia solanacearum* strains from Japan *Hort. Science*, **44**(7): 2038–2040.

Mimura, Y., Matsunaga, H., Yoshida,T., Sato, T. (2000). Pathogenicity of various isolates of *Ralstonia solanacearum* on tomato, eggplant and pepper varieties *J. Jpn. Soc. Hort. Sci.* **69**(*(suppl 1):* 231.

Mindrinos, M., Katagiri, F., Yu, G.-L. and Ausubel, F.M. (1994). The *Arabidopsis thaliana*disease resistance gene RPS2 encodes a protein containing a nucleotide binding site and leucine-rich repeats. *Cell,* **78**: 1089-1099.

Mishra, A., Mishra, S.K., Karmakar, S.K. Sarangi, C.R. and Sahu, G.S. (1995). Assessment of ytield loss due to wilting some popular tomato cultivars. *Environ. and Ecol.* **13**(2): 287-290.

Mishra, A., Mohanpatra, K.B. and Sinha S.K. (1999). Screening of tomato varieties against bacterial wilt with aid of streptomycin. *Trop. Sci.* **39**(1): 11-15.

Mitchell, B.A. (1962): Bacterial wilt in teak, *Tectona grandish* Linn. *Malayan Forester,* **25**: 164-166.

Mitra, M. (1937). India: Now plant disease recorded in 1936. *Internal. Bull, Plant. Prot.* **11:** 85-87.

Mochizuki, H., Yamakawa, K. (1979b). Resistance of selected eggplant cultivars and related wild *Solanum* species to bacterial wilt (*Pseudomonas solanacearum*). *Bull. Veg. Ornamental Crop Res. Stn.* A. 6: p. 10.

Moffett, M.C. and Hayward, A.C. (1980).The role of weed species in the survival of *Pseudomonas solanacearum* in tomato cropping. *Aust. Plant Path.* **9**: 6.

Mohamed, M.E.S., Pathmanothan, U. and Phelps, R.H. (1997).Genetic nature of bacterial wilt resistance in tomato (*Lycopersicon Esculentum* Mill.) accession LA 1421. *Euphytica*, **96**(3): 323-326.

Momol, M. T., Momol, E. A., Dankers, W. A., Olson, S. M., Simmons, J. A., and Rich, J. R. (1999). Evaluation of selected plant essential oils for suppression of *Ralstonia solanacearum* and *Meloidogyne arenaria* on tomato. (Abstr.) *Phytopathology,* **89**: S54.

Momol, M.T., Pradhanang, P. and Lopes, C.A. (2001). Bacterial wilt of pepper. University of Florida/IFAS, EPIS.Extesion Fact Sheet, 103-104.

Momol, T., J. Jones, and S. Olson. (2003). New outbreak of *Ralstonia solanacearum* (race 3 biovar 2) in geraniums in United States and effects of biofumigants on *Ralstonia solanacearum* (race 1 biovar 1). University of Florida Pest Alert. http://extlab7.entnem.ufl.edu/PestAlert/tmm 0303.htm.

Momol, T., Pingsheng Ji, Ken, P. McGovern, R. and Steve, O. (2005). Soil borne tomato diseases caused by *Ralstonia* and *Fusarium* species and their field diagnostics (A document is Fact Sheet pp. 205, one of a series of the Plant Pathology Department, Florida Cooperative Extension Service, Institute of Food and Agricultural Sciences, University of Florida).

Momol, T., Pradhanang, P. and Lopes, C.A. (2008). Bacterial wilt of pepper. This document is PP 189, one of a series of the Plant Pathology Department, Florida Cooperative Extension Service, Institute of Food and AgriculturalSciences, University of Florida. Original publication date September 2001. Reviewed May 2008. Visit the EDIS Website at http://edis.ifas.ufl.edu.

Momol, T., Pradhanang, P. and Lopes, C.A. (2002). Bacterial Wilt of Pepper.University of Florida, Gainesville, Fact Sheet 189, 4p.

Moraes, A.D.M. (1947). A Bacterial Wilt of Potato due to *Bacterium solanacearum. AgronomiaLusitana,* **9**: 277-328.

Morita, R.Y. (1997). In *Bacteria in Oligotrophic Environments. Starvation- Survival Lifestyle*, pp. 368-385. Edited by C.A. Reddy, A.M. Chakrabarty, A.L. Demain & J.M. Tiedje. New York: Chapman & Hall.

Morton, D.J., Duke, P.D. and Jenkin, S.F.Jr. (1965). Serological relationship among races 1, 2 and 3 of *Pseudomonas solanacearum. Plant Dis. Repor.* **50**: 275-277.

Mosse B., Stribley D.P., Letacon F. (1981) Ecology of mycorrhizae and mycorrhizal fungi.*Adv. Microbiol. Ecol.* **5**: 137-210

Moura, A.B., Romeiro, R.S.D. aand Neves, M.C.P. (1998). Bioaassay for mass assessment of antagonistic actinomycetes against *Ralstonia solanacearum* on tomato crops. *Psequisa Agrope. Brasil.* **33**(12): 2065-2072.

Muthiah, R., Zondlo, A., Zondlo, J. and Kagen, S. (1996). Eggplant anaphylaxis: Isolation of the major allergens and demonstration of cross-reactivity with other solanaceae. *J. Allergy Clin. Immunol.* **97**: 334

Muthoni, J., Shimelis, H, & Melis, R., 2012. Management of Bacterial Wilt [*Ralstonia solanacearum* Yabuuchi et al., 1995] of Potatoes: Opportunity for host resistance in Kenya. *J. Agri. Sci.* **4**: 64 - 78.

Myint, L. and Ranamukhaarachchi, S.L. (2006). Development of biological control of *Ralstonia solanacearum* through antagonistic microbial populations. *Inter. J. Agricul. Biol.* **8**(5): 657-660.

Nagamine, K., Hase, T. and Notomi, T. (2002a). Accelerated reaction by loop-mediated isothermal amplification using loop primers. *Mol. Cell. Probes*, **16**: 223-229

Nagamine, K., Kuzuhara, Y. and Notomi, T. (2002b). Isolation of single-stranded DNA from loop-mediatedisothermal amplification products. *Biochem. Biophys. Res. Commun.* **290**: 1195-1198.

Nagamine, K., Watanabe, K., Ohstuka, K., Hase, T. and Notomi, T. (2001). Loop-mediated isothermal amplification reaction using a nondenatured template. *Clin. Chem.* **47**: 1742–1743.

Nagrajan, S. (1983). *Plant Disease Epidemiology*. Oxford & IBH Publishing Co. New Delhi. 267p.

Naik, M.K. (2009). Integrated disease management strategies in chilli (*Capsicum annuum* L.) *5th Int. Conf. on Plant Pathol. in the Globalized Era,* Nov. 10-13, 2009, New Delhi, India 26 035 (S-8.I).

Nakaho, K., Takaya, S. and Sumida, Y. (1996). Condition that increase lateny infection of grafted on non-grafted tomatoes with *Pseudomonas solanacearum. Ann. Phytopathol. Soc. Japan,* **62**(3): 234-239.

Nakaho, K. (1999). Studies on mechanism of resistance in tomato rootstock cultivars to bacterial caused by *Ralstonia solanacearum. Ann. Phytopathol. Soc. Japan,* **65**(3): 245.

Nakaho, K., Hibino, H. and Miyagawa, H., (2000). Possible mechanisms limiting movement of *Ralstonia solanacearum* in resistant tomato tissues. *J. Phytopathol.* **148:** 181-190.

Nakaho, K., Inoue, H., Takayama, T., and Miyagawa, H. (2004). Distribution and multiplication of *Ralstonia solanacearum* in tomato plants with resistance dirived from different origins. *J. Gen. Plant Pathol.* **70**: 115-119.

Nakaho, K., Takaya, S. and Sumida, Y. (1996). Condition that increase lateny infection of grafted on non-grafted tomatoes with *Pseudomonas solanacearum. Ann. Phytopathol. Soc. Japan,* **62**(3): 234-239.

Nakayama, T., Homma, Y., Hashidoko, Y., Mizutani, J. and Tahara, S. (1999). Possible role of xanthobaccins produced by *Stenotrophomonas* sp. strain SB-K88 in suppression of sugar beet damping-off disease. *Appl. Environ. Microbiol.* **65**: 4334-4339

Narsimha Rao, N. (1979). The barriers to hybridization between *Solanum melongena* and some other species of*Solanum*. In: Hawkes J.G., Lester R.N., Skelding A.D. (Eds) the Biology and Taxonomy of the Solanaceae. Academic Press, London, pp. 605-614

Nath, G., Bora, L.C. Saikia, B.K. and Gogoi, P.K. (1996). Prevalence of three races of *Pseudomonas solanacearum* E.F.Smith. *J. Agri. Sci. Soc. North East India,* **9**(1): 21-27.

Natori, S. (1994). Function of antimicrobial proteins in insects. *Ciba Found. Symp.* **186**: 123-134.

Nautiyal, C.S.(1999). An efficient microbiological growth medium for screening phosphate solubilising microorganisms. *FEMC Microbiol. Lett.* **170**: 265-270.

Nazareno, N.R.X.de. (2009). Organic production of potato: Potential and challenges. *Producao organica de batata: potencialidades e desafios*. 249 pp.

Nesmith, W.C. and Jenkin, S.F. (1983). Survival of *Pseudomonas solanacearum* in selected North Carolina. *Phytopathology,* **73**: 1300-1304.

Nesmith, W.C.and Jenkin, S.F.(1985). Influence of antagonists and controlled matric potential on the survival of *Pseudomonas solanacearum* in four North Carolina. P*hytopathology,* **75**: 1182-1187.

Neto, A.F.L., da Silveira, M.A., de Souza, R.M., Nogueira, S.R. and André, C.M.G. (2002). Inheritance of bacterial wilt resistance in tomato plants cropped in naturally infested soils of the State of Tocantins. *Crop Breeding and Appl. Biotechnol.* **2**(1): 25-33.

Ngoc H. T. and Kim B. S. (2010). Influence of temperature, pathogen strain, inoculum density, seedling age, inoculation method and varietal resistance on infection of pepper seedlings by *Ralstonia solanacearum. Hort. Environ. Biotechnol.* **51**(2): 95-100.

NIAST. (2003). National. Institute of Agricultural Sciences and Technology Research highlights. http://www. niast.go.kv/eglish/sub/highlights. htm last accessed 15 May, 2002.

Nielsen, L.W. and Todd, F.A. (1945). Prelimary evaluatution of disinfectants for controlling southern bacteral wilt of potatoes. *Amer. Potato J.* **22**: 197-202.

Nimchuk, Z., Eulgem, T., Holt, B.E. and Dangl, J.L. (2003). Recognition and response in the plant immune system. *Annu. Rev. Genet.* **37**: 579-609.

Nirmala Devi, S. and Tikoo, S.K. (1989). Studies on genetic resistance to bacterial wilt and root-knot nematode in tomato. *Tomato Genet. Cooperative Repor.* **39**: 27.IN pp. 163-169.

Nirmala Devi, S. and Tikoo, S.K. (1992). Studies on the reaction of certain tomato genotypes and their F1 to combined infection by *Meloidogyne incognita* and *Pseudomonas solanacearum. Indian J. Genet. and Plant Breeding,* **52**(2): 118-125.

Nishi, T., Tajima, T., Noguchi, S., Ajisaka, H. and Negishi, H. (2003). Identification of DNA markers of tobacco linked to bacterial wilt resistance. *Theor. and Appl. Genet.* **106**(4): 765-770.

Nishiyama, K. Achmad, N. H., Wirtono, S., and Yamaguchi,T. 1980. Causal agents of cassava bacterial wiltin Indonesia. Contributions Central Research Institutefor Agriculture, Bogor, 59, 19 p.

Nishiyama, M., Shiomi,Y. Suzuki, S. and Marumoto, T. (1999). Suppression of growth of *Ralstonia solanacearum*, tomato bacterial wilt agent, on/in tomato seedlings cultivated in a suppressive soil. *Soil Sci. Plant Nutr.* **45** (I): 79-87.

Noda, H. and Machado, F.M. (1993). Selection progress of resistant tomato progenies to bacterial wilt disease by epidemiological evaluation. *Acta Amazo.* **23**(2/3): 107-114.

Nordeen, R.O., Sinden, S.L., Jaynes, J.M. and Owens, L.D. (1992). Activity of cecropin SB37 against protoplasts from several plant species and their bacterial pathogens. *Plant Sci.* **82**: 101-107.

Norman, D.J. and Yuen, J.M.F. (1998). A distinct pathotype of *Ralstonia solanacearum* race 1, biovar 1 entering Florida in pothos (*Epipremnum aureum*) cutting. *Can. J. Plant Pathol.* **20**(2):171-175.

Notomi, T., Okayama, H., Masubuchi, H., Yonekawa, T., Watanabe K, et al. (2000). Loop-mediated isothermal amplification of DNA. *Nucleic Acids Res.* **28**: e63.

Novitsky, J.A. and Morita, R.Y. (1976). Morphological characterization of small cells resulting from nutrient starvation of a pyschrophilic marine *Vibrio*. *Appl. Environ. Microbiol.* **32**: 617-622.

Obrero, P.P. (1969). Grafting tomato to control tomato bacterial wilt. *Hawaii Fm. Sci.* **18**(3): 1-4.

Obrero, P.P., Aragaki, M. and Trujillo, E.E. (1971). Tomato bacterial wilt: Inoculation of susceptible scions grafted to resistance rootstock. *Plant Dis. Repor.* **55**(6): 521-522.

Occhialini, A., Cunnac, S., Reymond, N., Genin, S. and Boucher, C. (2005). Genome-wide analysis of gene expression in *Ralstonia solanacearum* reveals that the *HrpB* Gene acts as a regulatory switch controlling multiple virulence pathways. *Mol. Plant- Microbe Intera.* **18**: 938-949.

Ohshima, M., Mitsuhara, I., Okamoto, M., Sawano, S., Nishiyama, K., Kaku, H., Natori, S. and Ohashi, Y. (1999). Enhanced resistance to bacterial diseases of transgenic tobacco plants overexpressing sarcotoxin IA, A bactericidal peptide of insect. *J. Biochem.* **125**: 431-435.

Okabe, N. and Goto, M. (1963). Bacteriophages of plant pathogens. *Annu. Rev. Phytopathol.* **1**: 397-418.

Okamoto, M., Mitsuhara, I., Ohshima, M., Natori, S. and Ohashi, Y. (1998). Enhanced expression of an antimicrobial peptides sarcotoxin IA by GUS fusion in transgenic tobacco plants. *Plant and Cell Physiol.* **39**: 57-63.

Oliveira, W.F., Giordano, L.B. and Lopes, C.A. (1999). Inheritance of resistance to bacterial wilt in tomato. *Fitopatologia*, **24**(1): 49-53.

Oliver, J.D. (2005). The viable but nonculturable state in bacteria. *J. Microbiol.* **43**: 93-100.

Oliver, J.D. (1993). Formation of viable but nonculturable cells. In: Kjelleberg S., ed. Starvation in bacteria. New York: Plenum Press, 239.

Olson, H.A. (2005). *Ralstonia solanacearum,* pathogen profile. pp 728, *Soilborne Plant Pathogens*, *Spring 2005.*

Olsson, K. (1976a). Experience of brown rot caused by *Pseudomonas solanacearum*. *EPPO Bull.* **6**: 199-207.

Olsson, K. (1976b). Overwintering of *Pseudomonas solanacearum* in Sweden. In: Sequiria, L. and Kelman, A.ed. Proc. First Intl. Planning Conf. and Workshop on Ecol. and Control of Bacterial Wilt caused by *Pseudomonas solancearum*.Reigh .N.C.18-24 July 1976. 105-109.

Opena, R.T. and TSchanz, A. T. (1987). Bacterial wilt resistance program on tomato at AVRDC. *ACIAR Bacterial Wilt Nwsl.* **2**: 1-2.

Opina, N. L. and Miller, S. A. (2005). Evaluation of immunoassays for detection of *Ralstonia solanacearum,* causal agent of bacterial wilt of tomato and eggplant in Philippines. *Acta Hort.* **695:** 353-356.

Opina, N., Tavner, F., Hollway, G., Wang, 1. F., Li, T.H., Maghirang, R., Fagan, M., Hayward, A. C., Krishnapillai, V., Hong, W.F., Holloway, B.W. and Timmis, J.N. (1997). A novel method for development of species and strain-specific DNA probes and peR primers for identifying *Burkholderia solanacearum* (formerly *Pseudomonas solanacearum*). *Asia-Pacific J. Molec. Biol. Biotechnol.* **5**: 19-30.

Orgambide, G., Montrozier, H., Servin, P., Roussel, J., Trigalet-Demery, D.and Trigalet, A. (1991). High heterogeneity of the exopolysaccharides of *Pseudomonas solanacearum* strain GMI 1000 and the complete structure of the major polysaccharide. *J. Biol. Chem.* **266**: 8312-8321.

Osiru, M.O., Rubaihayo, P.R. and Opio, A.F. (2001). Inheritance of resistance of tomato to bacterial wilt and its implication for tomato improvement in Uganda. *Afr. Crop Sci. J.* **9**: 9-16.

Oudhia, P. (2001, 2002, and 2003) Research Note: Traditional Medicinal Knowledge about Common Herb Dhaincha (*Sesbania* sp.) in Chhattisgarh, India. *The Tribune* Dated 30th June 2003.

Ozaki, K. and Kimura, T.T. (1989). Methods for evalutig resistance of *Solanum* plants to bacterial wilt caused by *Pseudomonas solanacearum*. *Bull. Chugoku Nati. Agri. Expe. Station* No. **4**: 103-117.

Ozakman, M. and Schaad, N. W. (2003). A real-time BIO-PCR assay for detection of *Ralstonia solanacearum* race 3, biovar 2, in asymptomatic potato tubers *Can. J. Plant Pathol.* **25**: 232-239.

Ozawa, H., Tanaka, H., Ichinose, Y., Shiraishi, T. and Yamada, T. (2001). Bacteriophage P4282, a parasite of *Ralstonia solanacearum*, encodes a bacteriolytic protein important for lytic infection of its host. *Mol. Genet. and Genom.* **265**: 95-101.

Pacumbaba, R.P. Beyl, C.A. and Pacumbaba, R.O.J. (1999). Shiitake myceial leachate suppresses growth of some bacterial species and symptom of bacterial wilt of tomato and lima bean. *Plant Dis.* **82**(1): 20-23

Paily, P.V. (1964). Control of bacterial wilts of tomato and brinjal by grafting on *Solanum torvum*. *Sci and Cult.* **30**: 295-296.

Palleroni, N.J. (1984). Pseudomonadaceae. In Bergey's Manual of Systematic Bacteriology, Vol. 1, pp. 141-219. Edited by N.R. Krieg & J.G. Holt. Baltimore: Williams & Wilkins.

Palleroni, N.J. *and* Doudoroff, M.1971.Phenotypic characterization and deoxyribonucleic acid homologies of *Pseudomonas solanacearum*. *J. Bacteriol.* **107***:* 690–696.

Panthee, D.R. and Bhattrai, B.P. (1999). Identification of the parents for the production of bacterial wilt resistant and high yielding F1 hybrids of tomato (*Lycopersocon esculentum).Working paper Lumie Agri. Res. Centre* 1996, No. 98/23.

Paret, M., L., Cabos, R., Kratky, B.A. and Alvarez, A.M. (2010). Effect of plant essential oils on *Ralstonia solanacearum* race 4 and bacterial wilt of edible ginger. *Plant Dis.* **94**(5): 521-527.

Paret, M.L., A.S. deSilva, and Alvarez, A.M. (2008). Detection of *Ralstonia solanacearum* with an immunostrip assay; Its specificity and sensitivity. *Indian Phytopath.* **61**: 518–522

Parham, E.V. (1934). Tomato bacterial wilt observation Part 2. Notes on wilts of tomato. *Fiji Agri. J.* **7**: 23-25.

Park, K.S., Ahn, I.P. and Kim, C.H. (2001). Systemic resistance and expression of the pathogenesis-related genes mediated by the plant growth-promoting rhizobacterium *Bacillus amyloliquefaciens* EXTN-1 against anthracnose disease in cucumber. *Mycobiology,* **29**: 48-53.

Park, K.S., Paul, D., Kim,Y. K., Nam, K. W., Lee, Y. K., Choi, H. W. and Lee, S. Y. (2007). Induced systemic resistance by *Bacillus vallismortis* EXTN-1suppressed bacterial wilt in tomato caused by *Ralstonia solanacearum*. *Plant Pathol. J.* **23**(1): 22-25.

Park, M. and Fernando, M. (1938). The Relative resistance of some tomato varieties to bacterial wilt (*Bacterium solanacearum* E.F. Smith). *Trop. Agri. J.* **7**: 23-25.

Partyka, R.E., White, J.E. and Miller, T.D. (1973). Controlling bacterial transmission when bare root tomato plants are dipped in water prior to transplanting. *Res. Summary*, *Ohio Agri. Res. and Develop. Centre* , No. **65**: 21-22.

Pastrik, K.H. and Maiss, E. (2000). Detection of *Ralstonia solanacearum* in potato tubers by polymerase chain reaction. *J. Phytopathol.* **148**(11-12): 619-626.

Pastrik, K.H., (2000). Detection of *Clavibacter michiganensis* subsp. *sepedonicus* in potato tubers by multiplex PCR with coamplification of host DNA. *Eur. J. Plant Pathol.* **106**: 155-165.

Pastrik, K.H., Elphinstone, J.G. and Pukall, R. (2002). Sequence analysis and detection of *Ralstonia solanacearum* by multiplex PCR amplification of 16S–23S ribosomal intergenic spacer region with internal positive control. *Eur. J. Plant Pathol.* **108**: 831-842.

Patel, M.K., Kulkami, N.B. and Kulkami, Y.S. (1952). Ring disease of potato in India. *Curr. Sci.* **21**:47-48.

Patel, P.N. (1972). *Bacterial wilt Disease of Plants in India.* ICAR, New Delhi.

Patil, B.P., Rajput, J.C., Palve, S.B., Jambhale, A.S. and Diwakar, M.P. (1990). New tomato, "Sonali" high yielding variety and wilt resistant. *Indian Hort.* **35**(3): 6-9.

Patil, R. (1989). Interspecific hybridization in eggplant. M.Sc. Thesis. University of Agricultural Sciences, Dharwad, India.

Patil, R. B., Siddaramehowda, R.K. and Shetty, K.S. (1973). Soil fumigtion tube for studying the root inhibition caused by *Pseudomonas solanacearum*, infecting vegetable crops. *Curr. Res.* **2**(2): 10-11.

Patil, V. U., Gopal, J. and Singh, B.P. (2012). Improvement for bacterial wilt resistance in potato by conventional and biotechnological approaches. *Agric. Res.* **1**(4): 299-316

Pawanexh, K. (2009). Potato storage and value preservation: The Basics. Crosstree Techno-visors.

Pegg, G.F. and Cronshaw, D.K. (1976). The Relationship of *in vitro in vivo* ethylene production in *Pseudomonas solanacearum*infection of tomato. *Physiol. Plant Pathol.* **9**(2): 145-154.

Pegg, K.G., Maftuff, M.L. and Coibran, R.C. (1974). Diseases of ginger in Queensland. *Queensland Agri. J.***100**: 611-618.

Peeters N, Guidot A, Vailleau F, Valls M (2013). *Ralstonia solanacearum*, a widespread bacterial plant pathogen in the post-genomic era..*Mol Plant Pathol.* **14**(7):651-62. Epub 2013 May 30.

Peixoto, AIR (1997). Biological control of bacterial wilt of tomato by fluorescent *Pseudomonas* spp. Literature review. *Cinencia Rural,* **27**(1): 153-160.

Peregrine, W.J.H. and Kassim B.A. (1982).Grafting a simple technique for overcoming bacterial wilt in tomato. *Tropical Pest Manage* **28**(1): 1-6.

Perez, J.E. (1961). Control of bacterial wilt of tomato by streptomycin in a hydtoponic farm. *J. Agric. Univ. P.R.* **45**: 193-194.

Perez, J.E. (1962). The use of Fuller's formide method in the serological identification of *Pseudomonas solanacearum. J. Agr. Univ. Puerto Rico,* **46**: 144-153.

Perez, J.E. (1967). Hemagglutination by bacterial phytopathogens. *J. Agr. Univ. Puerto Rico,* **51**: 337.

Perry, L. *et al.* (2007). Starch fossils and the domestication and dispersal of chili peppers (*Capsicum* spp.) in the Americas. *Science,* **315**: 986–988.

Perlsey, G.J. (1985). Bacterial wilt disease in Asia and the South Pacific. *ACIAR Proceeding No.* 13 pp. 45.

Persson, P. (1998). Successful Eradication of *Ralstonia solanacearum* from Sweden. OEPP/*EPPO Bull.* **28**: 113-119.

Peter, K. V. (2001). *Handbook of Herbs and Spices* Vol 1, CRC Press, ISBN 0-8493-1217-5 page 120

Peter, K. V., Goth, R.W. and Webb, R. E. (1984). Indian hot peppers as new sources of resistance to bacterial wilt, *Phytophthora* root rot and root rot nematodes. *Hort. Science,* **19**: 277-278.

Peter, K.V. Gopalakrishnan, T.R., Rajan, S. and Kumar, P.G.S. (1992). Breeding for resistance to bacterial wilt in tomato, eggplant and pepper. *ACIAR Proc.* **45**: 183-190.

Peterson, R.A., Inch, A.J., Herrington, M.E. and Saranath, J. (1983). Scorpio– a tomato resistant to bacterial wilt biovar III. *Australion Plant Pathol.* **12**(1): 8-10.

Pflieger, S., Lefebvre, V., Caranta, C., Blattes, A., Goffinet, P., and Palloix, A. (1999). Disease resistance gene analogs as candidates for Qtls involved in pepper/pathogen interactions. *Genome,* **42**: 1100-1110.

Phae, C.G., Shoda, M., Kita, N., Nakano, M. and Ushiyama, K.(1992). Biological control of crown and root rot and bacteria L; wilt of tomato by *Bacillus substilis. Ann. Phytopathol. Soc. Japan,* **58**(3): 329-339.

Pieterse, C.M.J., van Wees, S.C.M., Hoffland, E., van Pelt, J.A., van Loon, L.C. (1996). Systemic resistance in arabidopsis induced by biocontrol bacteria is independent of salicylic acid accumulation and pathogenesis-related gene expression. *Plant Cell,* **8**: 1225-1237.

Pochard, E., Palloix, A., and Daubèze, A.M. (1986). The use of androgenetic autodiploid lines for the analysis of complex resistance systems in the pepper. *VI Eucarpia Meeting on Genetics and Breeding on Capsicum and Eggplant,* Saragozza, Spain, 1986/10/21-24, 105-109.

Ponnuswami, V. (1997). Screening of bacterial wilt resistance in brinjal. *Capsicum and Eggplant Newslett. No.* **16**: 115-117.

Ponnuswami, V. (1999). Studies on bacterial wilt resistance of selected eggplant accessories inoculated with *Pseudomonas solanacearum* PSSS97 for 30 days under growth room conditions at Asian Vegetable Research and Development Centre. *Capsicum and Eggplant Newsl.* **12**: 91-93.

Porter-Jordan, K., Rosenberg, E. I., Keiser, J.F., Gross, J.D., Ross, A. M., Nasim, S., and Garrett, C. T. (1990). Nested polymerase chain reaction assay for the detection of cytomegalovirus overcomes false positives caused by contamination with fragmented DNA. *J. Med. Virol.* **30**: 85-91.

Posas, M. B. and Toyota, K. (2010). Mechanism of tomato bacterial wilt suppression in soil amended with lysine. *Microbes Environ.* **25**(2): 83–94.

Posas, M.B, Toyota, K. and IslamT.M.D. (2007). Inhibition of bacterial wilt of tomato caused by *Ralstonia solanacearum* by sugars and amino acids. *Microbes Environ.* **22** (3): 290-296.

Poussier, S., Vandewalle, P. and Luisetti, J. (1999). Genetic diversity of African and worldwide strains of *Ralstonia solanacearum* as determined by PCR-restriction fragment length polymorphism analysis of the hrp gene region. *Appl. Environ. Microbiol.* **65**: 2184-2194.

Poueymiro, M.and Genin, S. (2009). Secreted proteins from *Ralstonia solanacearum*: a hundred tricks to kill a plant. *Curr. Opinion Microbiol.* **12**: 44-52.

Poussier, S., Trigalet-Demery, D., Vandewalle, P., Goffiner, B., Luisetti, J. and Trigalet, A (2000). Genetic diversity of *Ralstonia solanacearum* as assessed by PCR-RFLP of the Hrp region, AFLP and 16S rRNA sequence analysis, and identification of an African subdivision. *Microbiology,* **146**: 1679-1692.

Poussier, S., Thoquet, P., Trigalet- D. *et al.* (2003). Host plant-dependent phenotypic reversion of *Ralstonia solanacearum* from non-pathogenic to pathogenic forms via alterations in the *Phca* Gene. *Mol Microbiol.* **49**: 991-1003.

Power, R.H. (1983). Relationship between the soil environment and tomato resistance to bacterial wilt (*Pseudomonas solanacearum*) control method. *Surinaamse Land Bouw,* **31**(2): 39-47.

Pradhanang, P.M., (1999). Transmission of *Ralstonia solanacearum* through drainage water. *Bacterial Wilt Newslet.* (16): 5-7.

Pradhanang, P.M., Momol, M.T., Dankers, H., Momol, E.A., Jones, J.B. (2002). First report of southern wilt caused by *Ralstonia solanacearum* on *Geranium* in Florida. *Plant Health Prog.* 2002 No. June pp. 1-2.

Pradhanang, P.M., Momol, M.T., Olson, S. M. and Jones, J. B. (2003). Effects of plant essential oils on *Ralstonia solanacearum* population density and bacterial wilt incidence in tomato. *Plant Dis.* **87**(4): 423-427.

Pradhanang, P.M. and Elphinstone, J.G. (1996). Identification of weed and crop hosts of *Pseudomonas solanacearum* race 3 in the hills of Nepal. In: Pradhanang, P.M., Elphinstone, J.G. (eds.) Integrated management of bacterial wilt of potato: Lessons from the hills of Nepal. Proceedings of a national workshop held at Lumle *Agricultural Research Centre*, Pokhara, pp. 39-49.

Pradhanang, P.M., Elphinstone, J.G. and Fox, R.T.V. (2000). Identification of crop and weed hosts of *Ralstonia solanacearum* biovar 2 in the hills of Nepal. *Plant Pathol.* **49**: 403-413.

Pradhanang, P.M., Ji, P., Momol, M.T., Olson, S.M., Mayfield, J.L., and Jones, J.B. (2005). Application of Acibenzolar-*S*-Methyl enhances host resistance in tomato against *Ralstonia solanacearum. Plant Dis.* **89**: 989-993.

Pradhanang, T. P. and Carlos, A. L. (2001). Bacterial wilt of pepper. University of Florida. IFAS Extension. PP 189.

Pramod, S.N. and Venkatesh, Y.P. (2008). Allergy to eggplant (*Solanum melongena*) caused by a putative secondary metabolite. *J. Investig Allergol. Clin. Immunol*: **18**(1): 59-62.

Principe, A., Alvarez, F.A., Castro, M.G., Zachi, L., Fischer, S.E., Mori, G.B. and Jofre, E. (2007). Biocontrol and PGPR features in native strain isolated from saline soils of Argentina. *Curr. Microbiol.* **55**: 314-322.

Prior, P. and Beramis, M. (1990). Induction de la resistance aufletrissement bacteriendfi 11 *Pseudomonas solanacearum* E.ES. chez un cultivar de tomate repute sensible. *Agronomie,* **10**: 391-401.

Prior, P. and Fegan, M. (2005a). Recent development in the phylogeny and classification of *Ralstonia solanacearum*. *In* T. Momol and J. B. Jones (ed.), *Proc. First Int. Symp.on tomato diseases*, 695. ISHS-Acta Horticulturae pp. 127-136.

Prior, P. and Fegan, M. (2005b). Diversity and molecular detection of *Ralstonia solanacearum* race 2 strains by multiplex PCR, p. 405-414. *In* C. Allen, P. Prior, and A. C. Hayward (ed.), *Bacterial wilt disease and the Ralstonia solanacearum* species complex. APS Press, Madison.

Prior, P. and Steva, H. (1990). Characteristics of strains of *Pseudomonas solanacearum* from the French West Indies. *Plant Dis.* **74**: 13-17.

Prior, P., Bart, S., Leclerco, S., Darrasse, A. and Axis, G. (1996). Resistance to bacterial wilt in tomato as discerned by spread of *Pseudomonas* (*Burkholderia*) *solanacearum* in the tissues. *Plant Pathol.* **45**(4): 720-726.

Prior, P., Grimault, V. and Schmit, J. (1994). Resistance to bacterial wilt (*Pseudomonas solanacearum*) in tomato: Present Status and Prospects. In: Hayward A.C., Hartman G.L., eds. *Bacterial wilt: the disease and its causative agent, Pseudomonas solanacearum.* Wallingford: CAB International, 209.

Prior, P., Steva, H. and Cadet, P. (1990). Aggressiveness of strains of *Pseudomonas solanacearum* from the French West Indies (Martinique and Guadeloupe) on tomato. *Plant Dis.* **74**: 962-65.

Priou S., Aley P., Chujoy E., Lemaga B. and French E. R. (1999). Integrated control of bacterial wilt of potato. agris.fao.org/agris-search/search/display.do?f=2008/QP/QP9901.

Priou, S., Gutarra, L. and Aley, P. (2006). Sensitive detection of *Ralstonia solanacearum* in soil: a comparison of different detection techniques. *Plant Pathol.* **49**: 414-422.

Priou, S., Gutarra, L., Aley, P., De Mendiburu, F. and Llique, R. (2010). Detection of *Ralstonia solanacearum* (biovar 2A) in stems of symptomless plants before harvest of the potato crop using post-enrichment DAS-ELISA.*Plant Pathol.* **59**: 59-67.

Priou, S., Gutarra, L. and and Aley, P. (2006). An improvement enrichment broth of the high sensitive and reliable detection of *Ralstonia solanacearum* (biovar 1 and 2A) in soil using double-antibody sandwich enzyme- linked immunosorbent assay. Plant Pathol. **55**(1): 36-45.

Prosen, D., Hatziloukas, E., schaad, N.W. and Panopoulos, N.J. (1993). Specific detection of *Pseudomonas syringaepv. phaseolicola* DNA in bean seed by PCR-based amplification of a phaseolotoxin gene region. *Phytopathology,* **83:** 965-970.

Pun, K.B. and Bhorali, P. (1997). Extracellular polysaccharides of *Pseudomonas solanacearum* Smith: method of extraction and biological properties. *Plant Dis. Res.* **12**(1): 15-19.

Pun, K.B. and Das, G.R. (1997). Management of bacterial wilt Asafetida and turmeric. *Bacterial Wilt Newsl.* No. **14**: 6.

Quezado-Soares, A.M. and Lopes, C.A. (1994). Resistance of tomato genotype *Pseudomonas solanacearum* biovar I and III. *Hort. Brasil.* **12**(2): 161-165.

Quezado-Soares, A.M. and Lopes, C.A. (1995). Stability of the resistance to bacterial wilt of thesweet pepper MC–4 challenged with strains of *Pseudomonas solanacearum. Fitopatol. Brasil.* **20**: 638-641.

Quimio, A.J. and Chan, H.H. (1979). Survival of *Pseudomonas solanacearum* E.F. Smith in the rhizosphere of some weed and economic plant species. *Phillipine Phytopath.* **15**: 108-121.

Raaijmakers, J.M. and Weller, D.M. (2001). Exploiting genotypic diversity of 2, 4-diacetylphloroglucinol producing *Pseudomonas* species: Characterization of superior root colonizing *Pseudomonas fluorescens* strain Q8r1-96. *Appl. Environ. Microbiol.* **67**: 2545-2554.

Raben, A., Tagliabue, A., Christensen, N.J., Madsen, J., Holst, J.J. and Astrup, A. (1994). Resistant starch: The effect on postprandial glycemia, hormonal response, and satiety. *Am. J. Clin. Nutr.* **60**(4): 544-551.

Rademaker, J.L. W., Hoste, B., Louws, F.J., Kersters, K., Swings, J., Vauterin, P. and de Bruijin, F.J. (2000). Comparison of AFLP and Rep-PCR genomic fingerprinting with DNA-DNA homology studies: *Xanthomonas* as model system. *J. Syst. Evolu. Microbiol.* **50**: 665-667.

Rahim, A. and Samraj, J. (1974). Comparative resistance of certain varieties of chillies to the bacterial wilt caused by *Pseudomonas solanacearum* Smith. *Agric Res. J. Kerala,* **12**(1): 105.

Rahman, M. A., Abullah, H., and Vanhaecke, M. (1999). Histopathology of susceptible and resistant *Capsicum annuum* cultivars infected with *Ralstonia solanacearum*. J. *Phytopathol.* **147**: 129-140.

Rahman, M. A., Ali, F., Hossain, K.M.A. and Laila, L. (2011). Screening of different eggplant cultivars against wilt disease caused by fungi, bacteria and nematodes. *J. Exp. Sci.* **2**(1): 6-10.

Rahman, M.A and Abdulah, H. (1997). Effect of calcium on *Ralstonia solanacearum*, pH and anatomical characters of *Capsicum annuum. Malays. Appl. Biol.* **26**(2): 29-36.

Rahman, M.A. and Hoque, M.O. (1986). Screening of eggplant Varieties/Lines against bacterial wilt. *Bangladesh Plant Pathol.* **2**(1): 15-18.

Rai, M., Gupta, P. N. and Agarwal, R. C. (1995). *Catalogue on eggplant* (*Solanum Melongena* L.) *germplasm Part - I.* National Bureau of Plant Genetic Resources, Pusa Campus, New Delhi. pp. 1-3.

Rajam, D.M. (1924). A new remedy for the potato ring disease. *Madras Agri. Students Union* J. **12:** 281-284.

Rajeshwari, N., Shydaja, M.D., Krishnappa, M., Shetty, H.S., Mortensen, C.N. and Mathur, S.B. (1998). Devlopment of ELISA for the detection of *Ralstonia solanacearum* in tomato: Its application in seed health testing. *World J. Microbiol. Biotechnol.* **14** (5): 697-704.

Ralston, E., Palleroni, N. J. and Doudoroff, M. (1973). Pseudomonas pickettii, a new species of clinical origin related to *Pseudomonas solanacearum. Inter. J. Sys. Bacteriol.* **23**: 15-19.

Ram, S. (1989). Diseases. In: Verma, S.C. and H.N. and (Eds.). Potato in Northeastern India. *Tech. Bull. No.* 18. Central Potato Research Institute (CPRI), Shimla, India. pp. 24-28.

Ramamoorthy, V., Viswanathanm R., Raguchanderm T., Prakasamm, V., Samiyappan, R. (2001). Induction of systemic resistance by plant growth promoting rhizobacteria in crop plants against pests and diseases. *Crop Prot.* **20**: 1-11.

Ramesh, C.R. and Bandyopadhyay, A.K. (1993a). Bacterial wilt potential of soil of Andaman and Nicobar Islands. In: Hartman, G.L. and A.C. Hayward (eds.). Bacterial wilt. Proc.Int.Conf. held at Kaoshioung, Taiwan, 28-31 Oct. 1992. *ACIAR Proceedings No.* 45. pp. 355-357.

Ramesh, C.R. and Bandyopadhyay, A.K. (1993b). Bacterial wilt of tomato in Andaman and Nicobar Islands. *ACIAR Proc.* - Australian Centre for International Agricultural Research Vol 45 *Bacterial wilt* in Hartman, G.L. Hayward, A.C. (eds.). - Canberra, A.C.T. (Australia): ISBN 1-86320-087-8. pp. 352-354.

Ramesh, R. (2006). Field evaluation of biological control agents for the management of *Ralstonia solanacearum* in brinjal. *J. Mycol. Plant Pathol.* **36**: 327-328

Ramesh, R., Anthony, J., Jaxon, T.C.D., Gaitonde, S. and Achari, G. (2011). PCR based sensitive detection of *Ralstonia solanacearum* from soil, eggplant, seeds and weeds. *Arch. Phytopathol. Plant Protec.* **44** (19): 1908-1919.

Ramesh, R., Joshi, A.A., and Ghanekar, A.M.P. (2009). Pseudomonads: Major antagonistic endophytic bacteria to suppress bacterial wilt pathogen, *Ralstonia solanacearumin* the eggplant (*Solanum melongena* L.). *World J. Microbiol. Biotechnol.* **25**: 47-55.

Ramesh R, Gaitonde S, Achari G, Asolkar T, Singh NP, Carrere S, Genin S, Peeters N. (2014). Genome sequencing of *Ralstonia solanacearum* biovar 3, phylotype I, strains Rs-09-161 and Rs-10-244, isolated from eggplant and chili in India. *Genome Announc.* 2(3):e00323-14. doi:10.1128/genomeA.00323-14

Ran, L. X., Liu, C. Y., Wu, G. J., van Loon L. C. and Bakker, P. A. H. M. (2005). Suppression of bacterial wilt in *Eucalyptus urophylla* by florescent *Pseudomonas* spp. in China. *Biol. Cont.* **32:** 111-120.

Rao, A. V. and Balachandran, B. (2002). Role of oxidative stress and antioxidants in neuro degenerative diseases. *Nutri. Neurosci.* **5**(5): 291–309.

Rao, A.N.S. (1990). Protection of tomato from bacterial wilt (*Pseudomonas solanacearum* Smith) by bacterriol –100. *Curr. Res. – Univ. Agri. Sci.* **9**(6): 100-101.

Rao, M.V.B. (1976). Bacterial wilt of tomato and eggplant in India. In: *Planning Conf. and Workshop on the Ecology and Control of Bacterial Wilt caused* by *Pseudomonas solanacearum*, North Carolina (July 18-23).

Rao, M.V.B. and Kishun, R. (1986). In: Advances in Research on Pathogenic Bacteria Based on Proceeding of the National Symposium on Phytobacteriology held at the University of Madras March 14-15 (edited by Gnanamanickam, S.S. Mahadevan, A.) New Delhi. *Today and Tomorrow's Printer's & Publisher* (1988) 139-45.

Rao, M.V.B. and Sohi, H.S. (1976). Additional hosts for *Pseudomonas solanacearum* Smith *Curr. Sci.* **45** (2): 75-76.

Rao, M.V.B., Sohi, H.S. and Tikoo, S.K. (1975). Reaction of wilt resistance tomato varieties and lines to *Pseudomonas solanacearum* in India. *Plant Dis. Repor.* **59**(9): 734-736.

Rashid, M.A., Ahmad, S. Rahman, A. Goffar, M. A. Karim, A.N.M.R. and Rajotte, E.G. (2004). Grafting of eggplant and tomatoes: An effective technique to control bacterial wilt disease and achieve high yields. In: *Asian Regional IPM CRSP Workshop:Impact and Lession learned*, held June 29-30. Makati City, Manila, Philippines.

Rashid, M.A., S. Ahmad, A. Rahman, M.A. Goffar, A.N.M.R. Karim, and E.G. Rajotte (2004). Grafting of Eggplant and Tomatoes: An effective Technique to Control Bacterial Wilt Disease and Achieve High Yields. In: Asian Regional IPM CRSP Workshop:Impact and Lession learned, Held June 29-30. Makati City, Manila, Philippines

Rashid, M.M. (1976). Bangladesher Sabjilst Edn. Bangla Academy, Dhaka Bangladesh. pp. 119-136.

Rashid, M.M. (2000). A Guidebook of Plant Pathology. Dept. of Plant Pathology. HSTU. Dinajpur. pp. 58.

Rasmussen, O.F., and Wulff, B.S. (1991). Detection of *Pseudomonas syringaepv.*pisi using PCR. In: Proceedings 4th International Working Group on *Pseudomonas syringae* Pathovars. Kluwer Academic Publishers, Dordrecht, The Netherlands. 369–376.

Read, D.J., Leuis, D.H., Fitter, A.H. and Alexander, I.J. (1992). Mycorrhizas in ecosystem. CAB International, Walling Ford, London.

Reboul, J. (1981). Grafting tomato in french polyneria. *Agronomie Trop.* **36**(3): 2993-306.

Reddy, K.C., Soffes, A. R., Prine G. M. and Dunn R. A. (1983). Tropical legumes for green manure. II. Nematode populations and their effects on succeeding crop yields. *Agro. J.* **78**(1): 5-10.

Reiter, R.J., Tan, D.X., Kim, S.J. and Qi, W. (1998). Melatonin as a pharmacological agent against oxidative damage to lipids and DNA. *Proc West Pharmacol Soc.* **41**: 229-236.

Remenant, B., Coupat-Goutaland, B., Guidot, A., Cellier, G., Wicker, E., *et al.* (2010). Genomes of three tomato pathogens within the *Ralstonia solanacearum* species complex reveal significant evolutionary divergence. *BMC Genomics*, **11:** 379.

Remenant, B., de Cambiaire, J.C., Cellier, G., Jacobs, J.M., Mangenot, S., *et al.* (2011). *Ralstonia syzygii*, the blood disease bacterium and some Asian *R. solanacearum* strain form a single genomic species despite divergent lifestyles. *PLoS ONE.* **6:** e24356.

Reyes, J.R.D. (1997). Astudy to determine the tolerance of the graft combination *Solanum diversifolium* tomato to bacterial wilt. *Proc. Trop. Reg. Amer. Soc. Hort. Sci.* **11**: 61.

Richmond, C.S., Glasner, J.D., Mau, R., Jin, H., and Blattner, F.R. (1999). Genome-wide expression profiling in *Escherichia coli* K-12. *Nucleic Acids Res.* **27:** 3821-3835.

Rick, C.M., DeVerna, J.W. and Chetelat, R.T. (1990). Experimental introgression to the cultivated tomato from related wild nightshades. In: AB Bennett and SD O'Neill (eds), *Hort. Biotechnol.* New York. pp. 19-30

Rivard, C.L. and Louws, F.J. (2006). Grafting: An integrated approach for soilborne disease management. Proc. 22nd annual tomato disease workshop November 9-10, 2006 North Carolina State University Mountain Horticultural Crops Research & Extension Center Fletcher, NC: pp. 13-16

Roach, J. (2008). fungus puts the heat in chili peppers, study says. Discover Magazine. Retrieved 13 August 2008.

Roberts, D.P., Denny, T.P. and Schell, M.A. (1988). Cloning of the *Egl* gene of *Pseudomonas solanacearum* and analysis of its role in phytopathogenicity. *J Bacteriol.* **170**: 1445-1451.

Robinson- Smith, A., Elphinstone, J. G. and Forde, S.M.D.O. (1995). Production of antibodies of *Pseudomonas solanacearum*, the causative agent of bacterial wilt. *Food and Agric. Immunol.* **7**: 67-79.

Romantschuk, M. (1992). Attachment of plant pathogenic bacteria to plant surfaces. *Annu. Rev. Phytopathol.* **30**: 225-243.

Rommens, C.M. and Kishore, G.M. (2000). Exploiting the full potential disease resistance genes for agriculture use. *Curr. Opin Biotechnol.* **11**: 120-125.

Roopali, S. (1994). Seed transmission of *Pseudomonas solanacearum* in tomato seed and eggplant. *ACIAR Bacterial Wilt Newsl.* No **11**: 12-13.

Rosqvist, R., Hakansson, S., Forsberg, A. and Wolf-Watz, H. (1995). Functional conservation of the secretion and translocation machinery for virulence proteins of Yersiniae, Salmonellae and Shigellae. *EMBO J.* **14**: 4187-4195.

Roszak, D. B. and Colwell, R.R. (1987). Survival strategies of bacteria in the natural environment. *Microbiol. Rev.* **51:** 365–379.

Rowe, P.R., Sequeira, L., and Gonzalez, L.C. (1972). Additionnal genes for resistance to *Pseudomonas solanacearum* in *Solanum phureja. Phytopathology,* **62**: 1093-1094.

Roy, A.K. Medhi, G. and Ramaih, Y. (1992). VC-48 tomato for disease prone area. *Indian Hort.* **37**: 6-7.

Ruiz-García, A.B., Olmos, A., Arahal, D.R., Antúnez, O., Llop, P., Pérez-Ortín, J.E., López, M.M., and Cambra, M. (2004). Biochip electrónicopara la detección y caracterizaciónsimultánea de los principales virus y bacteriaspatógenos de la patata. XII Congreso de la Sociedad Española de Fitopatología. Lloret de Mar. p. 12.

Sabita, J.N., Boruah, B.M. and Rachid, H.A. (2000). Yield potentiality of some brinjal cultivars in severely bacterial wilt infected condition. *Vege. Sci.* **27**:76-77.

Sadashiva, A.T., Madhavireddy, K., Deshpande, A. A. and Singh, R., (1994). Yield performance of eggplant lines resistance to bacterial wilt. *Capsicum and Eggplant Newslett.* **13**: 104-106.

Sadashiva, A.T. and Reddy, M. (1995-96). Breeding brinjal for yield, quality and resistance to biotic and abiotic stresses. *Annual report* (1995-96). Indian Institute of Horticulture Research. pp. 39-40.

Saddler, G.S. (2005). Management of bacterial wilt disease. Pages 121-132 in: Bacterial Wilt Disease and the *Ralstonia solanacearum* species complex. Allen, C., Prior, P., and Hayward, A.C., eds. APS press, St. Paul, M.N.

Sagar, V., Jeevalatha, A., Mian, S., Chakrabarti, S.K., Gurjar, M.S., Arora, R.K., Sharma, S., Bakade, R.R., Singh, B.P. (2014). Potato bacterial wilt in India caused by strains of phylotype I, II and IV of *Ralstonia solanacearum. European Journal of Plant Pathology* **138**(1): 51-65.

Sági, L. (2003). Engineering resistance to pathogenic bacteria. In: Atkinson, H., Dale, J., Harding, R., Kiggundu, A., Kunert, K., Muchwezi, J.M., Sagiand, L.and Viljoen A. (eds.). *Genetic transformation strategies to address the major constraints to banana and plantainproduction in Africa. IPGRI, Future Harvest.* pp. 39-73.

Sahambi, H.S. and Maiti, S. (1977). *Annual Report*, ICAR Research Complex for NEH Region, pp. 140.

Saile, E., McGarvey, J.A., Schell, M.A. and Denny, T.P. (1997). Role of extracellular polysaccharide and endoglucanase in root invasion and colonization of tomato plants by *Ralstonia solanacearum. Phytopathology,* **87**: 1264-1271.

Salanoubat, M., Genin, S., Artiguenave, F., Gouzy, J., Mangenot, S., et al. (2002). Genome sequence of the plant pathogen *Ralstonia solanacearum. Nature,* **415:** 497-502.

Salanoubat, M., Genin, S., Artiguenave, F., Gouzy, J., Mangenot, S., Arlat, M., Billault, A., Brottier, P. J., Camus, C., Cattolico, L., Chandler, M., Choisne, N., Claudel-Renard, C., Cunnac, S., Demange, N., Gaspin, C., Lavie, M., Moisan, A., Robert, C., Saurin,W., Schiex,T., Siguier, P., Thébault, Whalen, P. M., Wincker, P., Levy, M., Weissenbach J. and Boucher, C.A. (2001). Genome sequence of the plant pathogen *Ralstonia solanacearum. Nature,* **415**: 497-502.

Salmeron, J.M., Oldroyd, G.E.D., Rommens, C.M.T., Scofield, S.R., Kim, H.S., Lavelle, D.T., Dahlbeck, D. and Staskawicz, B.J. (1996). Tomato Prf is a member of the leucine-rich repeat class of plant disease resistance genes and lies embedded with the pto kinase gene cluster. *Cell,* **86**: 123-133.

Samaddar, K.R., Chakraborty, M. and Kanjilal, S. (1998). Identification of the race 1 of *Pseudomonas solanacearum* causing solanaceous vegetables in West Bengal and its survival. *J. Mycopathol. Res.* **36**: 51-58.

Sarowar, S., Kim, Y.J., Kim, E.N., Kim, K.D., Hwang, B.K., Islam, R. and Shin, J.S. (2005). Over-expression of a pepper basic pathogenesis-related protein 1 Gene in tobacco plants enhances resistance to heavy metal and pathogen stresses. *Plant Cell Rep.* **4**: 216-224.

Sasser, M. (1990). Identification of Bacteria by Gas Chromatography of Cellular Fatty Acids. *Technical Note* **101**: 1-6.

Sawant, D.M. (2002). *Trchoderma*: A potent plant disease control agent. In ICAR Sponsored Winter School on "*Frontiers in ecofriendly management of crop diseases and pests and utility of biofertilizers in improving soil fertility*" (Aug. 6 to Sept. 4). Department of Plant Pathology and Agriculture Microbiology. Mahatma Phule Krishi Vidyapeeth, Rahuri. pp. 14-16.

Schaad, M.C., Haldeman-Cahill, R., Cronin, S. and Carrington, J.C. (1996). Analysis of the VPg-proteinase (NIa) encoded by tobacco etch potyvirus: effects of mutations on subcellular transport, proteolytic processing, and genome amplification. *J. Virol.* **70**: 7039-7048.

Schaad, N.W. (1988). Laboratory guide for identification of plant pathogenic bacteria. American Phytopathological Society, St. Paul, MN.

Schaad, N.W., Berthier-Schaad, Y., Sechler, A. and Knorr, D. (1999). Detection of *Clavibacter michiganensis* subsp. *sepedonicus* in potato tubers by BIO-PCR and an automated real-time fluorescence detection system. *Plant Dis.* **83**: 1095-1100.

Schaad, N.W., Jones, J.B. and Chun, W. (2001). Laboratory Guide for Identification of Plant Pathogenic Bacteria. St. Paul, MN: APS Press

Schaad, N.W., Shaw, J.J., Vidaver, A., Leach, J., Erlick, B.J. (1999). Crop Biosecurity. APSnet Feature, 15 September–31 October 1999. (14 November 2001; *www.apsnet.org/online/feature/ BioSecurity /Top.html*).

Schans, J.A. and Steeghs, M.H.C.G. (1998). Strategy and results of eradication of brown rot in the Netherlands. *OEPP/EPPO Bull.* **28**: 121-133

Schell, M. (1987). Purification and characterization of an endoglucanase from *Pseudomonas solanacearum. Appl. Environ. Microbiol.* **53**: 2237–2241.

Schell, M.A. (2000a). Control of virulence and pathogenicity genes of Ralstonia solanacearum by an elaborate sensory array. *Annu. Rev. Phytopathol.* **38**: 263-292.

Schell, M.A. (2000b). Regulation of virulence and pathogenicity genes in *Ralstonia solanacearum* by a complex network. *Annu. Rev. Phytopathol.* **38**: 263–292.

Schell, M.A., Roberts, D.P. and Denny, T.P. (1988). Analysis of the *Pseudomonas solanacearum* polygalacturonase encoded by *Pgla* and its involvement in phytopathogenicity. *J. Bacteriol.* **170**: 4501-4508.

Schneider, M., Schweizer, P., Meuwly, P. and Métraux, J.P. (1996). Systemic acquired resistance in plants. *Intern. Rev. Cytol.* **168**: 303-340.

Schoen, C., De Weerdt, M., Hillhorst, R., Boender, P., Szemes, M., and Bonants, P. (2003). Multiple detection of plant (quarantine) pathogens by microarrays: innovative tool for plant health management. In Abstracts of the 19th International Symposium on Virus and Virus-like Diseases of Temperate Fruit Crops, Valencia, p. 108.

Scott , J. W., Somodi, C. and Jones, J. B. (1988). Bacterial spot resistance is not associated with bacterial wilt resistance in tomato. *Proc. Fla State Hort. Soc.* **101**: 390-392.

Scott, J.K. and Smith, G.. (1990). Searching for peptide ligands with an epitope library. *Science,* **249**: 386-390.

Seal S.E., Taghavi M., Fegan N., Hayward A.C., Fegan M., (1999). Determination of *Ralstonia* (*Pseudomonas*) *solanacearum* rDNA subgroups by PCR tests. *Plant Pathol.* **48**: 115-120.

Seal, S.E., Jackson, L.A. and Daniels, M.J. (1992). Isolation of a *Pseudomonas solanacearum*- specific DNA probe by substraction hybridization and construction of species-specific oligonucleotide primers for senstitive detection by polymerase chain reaction. *Appl. Environ. Microbiol.* **58**: 3751-3758.

Seal, S.E., Jackson, L.A., Young, J.P.W. and Daniels, M.J. (1993). Differentiation of *Pseudomonas solanacearum, Pseudomonas syzigii, Pseudomonas pickettii* and the blood disease bacterium by partial 16S rRNA sequencing: construction of oligonucleotide primers for sensitive detection by polymerase chain reaction. *J. Gen. Microbial.* **139**: 1587-1594.

Seleim, M.A.A., Saead, F.A., Abd-El-Moneem, K.M.H. and Abo-ELyousr, K.A.M. (2011). Biological control of bacterial wilt of tomato by plant growth promoting rhizobacteria. *Plant Pathol. J.* **10**:146-153.

Selsted, M.E., Novotny, M.J., Morris, W.L., Tang, Y.Q., Smith, W. and Cullor, J. S. (1992). Indolicidin, a Novel bactericidal tridecapeptide amide from neutrophils. *J. Biol. Chem.* **267**: 4292–4295.

Seneviratne, S.N. DeS. (1969). On the occurrence of *Pseudomonas solanacearum* in the hill country of Ceylon. *J. Hort. Sci.* **44**(4): 393-402.

Seneviratne, S.N. DeS. (1988). Soil survival of *Pseudomonas solanacearum*. In: Bacterial Diseases of Potato. Report on the planning conference on Bacterial Diseases of Pototo. Internatioal Potato Centre CIP. 85-91.

Sequeira, L. (1985). Surface components involved in bacterial pathogen-plant host recognition. *J. Cell Sci. Suppl.* **2**: 301-316.

Sequeira, L. and Graham, T.L. (1977). Agglutination of avirulent strains of *Pseudomonas solanacearum* by potato lectin. *Physiol. Plant Pathol.* **11**: 43-54.

Sequeira, L. and Kelman, A. (1976). *Proc. First Int. Planning Conf. and Workshop on the Ecology and control of Bacterial wilt caused by Pseudomonas solanacearum.* North Carolina State University, Raleigh, North Carolina, July 18-23, pp. 166.

Sequira, L. (1979). Development of resistance to bacterial wilt derived from *Solanum phureja*. In: Developments in control of potato bacterial diseases, CIP, Lima, Peru. 55-62.

Seshadri, K., Usman, K.M., Kandaswamy, T.K. and Seetharaman, K. (1977). Bacterial wilt of papaya caused by *Pseudomonas solanacearum. Madras Agri. J.* **64**: 181-182.

Shamsuddin, N., Lloyd, A.B. and Graham, J. (1979). Survival of the potato strain of *Pseudomonas solanacearum* in soil. *J. Aust. Inst. Agric. Sci.* **44**: 212-215.

Shapiro, S., and Guggenheim, B. (1995). The action of thymol on oral bacteria. *Oral Microbiol. Immunol.* **10**: 241-246.

Sharma, B.D. and Mukherjee, S.K. (1970). A bacterial wilt of jute (*Corchorus capsularis* L.,*C. olitorius* L.) caused by *Pseudomonas solanacearum* E.F. Smith. *Phytopathology,* **67**: 93-94.

Sharma, J.P., Jha, A.K., Singh, A.K., Pan, R.S. Rai, M. and Kumar, S. (2005). Screening of parental lines and their F_2 crosses of brinjal (*Solanurn melongena*) to *Ralstonia* wilt. *Indian J. Agri. Sci.* **75**(4): 197-199.

Sharma, J.P. and Kumar, S. (1997). Performance of elite lines of tomato (*Lycopersicon esculentum)* for resistance to wilt caused by *Pseudomonas solanacearum. Indian J. Agri. Sci.* **67**(4):175-176.

Sharma, J.P. and Kumar, S. (2000). Management of *Ralstonia* wilt through soil disinfectant, mulch, lime and cakes in tomato (*Lycopersicon esculentum*). *Indian J. Agri. Sci.* **70**(1): 17-19.

Sharma, J.P. and Kumar, S. (2005). Occurrence of biovar V of race I of *Ralstonia solanacerum* in Jharkhand (abstract). *Indian Phytopathol.* **58**(3): 362 (Abstract).

Sharma, J.P. and Kumar, S. (2009a). Linear reduction of propagules of *Ralstonia solanacearum* in soil by cake and chemicals. *Indian Phytopath.* **62**(1): 49-53.

Sharma, J.P. and Kumar, S. (2009b). Management of *Ralstonia* wilt of tomato through microbes, plant extract and combination of cake and chemicals. *Indian Phytopath.* **62**(4): 417-423.

Sharma, J.P. and Kumar, S. (1995). Evaluations of host resistance in brinjal to *Pseudomonas* wilt in Chotanagpur. *Indian Phytopath.* **48**(1): 81-83.

Sharma, J.P. and Kumar, S. (1995-96). Evaluation of bacterial wilt in resistant varietal trial in brinjal and tomato. *Annual report* (1995-96). Indian Institute of Horticulture Research. pp. 140-141

Sharma, J.P. and Kumar, S. (1996-97). *In Annual Report*. Indian Institute of Horticulture Research, Bangalore. pp. 178.

Sharma, J.P. and Kumar, S. (2004). Effects of crop rotation on population dynamics of *Ralstonia solanacearum* in tomato wilt sick soil. *Indian Phytopath.* **51** (1): 80-81.

Sharma, J.P. and Kumar, S. (2011). Annual Report. ICAR RCER

Sharma, J.P. Kumar, S. and Das, B. (2009). Biochemical changes in response to interaction with antagonists (*Trichoderma harzianum* and *T. viride*) in resistant and susceptible culivars. (S-02) *5th Int. Conf. on Plant Pathol. in the Globalized Era*, Nov. 10-13, 2009, New Delhi, India p. 86.

Sharma, J.P., Jha, A.K., Singh, A.K., Pan, R.S., Rai, M. and Kumar, S. (2006). Evaluation of tomato against bacterial wilt (*Ralstonia solanacearum*) in Jharkhand. *Indian Phytopath.* **59** (4): 405-409.

Sharma, J.P., Krishna Prasad, V.S.R., Pan, R.S. and Rai, M. (1999). Screening of exotic collection of tomato (*Lycopersicon esculentum*) for resistance to bacterial wilt (*Ralstonia solanacearum*). *Recent Trend in Horticulture Research, 188*-192.

Sharma, J.P., Kumar S. and Das, B. (2012). Soil application of *Trichoderma harzianum* and *T. viride* onbiochemical constituents in bacterial wilt resistant and susceptible cultivars of tomato**.** *Indian Phytopath.* **65**(3): 264-267 .

Sharma, J.P., Singh, D.P. and Krishna Prasad, V.S.R. (1995). Differential reaction and source of resistance to*Pseudomonas* wilt in eggplant (*Solanum melongena* L.). *Indian J. Agri. Sci.* **65**(9): 686-690.

Sharma, T.V. Rathore, R.S. and Kumar, S. (1998). Screening Tomato for bacterial wilt resistance in Andaman. *J. Andaman Sci. Asso.* **13**(1/2):32-36.

Shaw, D.E. (1984). Microorganism in Papua New Guinea Port Mores by Papua New Guinea, Department of Primary Industry. *Res. Bull.* **33**: 344p.

Shekhawat, G.S., Chakrabarti, S.K. and Gadewar, A.V. (1992a). Potato bacterial wilt in India. *Technical Bulletin No*. 38. Central Potato Research Institute (CPRIJ), Shimla, India. 52 p.

Shekhawat, G.S., Chakrabarti, S.K., Kishore, V., Sunaina, V. and Gadewar, A.V. (1992b). Possibilities of biological management of potato bacterial wilt with strains of *Bacillus* sp. *Bacillus subtilis*, *Pseudomonas fluorescens* and *Actinomycetes*. In: Hartman, G.L. and A.C. Hayward (eds.). Bacterial wilt. pp. 327-30. *ACIAR Proceedings*, No. 45: *Australian Centre for International Agricultural Research*, Camera.

Shekhawat, G.S., Singh, R. and Kishore,V. (1978). Distribution of bacterial wilt and races and biotypes of the pathogen in India. *J. Indian Potato Assoc.* **5**: 155-165.

Shekhwat, G.S. (1976). Bacterial wilt of the potato in India. *In Proc. Int. Planning Conf. and Workshop on Ecology and Control of Bacterial wilt caused by Pseudomonas solanacearum* (eds. Sequeira, L. and Kelman, A.) p. 89.

Sheoraj and Srivastava, L.S. (1979). *In Annual Report*, ICAR Research Complex for NEH Region Shillong, pp. 143.

Shidfar, F., Froghifar, N., Vafa, M., Rajab, A., Hosseini, S., Shidfar, S., Gohari, M. (2011). The effects of tomato consumption on serum glucose, apolipoprotein B, apolipoprotein A-I, homocysteine and blood pressure in type 2 diabetic patients. *Int. J. Food Sci. Nutr.* **62**(3): 289-94.

Shiomi, Y., Nishiyama, M., Onizuka, T. and Marumoto, T. (1999). Comparison of bacterial community structures in the rhizoplane of tomato plants grown in soils suppressive and conducive towards bacterial wilt. *Appl. Environ. Microbiol.* **65**: 3996-4001.

Shoda, M. (2000).Bacterial control of plant diseases *J. Biosci. Bioeng.* **89** (6): 515–521.

Shou, S. Y., Feng, Z. Z., Miao, L. X., Liao, F. B. (2006). Identification of AFLP markers linked to bacterial wilt resistance gene in tomato. *Hereditas,* **28**(2): 195-199.

Shuai, Z., Sujialic and Luo, Xiao B. (1997). Physiological specialization of tomato bacterial pathogen (*Pseudomonas solanacearum* Smith) in Chengalu and identification of the resistance of tomato varieties. *China Vege.* No. **3**: 16-18.

Sikirou, R., Beed, F., Ezin, V., Gbèhounou, G., Miller, S.A. and Wydra, K. (2009). First report of bacterial wilt of tomato (*Solanum lycopersicum*) caused by *Ralstonia solanacearum* in Benin. *Plant Dis.* **93**(5): 549.

Silveira, E.B., Gomes, A.M.A., Michereff, S.J. and Mariano, R.L.R. (1998). Variability of *Ralstonia solanacearum* populations causing wilt on tomato in Agreste of Pernambuco, Brazil. *Bacterial Wilt Newsl.* **15**: 8-10.

Silveira, E.B., Gomes, A.M.A., Ferraz, E., Maranhao, E.A.A. and Mariano, R.L.P. (1999). Identification of tomato progenies resistant to bacterial wilt. *Horti. Brasil.* **17**(1): 6-10.

Silverira, E.B. DA, Mariano, R., DE, L.R., Michereff, S.J. and Menezes, M. (1995). Antagonism of *Bacillus* spp. against *Pseudomonas solanacearum* and effect on tomato seedling growth. *Fitopathol. Brasil.* **20**(4): 605-612.

Simmaco, M., Barra, D., Chiarini, F., Noviello, L., Melchiorri, P., Kreil, G. and Richter, K. (1991). A family of bombinin-related peptides from the skin of *Bombina variegata. Eur. J. Biochem.* **199**: 217-222.

Simmaco, M., Mignogna, G., Barra, D. and Bossa, F. (1993). Novel antimicrobial peptides from skin secretion of the european frog *Rana esculenta. FEBS Lett.* **324**: 159-161.

Simmaco, M., Mignogna, G., Barra, D. and Bossa, F. (1994). Antimicrobial peptides from skin secretions of *Rana Esculenta.* Molecular cloning of cdnas encoding esculentin and brevinins and isolation of new active peptides. *J. Biol. Chem.* **269**: 11956-11961.

Simmaco, M., Mignogna, G., Canofeni, S., Miele, R., Mangoni, M.L. and Barra, D. (1996). Temporins, antimicrobial peptides from the european red frog *Rana temporaria. Eur. J. Biochem.* **242**: 788-792.

Simmonds, P., Balfe, P., Peutherer, J.F., Ludlam, C.A., Bishop, J.O., and Brown A.J. (1990). Human immunodeficiency virus-infected individuals contain provirus in small numbers of peripheral mononuclear cells and at low copy numbers. *J. Virol.* **64**: 864-72.

Simpson, B.B. and Ogorzaly, M.C. (1986). *Economic Botany*: Plants in Our World. *McGraw-Hill Science/ Engineering/Math; 3 edition* (December 20, 2000)

Singh A.K., Bhavana, P.S. Kumar, S. and Bhatt, B.P. (2013) Tomato-plant grafting for resistance to bacterial wilt. ICAR News **19**(3): 7

Singh, D. and Sharma, P. (2012). Current status of bacterial and fungal diseases of solanaceous fruit vegetables in India, Division of Plant Pathology, Indian Agricultural Research Institute, New Delhi-110012.

Singh, D. and Sharma, P. (2014). Diseases of tomato and brinjal. In Diseases of vegetable crops: Diagnosis and management edited by Singh, D., Chowdapa, P. and Sharma, P., Indian Phytopathological Society, IARI, New Delhi pp. 60-142.

Singh, D., Sinha, S., Chaudhary, G., Yadav, D.K. and Mondal, K.K. (2014a). Genetic diversity of biovar 3 and 4 of *Ralstonia solanacearum* causing bacterial wilt of tomato using BOX-PCR, RAPD and *hrp* gene sequences. *Indian J. Agri. Sci.* **84**(3): 391-395.

Singh, D., Sinha, S., Yadav, D.K. and Chaudhary, G. (2014b). Detection of *Ralstonia solanacearum* from asymptomatic tomato plants, irrigation water, and soil through non-selective enrichment medium with *hrp* gene-based bio-PCR. *Curr Microbiol.* **69**(2): 127-134.

Singh, B. (1993). Occurrence of a New Biotype of *P sthanacearion*in India. *Indian J. Agric. Sci.* **63**: 190-192.

Singh, B., Anita, K., Sharma, K.D., Agrawal, P.C. and Ramnath (1995). World distribution of phytopathogenic bacteria, NBPGR, New Delhi. p. 147.

Singh, D., Sinha, S., Yadav, D.K.; Sharma, J.P., Srivastava , D.K., Lal, H.C., Mondal, K.K., Jaiswal, R. K. (2010). Characterization of biovar/races of *Ralstonia solanacearum*, the incitant of bacterial wilt of solanaceous crops. *Indian Phytopath.* **63**(3): 261-265.

Singh, D., Yadav, D.K., Sinha, S. and Upadhyay, B.K. (2012). Utilization of plant growth promoting *Bacillus Subtilis* isolates for the management of bacterial wilt incidence in tomato caused by *Ralstonia solanacearum* race 1 biovar 3. *Indian Phytopath.* **65** (1): 18-24.

Singh, D., Yadav, D.K., Sinha, S., Mondal, K.K., Singh, G., Pandy, K.K. and Singh, R. (2013). Genetic diversity of iturin producing strain of *Bacillus* species antagonistic to *Ralstonia solanacerarum* causing bacterial wilt disease in tomato. *African J. Microbiol. Res.* **7**(48): 5459-5470.

Singh, K. (1961). Inheritance of North Carolina type of bacterial wilt resistance in tomato *Lycopersion esculentum* L. (Master Thesis). University of Hawaii, Honolulu.

Singh, K.P.(1991). Effect of continuous use of manureand lime on crop yield and fertility status in redloam soil of Chota Nagpur (Bihar). *New Agriculturist* **1**: 133-138.

Singh, R. (1994-95). Bacterial diseases of vegetable crops. *Annual Report* (1994-95). Indian Institute of Horticulture Research.

Singh, R. (1995-96). Bacterial diseases of vegetable crops. *Annual Report* (1995-96). Indian Institute of Horticulture Research. pp. 69

Singh, U.S., Zaidi, N.W., Joshi, D., Khan, T., John, D. and Bajpai, A. (2004). *Trichoderma:* Microbeswith Multifaceted Activity. In *Annu. Rev. Pl. Pathol. vol.* 3. pp. 33-75.

Sinha, S.K. (1985). Bacterial wilt in India. pp. 28-29. In *Bacterial wilt disease in Asia and South Pacific. ACIAR Proceeding No.* 13. (ed. Persley, G.J.) 145 p.

Sly, L. I. (1983). Preservation of microbial cultures, In: Plant Bacterial Diseases: Diagnostic Guide, Fahy, P.C. and Persley, G.J. eds. Academy Presss Sydney, Chapter 12, 275-278.

Smith E.F. (1920). The brown rot of *Solanaceae*. Bacterial Diseases of Plants. U.S.A.: Saunders Company, 1920:177.

Smith, A.F. (1994). *The tomato on America*: University of Ilorin Press

Smith, E.F. (1994). *The tomato in America: early history, culture, and cookery*. Columbia, S.C, USA: University of South Carolina Press. ISBN 1-5700-3000-6 .

Smith, E.F. (1896). A bacterial disease of the tomato, eggplant and Irish potato (*Bacillus solanacearum*) Bulletin, Division of vegetable physiology and pathology. United States, Department of Agriculture 12, 1-28.

Smith, E.F. (1908). The Granville tobacco wilt. *Bur. Plant Ind. Bul.* 141: 17 (Part II). U.S. Dept. Agr.

Smith, E.F. (1914). Bacteria in relation to plant diseases. Vol 3. Washington. Corneqie Institute.

Smith, J.J., Offord, L., Holderness, C.M. and Saddler, G.S. (1995). Genetic diversity of *Burkholderia solanacearum* (synonym *Pseudomonas solanacearum*) race 3 in Kenya. *Appl. Environ. Microbiol.* **61**: 4263-4268.

Sohi, H.S., Rao, M.V.B., Rawal, R.D. and Kishun, R. (1981). Effect of crop rotation on bacterial wilt of tomato and brinjal. *Ind. J. Agril. Sci.* **51**(8): 572-573.

Song, W.Y., Wang, G.L., Chen, L.L., Kim, H.S., Pi, L.Y., Holsten, T. Gardner, J., Wang, B., Zhai, W.X., Zhu, L.H., Fauquet, C. and Ronald, P. (1995). A receptor kinase-like protein encoded by the rice resistance gene, Xa21. *Science,* **270**: 1804- 1806.

Sonoda, R.M. (1975). Effect of time of planting on incidence of bacterial wilt in tolerant and susceptible lines of tomato. *Res. Repor. Agri. Res. Centre, Fort Pierce* No RL –1975, 25p.

Sonoda, R.M. (1979). Effect of fumigation and partial resistance on survival of tomato lines to *Pseudomonas solanacearum* infected soil in late spring. *Fort Pierce, Florida, USA Agricultural Research Centre. RL-1979-8, 4pp.* RL –1977-6, 5p.

Sood, A.K., Kalha, C.S. and Ambardar, V.K. (1997). Soil solarization –experience with *Burkholderia solanacearum* causing wilt of tomato (Abstr.) p. 89. *Indian Phytopathol. Soc. – Golden Jubilee-Int. Conf. Integrated Plant Disease Management for Sustainable Agriculture* 10-15 Nov.

Sood, A.K., Kalha, C.S. and Parasheor, A. (1998). Ecofriendly methods for the management of bacterial wilt caused by *Ralstonia solanacearum. Bacterial Wilt Newslett.* No. **15**: 7.

Soohee, C., Hyesuk, K., Jeffrey, S.B., Dilip, K.L., John, L., Sang, D.K. and Daniel, P.R. (2008). Isolation and partial characterization of *Bacillus subtilis* ME488 for suppression of soilborne pathogens of cucumber and pepper. *Appl. Microbiol. Biotechnol.* **80**: 115-121.

Sosnowski, R.G., Tu, E., Butler, W.F., O'Connell, J.P., and Heller, M.J. (1997). Rapid determination of single base mismatch mutations in DNA hybrids bydirect electric field control. *Proc. Natl. Acad. Sci. U.S.A.* **94**: 1119-1123.

Spadaro, D. and Gullino, M. L. (2005). Improving the efficacy of biocontrol agents against soilborne pathogens. *Crop Protec.* **24**(7): 601-613.

Spooner, D.M., McLean, K., Ramsay, G., Waugh, R. and Bryan G.J. (2005). A Single domestication for potato based on multilocus amplified fragment length polymorphism genotyping. *PNAS* **102**(41):14694-14699.

Srinivasan, K. and Gopimony, R. (1969). On the resistance of wild brinjal variety to bacterial wilt. *Agri. Res. J.*, Kerala, **7**(1): 39-40.

Sriram, S. and Rangeshwaran, R. (2010-11). In Annual Report. *Phytophthora, Fusarium* and *Ralstonia* Diseases of Horticultural Crops. *Indian Institute of Spices Research*: 93-94.

Stackebrandt, E., Murray, R.G.E. and Truper, H.G. (1988). Proteobacteria Classis Nov., A name for the phylogenetic taxon that includes the purple bacteria and their relatives. *Int. J. Syste. Bacteriol.* **38**: 321-325.

Stapp, C. (1965). Bacterial wilts and its causal organism. *Zbl.* **119**: 166-190.

Staskawicz, B.J., Dahlbeck, D. and Keen, N.T. (1984). Cloned avirulence gene of *Pseudomonas syringae* pv. *glycinea* determines race-specific incompatibility on *Glycine max* (L.) Merr. *Proc. Nati. Acad. Sci.* USA, **81**: 6024-6028.

Stefani, E., Giosuè, S. and Mazzucchi, U. (2005). Detection of Latent Infections of *Ralstonia solanacearum*biovar 2, Race 3 in Tomato Crops. *J. Plant Path.* **87**(3): 167-171.

Stevens, R.B. (1960). Plant Pathology, an Advanced Treatise, Vol. 3. p. 357-429. In J.G. Horsfall and A.E. Dimond (eds.). Academic Press, NY.

Stevenson, W.R., R. Loria, G.D. Franc, and D.P. Weingartner, eds. (2001). Compendium of potato diseases, 2nd edn. APS Press, St. Paul, MN.

Sudakevitz, D., Kostlanova, N. and Blatman-Jan, G. *et al.* (2004). A new *Ralstonia solanacearum* high-affinity mannose-binding lectin rsiil structurally resembling the *Pseudomonas aeruginosa* fucose-specific lectin PA-IIL. *Mol. Microbiol.* **52**: 691-700.

Suganuma, H., Hirano, T., Arimoto, Y., Inakuma, T. (2002). Effect of tomato intake on striatal monoamine level in a mouse model of experimental parkinson's disease. *J. Nutri. Sci. and Vitaminol.* **48** (3): 251–254.

Sun, B., Liao, Yi, Li Zhiliang, Li Zhenxing, Sun, G. (2008). AFLP markers linked to genes related to bacterial wilt resistance of eggplant. *Mol. Plant Breed.* **6**: 929-934.

Sun, S.K. and Huang, J.W. (1985). Formulated soil amemndment for contolling *Fusarium* and other soilborne disease. *Plant Dis.* **69**: 917-920.

Sun, S.K. and Huang,J.W. (1985b). Mechanisms of control of Fusarium wilt disease by amendment of soil with S-H mixture. *Plant Prot. Bull.* **27:** 159-169.

Sunaina, V., Gupta, J.P.K., Shekhawat, G.S. and Kishore, V. (1998). Multiplication of *Pseudomonas solanacearum* in pure water. *Indian Phytopath.* **51**(1): 95-96.

Sunaina, V., Pandy, S. K. and Singh, B. P. (2006). Bio-B5, an ecofriendly biopesticide- cum biofertilizer for management of soil and tuber borne diseases of potato and other crops. Central Potato Research Institute Campus, Modipuram, Meerut, Extension. Bull. pp.10.

Sunaina, V., V. Kishore, and Shekhawat, G.S. (1989). Latent survival of *Pseudoinonas solanacearum* in potato tubers and weeds. *J. Plant Dis. Prot.* **96**: 361-364.

Suresh, A.K. and Mody, K. (2009). *Microbial exopolysaccharides*: Variety and potential applications. *Microbial Production of Biopolymers and Polymer Precursors*. Caister Academic Press. ISBN 978-1-904455-36-3 en.wikipedia.org/wiki/Exopolysaccharide.

Suresh, C.K. and Rai, P.V. (1991). Interaction of *Pseudomonas solanacearum* with antagonistic bacterium and VAM mycorrhiza. *Current Res. Univ. Agri. Sci.* **20**(3): 36-37.

Suzuki, I. and Others. (1964). Studies on breeding egg plant and tomato lines for resistance to bacterial wilt. I. Investigation on methods of evauating resistance and on the Ssource of resistance in eggplants and tomatoes. *bull. Hort. Res. Statn. Hiratsuka, Ser* A No. 3 pp. 77-106.

Suzuki, S., Ohe, Y., Okube, T., Kakegawa, T. and Tatemoto, K. (1995). Isolation and characterization of novel antimicrobial peptides, rugosins A, B and C, from the skin of the frog, *Rana rugosa. Biochem. Biophysi. Res. Commun.* **212**: 249-254.

Swaminathan, M. and Srinivasan, K. (1972). Studies on brinjal hybridization II. transference of bacterial wilt resistance from wild brinjal variety. *Agric. Res. J.* Kerala. **9**: 11-13.

Swanson, J.K., Montes, L., Mejia, L., and Allen, C. (2007). Detection of latent infection of *Ralstonia solanacearum* race 3 biovar 2 in geranium. *Plant Dis.* **91**(7): 828-834.

Swaroop, K. and Suryanarayana, M.A. (2005). Evaluation of tomato varieties and lines for growth, yield, quality, and bacterial wilt resistance under coastal tropical condition of the Andaman Islands. *Trop. Agri.* 82 (3-4): 294-299.

Tahat, M.M. and Sijam, K. (2010). *Ralstonia solanacearum*: The bacterial wilt causal agent. *Asian J. Plant Sci.* **9**: 385-393.

Tainter, D.R. and Anthony, T.G. (2001). *Spices and Seasonings*. A Food Technology Handbook. *John Wiley & Sons,* Science - 249 pages.

Taiwan, AVRDC (1984). Semi tropical proceeding tomato developed centre point **3**(6): 2.

Talukdar, M.J. (1974). Plant diseases in Bangladesh. *Bangladesh J. Agric. Res.* **1**(1): 161-186.

Tan, S., Jiang, Y., Song. S., Huang, J., Ling, N., Xu, Yangchun and Shen Qirong. (2013). Two *Bacillus amyloliquefaciens* strains isolated using the competitive tomato root enrichment method and their effects on suppressing *Ralstonia solanacearum* and promoting tomato plant growth. *Crop Protec.* **43:** 134-140.

Tanaka, Y. and Noda, N. (1973). Studies on the factors affecting survival of *Pseudomonas solanacearum* E.F. Smith, the causal agent of tobacco wilt disease. *Bull. Okayama Tobacco Exp. Stn.* **32:** 81–94.

Tanigawa, M., Asao, H. and Okayama, K. (1991).Wilt including activity and components of exopolysaccharide produced by a virulent *Pseudomonas solonacearum. Bull. Nira Agri. Exp. Sta. No.* **22**: 23-27

Tans-Kersten, J., Guan, Y.F.and Allen, C. (1998). *Ralstonia solanacearum* pectin methylesterase is required for growth on methylated pectinbut not for bacterial wilt virulence. *Appl. Environ. Microbiol.* **64**: 4918-4923.

Tans-Kersten, J., Huang, H.Y. and Allen, C. (2001). *Ralstonia solanacearum* needs motility for invasive virulence on tomato. *J. Bacteriol.* **183**: 3597-3605.

Termorshuizen, A.J., Moolenaar, S.W., Veeken, A.H.M. and Blok, W.J. (2004). The value of compost. *Rev. Environ. Sci. Biotechnol.* **3**: 343-347

Thakur, R.P., Rao, V.P., Wu, B.M., Subbarao, K.V., Shetty, H.S., Singh, G., Lukose, C., Panwar, M.S., Sereme, P., Hess, D.E., Gupta, S.C., Dattar, V.V., Panicker, S., Pawar, N.B. Bhangale, G.T. and Panchbhai, S.D. (2004). Host resistance stability to downy mildew in pearl millet and pathogenic variability in Sclerospora graminicola. *Crop Protec.* **23** (10): 901-908.

Thanh, D.T., Tarn, L.T.T., Hanh, N.T., Tuyen, N.H., Bharatkumar, Srinivasan, Lee, S.Y. and Park, K.S. (2009). Biological control of soil borne diseases on tomato, potato and blackpepper by selected pgpr in green house and field in Vietnam. *Plant Pathol. J.* **25**(3): 261-269.

Thomas, C., Hill, D.J. and Mabey, M. (1999). Morphological changes of synchronized *Campylobacter Jejuni* populations during growth in single phase liquid culture. *Lett. Appl. Microbiol.* **28**: 194-198.

Thoquet, P. Stephens, S. and Grimsley, N. (1993). Mapping of bacterial wilt resistance genesin tomato variety Hawaii 7996. pp. 176.

Thoquet, P., Olivier, J., Sperisen, C., Rogowsky, P. Laterrot, H. and Grimsley, N. (1996a). Quantitative traits loci determing resistance to bacterial wilt in tomato cultivar Hawaii 7996. *Mol. Plant –Microbe Intera.* **9**(9): 826-836.

Thoquet, P., Olivier, J., Sperisen, C., Rogowsky, P., Prior, P., Anaïs, G., Mangin, B., Bazin, B., Nazer, R., and Grimsley, N. (1996b). Polygenic resistance of tomato plants to bacterial wilt in the French West Indies. *Mol. Plant– Microbe Interact.* **9**: 837-842.

Thorold, C.A. (1949). The effects of certain crop rotation on the incidence of tomato (in Trinidad). *Trop. Agri. Trin.* **26**: 28-32.

Thurston, H.D. (1976). Resistance to bacterial wilt *(Pseudomonas solanacearum*). In: Sequeira, L., Kelman A. (eds) Proc. First Int. Planning Conf. and Workshop on the Ecology and Control of Bacterial Wilt caused by *Pseudomonas solanacearum*. NC State University, Raleigh, pp. 58-62.

Tian, S.B., Wang, Y.Q., Luo, Z.Y., Li, M., Chen, L., Chen, Y.K., Hong, Y.J. (2007). Genetic analysis of resistance to bacterial wilt in eggplant *(Solanum melongena*), Southwest China. *J. Agric. Sci.* **20:** 642-645.

Tikoo, S.K., Anand, N., Kishun, R. and Reddy, P.P. (1988). In: Tomato and pepper production in the tropics. *Proc.e Int. Symp. Integr. Management Practices*, Tainan, Taiwan, 21-26 March (edited by Green, S.K.; Gripps, T.D.; MeClean, B.T.) Shanhua, Taiwan, AVRDC (1989) 99-106.

Tikoo, S.K., Mathal, P.J. and Kishun, R. (1979). Successful graft culture of tomato in bacterial wilt sick soils. *Curr. Sci.* **48**(6): 259-260.

Timm, R.W. and Amen, M. (1960). Nematodes associated with commercial crops in East Pakistan. *Agric Pak.* **11**(3): 355- 363.

Titaturn, V. and Tong Jeam, M. (1985a). Survival of *Pseudomonas solanacearum* in soil from potato growing area. *Ann. Repor. Departm. Agri.* Thailand.

Titaturn, V. and Tong Jeam, M. (1985b). Studies on weed hosts of bacterial wilt of potato in the fields. *Annu. Repor. Departm. Agri.* Thailand.

Tolba, A.F. (1998). Biological control of some potato bacterial diseases. Ph.D. Thesis, Faculty of Agriculture, Minufyia University Egypt.

Tomar, B. S. and Kalda, J. S. (1998). Production System. Is Eggplant Nutritious? *TVIS Newsl.* **1**(1): 26.

Tomer, D.K. and Sunaina, V. (2010). Antagonistic action of bacterial wilt of potato. *Progr. Agri.* **10**: 301-304.

Tör, M., Yemm, A. and Holub, E. (2003). The role of proteolysis in r gene mediated defence in plants. *Mol. Plant Pathol.* **4**: 297-305.

Toyoda, H., Nomura, T., Matsuda, Y., Dogo, M., Kato Y., Bigno, M. and Ouchis, S. (1998). Valuation of disease resistance expression in self-pollinated progenies of heat callus derived tomato regenerated resistance to bacterial wilt. *Bull. Insti. Compreh. Agri. Sci. Kinki Univ.* No. **6**: 67-72.

Toyoda, H., Shimizu, K., Chtani, K., Kita, N., Matsuda, Y. and Ouchis, S. (1989). Selection of bacterial wilt resistant tomato mate through tissue culture. *Plant Cell Repor.* **8**(6): 3173-20.

Tran, N.H. and Kim, B.S. (2010). Inheritance of resistance to bacterial wilt (*Ralstonia solancearum*) in pepper (*Capsicum annuum* L.). *Hort. Environ. Biotechnol.* **51**: 431-439.

Trigalet, A. and Demery, D. (1986). Invasiveness in tomato plants of tn5-induced avirulent mutants of *Pseudomonas solanacearum*. *Physiol. Mol. Plant Pathol.* **28**: 423-430.

Tsao and Lo (2006) in "Vegetables: Types and Biology". *Handbook of Food Science, Technology, and Engineering* by Yiu H. Hui (2006). CRC Press. ISBN 1-57444-551-0.

Tsuro, M., Minamiyama, Y. and Hirai, M. (2007). QTL Analysis for bacterial wilt resistance in Japanese pepper (*Capsicum annuum* L.). *Breeding Res.* **9**: 111–115.

Tung, P.X, Rasco, E.E., Vander Zaag, P. and Schmiediche, P. (1990b). Resistance to environment interactions. *Euphytica,* **46**: 211-215.

Tung, P.X. (1985). Bacterial wilt in Vietnam. In: *Bacterial wilt disease in Asia and the South Pacific* (ed. Persley, G.J.). *ACIAR Proceeding* No.13. pp. 68-70.

U N F AO (2009). International Year of the Potato 2008 – The potato. United Nations Food and Agricultural Organisation. 2009. Retrieved 26 October 2011.

Uehara, Y. (1988). Control of tomato wilt disease by using a sheet to interupt the root system. *Agri. and Hort.* **63**(11): 1305-1309.

Uematsu, T. *et al.* (1983). Pathogenecity and some bacteriological properties of six strains of *Pseudomonas solanacearum* in Thailand. *Repor. Plant Pathol.*, Bangkok, Thailand.

Uesugi, T., Froh, M., Artell, G., Bradford, B., Wheeler, M., Gabele, E., Isayama, F.,Thurman, R, (2002). Role of lipopolysaccharide-binding protein in early alcohol-induced injury in mice. *J. Immunol.* **168**: 2963–2969.

Valls, M., Genin, S. and Boucher, C. (2006). Integrated regulation of the type III secretion system and other virulence determinants in *Ralstonia solanacearum. Plos Pathogens,* **2**: 798-807.

van Bruggen, A.H.C. and Termorshuizen, A.J. (2003). Integrated approaches to root disease management in organic farming systems. *Australasian Plant Pathol.* **32**: 141-156.

van der Biezen, E.A. and Jones, J.D.G. (1998). Plant disease-resistance proteins and the gene-for-gene concept. *Trends in Biochem. Sci.* **23**: 454-456.

Van der Wolf, J. and Schoen, C. D. (2004). Bacterial pathogens: detection and identification methods. *Encyclopedia of Plant and Crop Sci.* 84-88.

Van der Wolf, J.M., Bonants, P.J.M., Smith, J.J., Hagenaar, M., Nijhuis, E., Van Beckhoven, R.C.M., Saddler, G.S., Trigalet, E., and Feuillade, R. (1998).Genetic diversity of *Ralstonia solanacearum* race 3 in Western Europe Determined by AFLP, RC-PFGE and Rep-PCR. pp. 44-49 in: Bacterial Wilt Disease, Molecular and Ecological Aspects. P. Prior, C. Allen, and J. Elphinstone, eds. Springer-Verlag, Berlin, Heidelberg, New York.

Van der Wolf, J. M., Van Beckhoven, J.R. C.M., De Haan, E.G., Van Den Bovenkamp, G.W. and Leone, G.O.M. (2004). Specific detection of *Ralstonia solanacearum* 16S rRNA sequences by AmpliDet RNA. *Eur. J. Plant Pathol.* 110: 25-33.

Van der Wolf, J. M., van Bekkum, P.J., van Elsas, J.D., Nijhuis, E.H., Vriend, S.G.C. and Ruissen, M.A. (1998). Immunofluorescence colony staining and selective enrichment in liquid medium for studying the population dynamics of *Ralstonia solanacearum* (race 3) in soil. *Bulletin OEPP/EPPO*, **28:** 71–79.

van der Wolf, J.M., Vriend, S.G.C., Kastelein, P., Nijhuis, E.H., van Bekkum, P.J. and van Vuurde, J.W.L. (2000) Immunofluorescence colony-staining (IFC) for detection and quantification of *Ralstonia (Pseudomonas) solanacearum* biovar 2 (race 3) in soil and verification of positive results by PCR and dilution plating. *Eur. J. Plant Pathol.* **106**: 123–133.

van Diepeningen, A.D., de Vos, O.J., Korthals, G.W. and van Bruggen, A.H.C. (2006). Effects of organic versus conventional management on chemical and biological parameters in agricultural soils. *Appl. Soil Ecol.* **31**: 120-135.

van Elsas, J.D., Kastelein, P., van Bekkum, P., van der Wolf, J.M., de Vries, P.M. and van Overbeek, L.S. (2000). Survival of *Ralstonia solanacearum* biovar 2, the causative agent of potato brown rot, in field and microcosm soils in temperate climates. *Phytopathology,* **90**: 1358-1366.

van Elsas, J. D., Kastelein,P., deVries, P. M. and van Overbeek, L.S. (2001). Effects of ecological factors on the survival and physiology of*Ralstonia solanacearum* bv 2 in irrigation water. *Can. J. Microbiol.* **47:** 842–854.

van Elsas, J.D., van Overbeek, L.S. and Trigalet, A. (2005a).The viable but non-culturable state in *Ralstonia solanacearum*: Is there a realistic threat to our strategic concepts? In: Bacterial Wilt Disease and the *Ralstonia solanacearum* Species Complex, pp. 103–119. Edited by C. Allen, P. Prior & A.C. Hayward. St. Paul, MN: APS Press.

van Elsas, J.D., van Overbeek, L.S., Bailey, M.J., Schonfeld, J. and Smalla, K.(2005b). Fate of *Ralstonia Solanacearum* biovar 2 as affected by conditions and soil treatments in temperate Climate Zones. In: Allen C., Prior P, Hayward A.C., eds. *Bacterial wilt disease and the Ralstonia solanacearum species complex.* St. Paul, MN: APS Press, 39.

van Gijsegem, F., Genin, S. and Boucher, C. (1993). Conservation of secretion pathways for pathogenicity determinants of plant and animal bacteria. *Trends Microbiol.* **1**: 175-180.

van Gijsegem, F., Vasse, J., Camus, J.C., Marenda, M.and Boucher, C. (2000). *Ralstonia solanacearum* produces Hrp-dependent pili that are required for popa Secretion but not for attachment of bacteria to plant cells. *Mol. Microbiol.* **36:** 249-260.

Van Overbeek M., Cassidy M., Kozdroj J., Trevors J.T., Van Elsas J.D. (2002). A polyphasic approach for studying the interaction between *Ralstonia solanacearum* and potential control agents in the tomato phytosphere. *J. Microbiol. Methods,* **48**: 69-86.

van Overbeek, L. S., Bergervoet, J. H. W., Jacobs, F. H. H. and van Elsas, J. D. (2004). The low-temperature-induced viable-but nonculturable state affects the virulence of *Ralstonia solanacearum* biovar 2. *Phytopathology,* **94**: 463–469.

Vandamme, P. and Coenye, T. (2004). Taxonomy of the genus Cupriavidus: a tale of lost and found. *Inter. J. Sys. andEvol. Microbiol.* **54**: 317- 327.

Vanderplank, J.E (1963). *The Disease: Epedemic and Control.* Academic Press, New York, 349 p. 4.

vanDoorn, R., Szemes, M., Bonants, P., Kowalchuk, G.A., Salles, J.F., Ortenberg, E., and Schoen, C.D. (2007). Quantitative multiplex detection of plant pathogens using a novel ligation probe-based system coupled with universal, high-throughput real time PCR on Open Arrays. *BMC Genomics,* **14**: 276.

Vaneechoutte, M., Kampfer, P., De Bacre, T., Falsen, E. and Verschraegen, G. (2004). *Wautersia* gen. nov., a novel genus accommodating the phylogenetic lineage including *Ralstonia eutropha* and related species, and proposal of *Ralstonia* [*Pseudomonas*] *syzygii* (Roberts et al., 1990) comb. Nov. *Inter. J. Sys. and Evol. Microbiol.* **54**: 317- 327.

vanOverbeek M., Cassidy M., Kozdroj J., Trevors J.T., Van Elsas J.D. (2002). A polyphasic approach for studying the interaction between *Ralstonia solanacearum* and potential control agents in the tomato phytosphere. *J. Microbiol. Methods,* **48**: 69-86.

Vasse, J., Danoun, S. and Trigalet, A. (2005). Microscopic studies of root infection in resistant tomato cultivar Hawaii 7996. In: Allen C., Prior P., Hayward A.C., eds. Bacterial wilt disease and the *Ralstonia solanacearum* species complex. St. Paul, MN: APS Press, (pp 285-291) St. Paul, MN: APS Press,

Vasse, J., Frey, P.and Trigalet, A. (1995). Microscopic studies of intercellular infection and protoxylem invasion of tomato roots by *Pseudomonas solanacearum. Mol. Plant-Microbe Intera.* **8**(2): 241-251.

Vavilov, N.I. (1931). Mexico and central america as the principal centre of origin of cultivated plants in the new World. *Bull. Appl. Bot. Genet. Plant Breed,* **26**: 135-199.

Velupillai, M. (1986). Bacterial wilt in Sri Lanka. In: Pers!ey, G.J. ed., Bacterial Wilt Disease in Asia and the South Pacific. *ACIAR Proce.* No. **13:** 57-64.

Venkatraman, S. and Sen Gupta, D.K. (1979). In *Annual Report.* ICAR Research Complex for NEH Region. Shillong, p. 14.

Verma, R.K. and Shekawat, G.S. (1991). Effect of crop rotation and chemical soil treatments on bacterial wilt of potato. *Indian. Phytopath.* **44**: 5-8.

Verma, R.N. and Gupta, B.P. (1979). In: *Annual Report.* ICAR Research Complex for NEH Region. Shillong, p. 142.

Vidhyasekaran, P. (1988). Physiology of disease resistance in plants. Vol. 1, CRC Press, I.C. Boca Raton, Florida, *R. solanacearum.* Ph.D. Thesis. University of Pertanian Malaysia p. 149.

Villa, J., Tsuchiya, K., Horita, M., Natural, M., Opina, M. and Hyakumachi, M. (2003). DNA analysis of *Ralstonia solanacearum* and related bacteria based on 282–bp PCR amplified fragment. *Plant Dis.* **87**: 1337-1343.

Vinitha, S.C., Niranjana, S. R., Mortensen, C. N. and Umesha, S. (2009). Bacterial wilt of tomato in Karnataka and its management by *Pseudomonas fluorescens. Bio Control,* **54**(5): 685-695.

Vishwadhar, T.S. and Hazarika, B.P. (1988). In: *Annual Report.* ICAR Research Complex for NEH Region. Shillong. p 190.

Viswanathan, R. and R. Samiyappan (2002). Induced systemic resistance by fluorescent pseudomonads against red rot disease of sugarcane caused by *Colletotrichum falcatum*. Crop Protec. **21**: 1-10.

Volcani, Z.and Palti, J. (1960). *Pseudomonas solanacearum* in Israel. *Plant Dis. Repor.* **44**: 448-449.

Vos P., Hogers R., Bleeker M., Reijans M., van de Lee T., Hornes M., Frijters A., Pot J., Peleman J., Kuiper M., Zabeau M., (1995). AFLP: a new technique for DNA fingerprinting. *Nucleic Acids Res.* **23**: 4407-4414.

Voss, A., Niersbach, M., Hain, R., Hirsch, H.J., Liao, Y.C., Kreuzaler, F. and Fischer, R. (1995). Reduced virusinfectivity in *N. tabacum*secreting a tmvspecific full size antibody. *Mol. Breeding,* **1:** 39-50.

Waggoner, P.E., Miller, P.M. and DeRoo, H.C. (1960). Plastic mulching principles and benefit. *Connec. Agri. Exp. Sta. Bull.* **634**: 44.

Wakimoto, S. (1987). Biological control of bacterial wilt of tomato by non-pathogenic strain of *Pseudomonas glumae* Korean J. Plant. Pathol. **3:** 300-303.

Wakimoto, S., Utatsu, I., Matsuo, N. and Hayashi, N. (1982). Multiplication of *Pseudomonas solanacearum* in pure water. *Nippon Shokubutsu Byori Gakkaiho,* **48**: 620–627.

Wang, J.F. (1998). Basic protocols for conducting research on tomato bacterial wilt caused by *Ralstonia solanacearum. Asian Vegetable Research and Development Centre*, p. 23.

Wang, J.F. and Ramasamy, S. (2011). Decades of IPM innovation recognized. AVRDC, The World Vegetable Centre. pp. 1-2.

Wang, J.F., Hanson, P. and Barnes, J.A. (1998). Worldwide evauation of an international set of resistance sources to bacterial wilt in tomato. pp. 269-75. In: *Bacterial wilt Disease: Molecular and Ecological Aspects* (Ed. Prior, P.; Allen, C.and Eliphinastone, J.). Sringer-Verag, Berlin; INRA, Paris

Wang, J.F., Olivier, J., Thoquet, P., Mangin, B., Sauviac, L., and Grimsley, N.H. (2000). Resistance to tomato line Hawaii 7996 to *Ralstonia solanacearum* PSS4 in Taiwan is controlled mainly by a major strain specific locus. *Mol. Plant-Microbe Intera.* **13**: 6-13.

Wang, J.F.and Lin, C.H. (2005). Integrated management of tomato bacterial wilt. Shanhua, Tainan: AVRDC - *The World Vegetable Center*. 12p.

Wani, P.V. (2002). Biofertilizers as bio-agents. In: ICAR Sponsored Winter School on "*Frontiers in ecofriendly management of crop diseases and pests and utility of biofertilizers in improving soil fertility*." (Aug. 6 to Sept. 4). Department of Plant Pathology and Agriculture Microbiology. Mahatma Phule Krishi Vidyapeeth, Rahuri. pp. 27-36.

Warren, R.F., Henk, A., Mowery, P., Holub, E. and Innes, R.W. (1998). A mutation within the leucine-rich repeat domain of the *Arabidopsis* disease resistance gene Rps5 partially suppresses multiple bacterial and downy mildew resistance genes. *Plant Cell,* **10**: 1439–1452.

Watcharachaivakup, J., Thaveechai, N., Chunwongse, J. and Kositratana, W. (2005). Mechanism of resistance to bacterial wilt disease in chilli pepper caused by *Ralstonia solanacearum. In Proceeding of 43rd* Kesetsart University Annual Conference Thialand, 1-4 February 2005 Subject: Plant pp. 262-26

Wei Chun, M., Zhang, C., Liu, Z.Z., Xu Suzheri and Zhao, J.H. (1994). Adsorption, penetration and movement of bacterial wilt antagonistic strain 90 B 4-2-2 in roots of tomato plants. *Acta Agri. Shanghai,* **10**(4): 48-52.

Weller S.A., Elphinstone J.G., Smith N., Stead D.E. and Boonham N. (1999). Detection of *Ralstonia solanacearum* strainsusing an automated and quantitative fluorogenic 5' nuclease TaqMan assay. *Appl. Environ. Microbiol.* **66**: 2853-2858.

Weller, S.A. and Stead, D.E. (2002). Detection of root mat associated Agrobacterium strains from plant material andother sample types by post-enrichment TaqMan PCR. *J. Appl. Microbiol.* **92**: 118-126.

Weller, S.A., Elphinstone, J.G., Smith, N., and Stead, D.E. (2000b). Detection of *Ralstonia solanacearum* from potato tissue by post enriched TaqMan™ PCR. *Bull. Organ. Eur. Mediterr. Prot. Plant,* **30**: 381-383.

Weller, S.A., Elphinstone, J.G., Smith, N.C., Boonham, N., and Stead, D.E. (2000a). Detection of *Ralstonia solanacearum* strains with a quantitative, multiplex, real-time, fluorogenic PCR (TaqMan) assay. *Appl. Environ. Microbiol.* **66**: 2853-2858.

Wenneker, M., Verdel, M. S. W., Groeneveld, R. M. W., Kempenaar, C., van Beuningen, A. R. and Janse, J. D. (1999). *Ralstonia (Pseudomonas) solanacearum* race 3 (biovar II) in surface water and natural weed hosts: First report on stinging nettle (Urtica dioica). *Eur. J. Plant Pathol.* **105**: 307-315.

Wicker, E., Grassart, L. Coranson-Beaudu, R., Mian, D., Guilbaud, C. Fegan, M. and P. Prior (2007). *Ralstonia solanacearum* strains from Martinique (French West Indies) exhibiting a new pathogenic potential. *Appl. Environ.Microbiol.* **73**(21): 6790-6801.

Wicker, E., Grassart, L., Coranson-Beaudu, R., Mian, D., Guilbaud, C. andFegan, M. (2007). *Ralstonia solanacearum* strains from Martinique (French West Indies) exhibiting a new pathogenic potential. *Appl. Environ. Microbiol.* **71**: 6790-6801.

Wicker, E., Lefeuvre, P., de Cambiaire, J.C., Lemaire, C., Poussier, S. and Prior, P. (2012). Contrasting recombination patterns and demographic histories of the plant pathogen *Ralstonia solanacearum* inferred from MLSA. *ISME J.* **6:** 961-974.

Winstead, N.N. and Kelman, A. (1952). Inoculation techniques for evaluation resistance to *Pseudomonas solanacearum. Phytopathology,* **42**: 628-634.

Xiao, R. andKisaalita W. S. (1998). Fluorescent pseudomonad pyoverdines bind and oxidize ferrous ion. *Appl. Environ. Microbiol.* **64:** 1472–1476.

Xue, Y.Q., Chen, Y., Li S.M., Chen, L.F., Ding, G.C., Guo, D.W. and Guo J.H. (2009). Evaluation of the strains of *Acinetobacter* and *Enterobacter* as potential biocontrol agents against *Ralstonia* wilt of tomato. *Biol. Contr.* **48**: 252-258.

Yabuuchi, E., Kosako, Y., Oyaizu, H. *et al* (1992). Proposal of *Burkholderia* gen. nov. and transfer of 7 species of the genus *Pseudomonas* homology group-II to the new genus, with the type species *Burkholderia cepacia* (Palleroni and Holmes, 1981) comb. nov. *Microbiol. Immunol.* **36**: 1251-1275.

Yabuuchi, E., Kosako, Y., Yano L., Hotta, H. and Nishiuchi, Y. (1996). Validation of the publication of new names and new combinations previously effectively published outsidethe IJSB. *Int. J. Syst. Bacteriol.* **46**: 625-626.

Yabuuchi, E., Kosako, Y., Yano, I., Hotia, H. and Nishiuchi, Y. (1995). Transfer of two *Burkholderia* and *Alcaligene* species to *Ralstonia*. Gen. Nov. propal of *Ralstonia picketti* (Ralston, Palleroni and Doudoroff 1973) Comp. Nov. and *Ralstonia eutropha* (Davis, 1969) *Com. Nov. Microbiol. Iummunol.* **39**: 897-904.

Yadav R.K., B.C. Deka and Sanwal, S.K. (2009). Genetic resources of vegetable crops of North Eastern himalayan Region Envis Bulletin: Himalayan Ecology 17 Division of Horticulture, ICAR Research Complex for NEH Region, Umiam-793 103, Meghalaya, India.

Yadessa, G.B. van Bruggen, A.H.C. and Ocho, F.L. (2010). Effects of different soil amendments on bacterial wilt caused by *Ralstonia solanacearum* and on the yield of tomato. *J. Plant Pathol.* **92**(2): 439-450.

Yamada, M., Nakazama, Y. and Kitamura, T. (1997). Control of tomato bacterial wilt by dazomet combined with soil solarization. *Proc. Kento-Tosan Plant Protec. Soc.* **44**: 75-78.

Yamazaki, H. and Hoshima, T. (1995). Calcium nutrition affected resistance of tomato seedling to bacterial wilt. *Hortscience,* **30**(1): 91-93.

Yamazaki, H., Ishizuka, O. and Hoshina, T., (1996). Relationship between resistance to bacterial wilt and nutrient uptake in tomato seedlings. *Soil Sci. Plant Nutri.* **42**: 203-208.

Yang, B., Zhu, W., Johnson, L.B., and White, F.F. (2000). The virulence factor AvrXa7 of *Xanthomonas oryzae* pv. *oryzae* is a type III secretionpathway-dependent nuclear-localized double-stranded DNA-bindingprotein. *Proc. Natl. Acad. Sci.* USA, **97**: 9807-9812.

Yang, J.G., Pi, X.H., Chen, H.M., Yao, Y.G., and Zou, Y. (2006). Resistance of ER300 eggplant (*Solanum melongena.*) against genetic bacterial wilt *(Ralastonia solanacearum*) and its application to breeding. *J. Hunan Agric. Univ.* (Natural Science Edition). **32**(3): 277-279.

Yao, G., Zhang, F. and Li, Z. (1994). Control of bacterial wilt with soil amendment. *Chinese J. Biol. Contr.* **10**(3): 106-109.

Yao, J. and Allen, C. (2007). The plant pathogen *Ralstonia solanacearum* needs aerotaxis for normal biofilm formation and interactions with its tomato host. *J. Bacteriol.* **189**: 6415-6424.

Yao, J. and Allen, C.(2006a). Chemotaxis is required for virulence and competitive fitness of the bacterial wilt pathogen *Ralstonia solanacearum. J. Bacteriol.* **188**: 3697-3708

Yao, J.and Allen, C. (2006b).The role of bacterial aerotaxis in *Ralstonia solanacearum* interactions with host plants. *Phytopathology,* **96**: 128-135.

Yawalkar, K. S. and Agarwal, J. P. (1962). Manures and fertilizers agri-horticulture, Pub. House, pp. 251.

Yik, C.P., Ong, A.K. and Ho, R. (1994). Characteristics of *Pseudomonas solanacearum* strain from Singapore. *J. Primary Indus.* **22**(2): 57-62.

Yim, W., Seshadri, S., Kim, K., Lee, G. and Sa, T. (2013). Ethylene emission and PR protein synthesis in ACC deaminase producing *Methylobacterium* spp. inoculated tomato plants (*Lycopersicon esculentum* Mill.) challenged with *Ralstonia solanacearum* under greenhouse conditions. *Plant Physiol. and Biochem.* **67:** 95-104.

Yoon, J.Y., Green, S.K., Tschanz, A.T., Tsou, S.C.S., Chang, L.C. (1989). Pepper improvement for the tropics: Problems and the AVRDC approach. In: Tomato and pepper production in the tropics. Shanhua, Tainan: AVRDC. pp. 86-98.

Yoshimura, S., Yamanouchi, U., Katayose, Y., Toki, S., Wang, Z.X., Kono, I., Kurata, N., Yano, M., Iwata, N. and Sasaki, T. (1998). Expression of Xa1, a bacterial blight-resistance gene in rice, is induced by bacterial inoculation. *Proc. Nat. Acad. Sci.* USA, **95**: 1663-1668.

Yossen, V.E., Michie, J.L.N., Loch, L.C. and Daarte, V. (1998). Infectivity titration of *Ralstonia solanacearum* strains resistance to antibiotics. *Fitopathol. Brasi.* **23**(4): 495-497.

Young, D.H. and Sequeira, L. (1986). Binding of *Pseudomonas solanacearum* fimbriae to tobacco leaf cell-walls and its inhibition by bacterial extracellular polysaccharides. *Physiol. Mol. Plant Pathol.* **28**: 393-402.

Yu, J.Q *et al.* (1995a). Effcets of sugi and hinoki bark fibre on the incidence of fusarium wilt and bacterial wilt in the soilless culture of of tomato. *Ann. Japanese Soc. Phytopathol.* **61**: 241-242.

Yu, J.Q *et al.* (1995b). Suppressive effects of sugi and hinoki bark fibre media on soil-bome diseases in soilless culture of tomato. *Plant Epidemic Prove*. **49**: 24-26.

Yu, J.Q. (1999a). Alleopathic suppression of *Pseudomonas solanacearum* infection of tomato (*Lycopersicon esculentum*) in a tomato Chinese Clive (*Allium tuberosum*) intercropping system. *J. Chem. Ecol.* **25**(11): 2409-2417.

Yu, J.Q. and Komada, H. (1999). Hinoki (*Chamaecypasis obtusa*) Bark, a substrate with antipathogen properties that suppress some root disease of tomato. *Scient. Hort.* **81**(1): 13-24.

Yuan, L.H., Su, O.L., Yi, Y. H. and Yuan, B. (1995). A preliminary study on bacterial wilt disease suppressive soil of the tomato in Jiangxi Province. *Acta Phytophyl. Sin.* **22**(1): 93-94.

Yuan, L.H., Su, O.L., Yi, Y.H. and Yuan, B. (1997). Study on tomato bacterial wilt suppressive soil of the tomato.*Acta Phytophyl. Sin.* **24**(2): 121-125.

Zachos, D.G. (1957). The Brown Rot of Potatoes in Greece. *Annales de l Institut Phytopathologique Benaki, New Series* **1**: 115-117.

Zdeňka, P., Petr Šmerák, H.D. and Milan, H. (2010). Antimutagenic effects of lycopene and tomato purée. *J. Medicinal Food.* **13**(6): 1443-1450.

Zehr, E.I. (1969). Studies of the distribution and economic importance of *Pseudomonas solanacearum* in certain crops in the Philippines. *Philippine Agri.* **53**: 218-223.

Zehr, E.I. (1970). Isolation of *Pseudomonas solanacearum* from abaca and banana in the Philippines. *Plant Dis. Repor.* **54**(6): 516-519.

Zhang, C.X., Ho, S.C., Chen, Y.M., Fu, J.H., Cheng, S.Z. and Lin, F.Y. (2009). Greater vegetable and fruit intake is associated with a lower risk of breast cancer among Chinese women. *Int. J. Cancer,* **125** (1): 181-188.

Zhang, L.P., Khan, A., Niño-Liu, D., Foolad, M.R. (2002). A molecular linkage map of tomato displaying chromosomal locations of resistance gene analogs based on a *Lycopersicon esculentum* × *Lycopersicon hirsutum* cross. *Genome*, **45**(1): 133-146.

Zhang, Z.H. and Liu, Y. (1998). Evaluation of bacterial wilt resistance of tomato variety Zao Kang 1. *Hunan Agri. Res. News Lett.* **59**(3): 11.

Zhu, H.H. and Yao, Q. (2004). Localized and systemic increase of phenols in tomato roots induced by *Glomus versiforme* inhibits *Ralstonia solanacearum*. *J. Phytopathol.* **152**: 537-542.

Zimmer, D.P., Soupene, E., Lee, H.L., Wendisch, V.F., Khodursky, A.B., Peterm B.J., Bender, R.A., and Kustu, S. (2000). Nitrogen regulatory protein controlledgenes of *Escherichia coli*: scavenging asa defense against nitrogen limitation. *Proc. Natl. Acad. Sci. USA*. **97**: 14674-14679.

Zinniel, D.K., Lambrecht, P., Harris, N.B., Feng, Z., Kuczmarski, D., Higley, P. *et al.* (2002). Isolation and characterization of endophytic colonizing bacteria from agronomic crops and prairie plants. *Appl. Environ. Microbiol.* **68**: 2198–2208.

Subject Index

Author Index

Q

R